Weyward *Macbeth*

SIGNS OF RACE
Series Editors: **Arthur L. Little, Jr. and Gary Taylor**

Writing Race across the Atlantic World: Medieval to Modern
Edited by Phillip D. Beidler and Gary Taylor (January 2005)

Buying Whiteness: Race, Culture, and Identity from Columbus to Hip-Hop
By Gary Taylor (January 2005)

English and Ethnicity
Edited by Janina Brutt-Griffler and Catherine Evans Davies (December 2006)

Women & Others: Perspectives on Race, Gender, and Empire
Edited by Celia R. Daileader, Rhoda E. Johnson, and Amilcar Shabazz (September 2007)

The Funk Era and Beyond: New Perspectives on Black Popular Culture
Edited by Tony Bolden (August 2008)

Race and Nature from Transcendentalism to the Harlem Renaissance
By Paul Outka (August 2008)

Weyward Macbeth: *Intersections of Race and Performance*
Edited by Scott L. Newstok and Ayanna Thompson (January 2010)

WEYWARD *MACBETH*

INTERSECTIONS OF RACE AND PERFORMANCE

Edited by
Scott L. Newstok and Ayanna Thompson

WEYWARD *MACBETH*
Copyright © Scott L. Newstok and Ayanna Thompson, 2010.

All rights reserved.

First published in 2010 by
PALGRAVE MACMILLAN®
in the United States—a division of St. Martin's Press LLC,
175 Fifth Avenue, New York, NY 10010.

Where this book is distributed in the UK, Europe and the rest of the world, this is by Palgrave Macmillan, a division of Macmillan Publishers Limited, registered in England, company number 785998, of Houndmills, Basingstoke, Hampshire RG21 6XS.

Palgrave Macmillan is the global academic imprint of the above companies and has companies and representatives throughout the world.

Palgrave® and Macmillan® are registered trademarks in the United States, the United Kingdom, Europe and other countries.

ISBN: 978–0–230–61642–4

Library of Congress Cataloging-in-Publication Data is available from the Library of Congress.

A catalogue record of the book is available from the British Library.

Design by Newgen Imaging Systems (P) Ltd., Chennai, India.

First edition: January 2010

10 9 8 7 6 5 4 3 2 1

Printed in the United States of America.

To our students

You put me in *Macbeth*...
And in everything but what's about me.

—Langston Hughes,
"Note on Commercial Theatre" (1940)

Contents

List of Figures	xiii
Series Editors' Preface	xv
Acknowledgments	xvii

1 Beginnings

1	What Is a "Weyward" *Macbeth*? Ayanna Thompson	3
2	Weird Brothers: What Thomas Middleton's *The Witch* Can Tell Us about Race, Sex, and Gender in *Macbeth* Celia R. Daileader	11

2 Early American Intersections

3	"Blood Will Have Blood": Violence, Slavery, and *Macbeth* in the Antebellum American Imagination Heather S. Nathans	23
4	The Exorcism of Macbeth: Frederick Douglass's Appropriation of Shakespeare John C. Briggs	35
5	Ira Aldridge as Macbeth Bernth Lindfors	45
6	Minstrel Show *Macbeth* Joyce Green MacDonald	55
7	Reading *Macbeth* in Texts by and about African Americans, 1903–1944: Race and the Problematics of Allusive Identification Nick Moschovakis	65

3 Federal Theatre Project(s)

8   Before Welles: A 1935 Boston Production 79
    Lisa N. Simmons

9   Black Cast Conjures White Genius: Unraveling
    the Mystique of Orson Welles's "Voodoo" *Macbeth* 83
    Marguerite Rippy

10 After Welles: Re-do Voodoo *Macbeth*s 91
    Scott L. Newstok

11 *The Vo-Du Macbeth!*: Travels and Travails of a
    Choreo-Drama Inspired by the FTP Production 101
    Lenwood Sloan

4 Further Stages

12 A Black Actor's Guide to the Scottish Play, or,
    Why *Macbeth* Matters 113
    Harry J. Lennix

13 Asian-American Theatre Reimagined:
    Shogun Macbeth in New York 121
    Alexander C. Y. Huang

14 The Tlingit Play: *Macbeth* and Native Americanism 127
    Anita Maynard-Losh

15 A Post-Apocalyptic *Macbeth*: Teatro LA TEA's *Macbeth 2029* 133
    José A. Esquea

16 Multicultural, Multilingual *Macbeth* 137
    William C. Carroll

5 Music

17 Reflections on Verdi, *Macbeth*, and
    Non-Traditional Casting in Opera 145
    Wallace McClain Cheatham

18 Ellington's Dark Lady 151
    Douglas Lanier

19 Hip-Hop *Macbeth*s, "Digitized Blackness," and the Millennial
    Minstrel: Illegal Culture Sharing in the Virtual Classroom 161
    Todd Landon Barnes

6 Screen

20 Riddling Whiteness, Riddling Certainty:
Roman Polanski's *Macbeth* — 173
Francesca Royster

21 Semper *Die*: Marines Incarnadine in Nina Menkes's
The Bloody Child: An Interior of Violence — 183
Courtney Lehmann

22 Shades of Shakespeare: Colorblind Casting and
Interracial Couples in *Macbeth in Manhattan*,
Grey's Anatomy, and Prison *Macbeth* — 193
Amy Scott-Douglass

7 Shakespearean (A)Versions

23 Three Weyward Sisters: African-American Female
Poets Conjure with *Macbeth* — 205
Charita Gainey-O'Toole and Elizabeth Alexander

24 "Black up again": Combating *Macbeth* in
Contemporary African-American Plays — 211
Philip C. Kolin

25 Black Characters in Search of an Author:
Black Plays on Black Performers of Shakespeare — 223
Peter Erickson

Epilogue

26 ObaMacbeth: National Transition as National Traumission — 235
Richard Burt

Appendix

27 Selected Productions of *Macbeth* Featuring
Non-Traditional Casting — 241
Brent Butgereit and Scott L. Newstok

References — 253

Notes on the Contributors and Editors — 273

Index of Passages from Macbeth — 279

General Index — 281

Figures

Cover: Premiere of the Federal Theatre Project *Macbeth*, Harlem, 1936. Library of Congress, Music Division.

S.1	Text from *Macbeth, Scena Tertia*, First Folio, 1623.	1
S.2	"A Proslavery Incantation Scene, or Shakespeare Improved." David Claypool Johnston, 1856. Courtesy, American Antiquarian Society.	21
3.1	"The Hurly-Burly Pot." *Harper's Weekly*, 1850. Library of Congress, Prints and Photographs Division.	24
5.1	Ira Aldridge as Macbeth. Courtesy of McCormick Library of Special Collections, Northwestern University.	49
S.3	Romare Bearden, poster for New Federal Theatre production of *Macbeth*, 1977. Art © Romare Bearden Foundation / Licensed by VAGA, New York, NY.	77
S.4	Weird Sisters from Teatro LA TEA's *Macbeth 2029*. Credit: Antoni Ruiz: aruizphotography.com.	111
S.5	Duke Ellington, sheet music for "Lady Mac," c. 1956. Courtesy of Duke Ellington Collection, Archives Center, National Museum of American History, Smithsonian Institution.	143
S.6	Harold Perrineau, *Macbeth in Manhattan*, 1999.	171
22.1	Sandra Oh and Isaiah Washington, "From a Whisper to a Scream," *Grey's Anatomy*, November 23, 2006.	197
S.7	E. McKnight Kauffer (1890–1954), dust jacket for Langston Hughes's *Shakespeare in Harlem* (1942). Copyright © Simon Rendall, Estate of E. McKnight Kauffer. Image courtesy of the Emory University Manuscript, Archives, & Rare Book Library.	203
S.8	Obamicon, Paste Media Group, 2009.	233
S.9	Playbill from the Los Angeles FTP production of *Macbeth* (1937). Library of Congress, Music Division, Finding Aid Box 1095.	239

Series Editors' Preface

The first thing you see when you enter the permanent exhibits at the Birmingham Civil Rights Institute is a pair of drinking fountains. Over one hangs a sign that says "White." Over the other hangs a sign that says "Colored."

To the extent that every social identity is to some degree local, the meanings of race in Birmingham, Alabama, necessarily differ, in some demographic and historical particulars, from the meanings of race in North Dakota and Northern Ireland, New York and New South Wales, Cape Town and Calcutta. But the same questions can be asked everywhere in the English-speaking world.

How do people signal a racial identity?

What does that racial identity signify?

This series examines the complex relationships between race, ethnicity, and culture in the English-speaking world from the early modern period (when the English language first began to move from its home island into the wider world) into the postcolonial present, when English has become the dominant language of an increasingly globalized culture. English is now the medium of a great variety of literatures, spoken and written by many ethnic groups. The racial and ethnic divisions between (and within) such groups are not only reflected in, but also shaped by, the language we share and contest.

Indeed, such conflicts in part determine what counts as "literature" or "culture."

Every volume in the series approaches race from a transracial, interdisciplinary, intercultural perspective. Each volume in the series focuses on one aspect of the cross-cultural performance of race, exploring the ways in which "race" remains stubbornly local, personal, and present.

We no longer hang racial signs over drinking fountains. But the fact that the signs of race have become less obvious does not mean that they have disappeared, or that we can or do ignore them. It is the purpose of this series to make us more conscious, and more critical, readers of the signs that have separated, and still separate, one group of human beings from another.

GARY TAYLOR

Acknowledgments

This volume has been deeply collaborative since its outset, emerging from ongoing conversations between us long before we even imagined such a collection.

Rhodes College supported the January 2008 symposium on *Macbeth* that provided the seed for *Weyward Macbeth*. This symposium brought together Ayanna Thompson, Peter Erickson, Marguerite Rippy, Amy Scott-Douglass, Wallace McClain Cheatham, and Harry J. Lennix, as well as Aleta Chappelle and performers and directors from the Hattiloo Theatre and Opera Memphis. The Rhodes College Center for Outreach in the Development of the Arts (CODA) was the main sponsor, with additional support coming from the Departments of Theatre, English, and African American Studies, as well as Provost Charlotte Borst.

We thank Dean Deborah Losse at Arizona State University and Dean Michael Drompp at Rhodes College for their support toward permissions fees. A grant from the Mike Curb Institute for Music at Rhodes College also provided support for research at the Center for Popular Music, Middle Tennessee State University.

We are grateful for the hundreds of artists and scholars with whom we have corresponded about this project over the past two years; they helped deepen our understanding of the play and its place within American racial discourses, and their comments have particularly enriched the Appendix.

We dedicate this collection to our students—those whom we have taught in the past as well as those whom we look forward to teaching in the future. For their work on helping us edit this book, we are particularly thankful for the assistance of Heather Ackerman, Valerie Fazel, and Julianne Mayer. Brent Butgereit's work on the Appendix was supported through the Rhodes College Student Research Assistant Program.

Finally, we are ever grateful to Derek, Dashiell, Sarah, and Ruth, who put up with spouses and parents whom this fascinating project made too often weyward.

Note: All citations from Shakespeare in this collection are taken from the Norton anthology. *The Norton Shakespeare: Based on the Oxford Edition.* 2nd edition, Ed. Stephen Greenblatt *et al.* (New York: W. W. Norton & Company, 2008). We thank the following poets for permission to reprint excerpts from their work: Al Young, "Identities," copyright © 1975, 1992 by Al Young; reprinted with permission of the author; Rita Dove, "In the Old Neighborhood," reprinted with permission of the author.

1

Beginnings

Looke what I haue.
  2. Shew me, shew me.
  1. Here I haue a Pilots Thumbe,
Wrackt, as homeward he did come.   *Drum within.*
  3. A Drumme, a Drumme:
Macbeth doth come.
  All. The weyward Sisters, hand in hand,
Posters of the Sea and Land,
Thus doe goe, about, about,
Thrice to thine, and thrice to mine,
And thrice againe, to make vp nine.
Peace, the Charme's wound vp.

      Enter Macbeth and Banquo.

Macb. So foule and faire a day I haue not seene.

Figure S.1 Text from *Macbeth, Scena Tertia*, First Folio, 1623.

1

WHAT IS A "WEYWARD" *MACBETH*?

Ayanna Thompson

Let's begin at the beginning, that is, at the first word—*weyward*. As Margreta de Grazia and Peter Stallybrass explain:

> In most editions of [*Macbeth*], the witches refer to themselves as the "weird sisters" [see Figure S.1], and editors provide a footnote associating the word with Old English *wyrd* or "fate." But we look in vain to the Folio for such creatures. Instead we encounter the "weyward Sisters" and the "weyard Sisters"; it is of "these weyward Sisters" that Lady Macbeth reads in the letter from Macbeth, and it is of "the three weyward sisters" that Banquo dreams. (263)

Wondering why editors choose "weird" instead of "wayward" as the modern gloss for "weyward," de Grazia and Stallybrass note that "a simple vowel shift" transposes "the sisters from the world of witchcraft and prophecy...to one of perversion and vagrancy" (263). In thinking about the roles *Macbeth* has played—and continues to play—in American constructions and performances of race, we want to maintain the multiplicity *and* instability of the original text's typography. Unlike other critical texts that seek to capitalize on the subversive potentials of perversity through an appropriation of "wayward," this collection recognizes the ambivalent nature of the racialized re-stagings, adaptations, and allusions to *Macbeth*.[1] "Weyward"—as weird, fated, fateful, perverse, intractable, willful, erratic, unlicensed, fugitive, troublesome, and wayward—is precisely the correct word for *Macbeth*'s role in American racial formations.[2]

Why *Macbeth* and race? What is "weyward" about the intersections of race and performance in *Macbeth*? In some respects the intersections seem arbitrary: just as easily, and perhaps more logically, *Othello*, *The Tempest*, or any of Shakespeare's so-called "race" plays could serve as the theatrical placeholder in this book. This collection itself grows out of the 2008 symposium that Scott L. Newstok organized at Rhodes College when he noticed the coincidence of the Hattiloo Theatre Company staging Shakespeare's *Macbeth* with a predominantly black cast in conjunction with Opera Memphis staging Verdi's *Macbeth* with black principals. So perhaps thinking through *Macbeth* and race together might seem something of a fluke, since these companies could have staged several other play-opera combinations. Would there be as much to say, or different things to say, about the intersections of race and performance for *Othello* and Verdi's *Otello*, or *Romeo and Juliet* and Gounod's *Roméo et Juliette*? Is *Macbeth*'s place in this volume thus capricious?

Yet there remains something unique about *Macbeth*: the haunted play, the play that shall not be named, in which "nothing is / But what is not" (1.3.140–41). Whether or not its very essence is unique, people have historically treated *Macbeth* as anomalous, different, and Other—as "weyward," in fact. At its very core, the "Scottish play" is about the very real distinctions between a king and one who dons "borrowed robes" (1.3.107), the English and the Scottish, the physical and the metaphysical. So should we be surprised that *Macbeth* is *not* the antithesis of a "race" play: that is, not an un-raced, non-raced, or normatively-raced play (you see how the English language bucks the sense that something is not raced!)? After all, the "weyward" qualities of the witches seem to stem in part from their Scottishness—a sort of seventeenth-century joke about King James's ethnicity. And it is these "weyward Sisters" who most clearly verbalize how racially inflected the play is in their famous (and famously censored) incantation[3]:

> Double, double, toil and trouble,
> Fire burn, and cauldron bubble.
> . . .
> Liver of blaspheming Jew,
> Gall of goat, and slips of yew
> Slivered in the moon's eclipse,
> Nose of Turk and Tartar's lips
>   (4.1.20–21, 26–29)

However, it is not simply the witches' brew, with bits of a Jew, Turk, and Tartar, nor even the play's consistent recourse to early modern figurative invocations of whiteness and blackness that makes *Macbeth* read as much more racially engaged than is conventionally assumed.[4] The play's very rhetoric of blood and staining informs—or seeps into—early American racial rhetoric as well. *Macbeth*'s focus on the indelible quality of blood, that staining and smelling substance that Lady Macbeth cannot fully wash from her hands, unnervingly coincides with early American debates about the nature—the essence—of race. On the one hand, the proponents of slavery (and later segregation) insist that the *blood* is the thing; the essential substance that identifies, divides, and classifies races, even down to a single drop. On the other hand, many opponents of slavery (and later segregation) appropriate *Macbeth*'s construction of blood into their protest speeches; the *mark* that stains America's formation and history. On January 7, 1828, for example, J. C. Clark, a Congressman from New York, argued on the House floor that slaves could not be considered property, and he did so by appropriating lines from *Macbeth*:

> Is the ghost of the Missouri question again to be marched, with solemn and terrific aspect, through these halls? Is it again to "shake its [*sic*] gory locks at

us [*sic*]," and, pointing with one hand to the North, and with the other to the South, and gazing its blood-shotten eye on slavery, written on the escutcheon of the Constitution, to proclaim, with unearthly voice, "out damned spot?" (917–18; *Macbeth* 3.4.49–50; 5.1.30)

Clark "imagined" that the debate about the status of slaves had been put to rest, but he conjures its ghostly re-vision both in terms of Banquo's ghost and in terms of a bloody stain on the Constitution—something that cannot be removed with ease (918).

Likewise, in his June 11, 1854 speech in Westfield, Massachusetts, protesting the North's lack of a stance against Southern slavery, the Methodist Reverend Gilbert Haven intoned,

> We were indifferent to the perils and defeats of freedom. We eagerly snatched and swallowed the few beggarly slops of office and enactments which our shrewd Southern masters tossed us.... Look on your hands! Blood! Cry, "Out damned spot! out, I say!" It flees not; it blears our eyes; it stains our souls; it smells to heaven. Not all the perfumes in Arabia can sweeten this Northern hand. (71)

Haven employs *Macbeth* not only because the play presents the visceral aspects of blood (it "stains" and "smells"), but also because it constructs blood in essentialist terms ("It flees not").

Even into the twentieth century, America's history is often described as analogous to the invisible yet seemingly permanent bloodstains on Lady Macbeth's hands. Writing about the structural nature of inequality in the United States in 1995, John Sedgwick notes, "Then there is the matter of racism in America, which, like bloodstains on the hands of Lady Macbeth, cannot be washed away" (158). The legal scholar Norman Redlich cites the "out, damned spot" passage directly in the title of his essay on the persistence of race in American law, and invokes the "spot" repeatedly as a governing metaphor throughout. And finally, Stephan Lesher, the biographer of the infamous segregationist George Wallace, writes, "Like the imagined indelible bloodstain on Lady Macbeth's hands, the stains of racism on Wallace's reputation will never be washed away" (xiii).

The convergence of Memphis's non-traditionally cast productions of *Macbeth*, then, is not as singular as it might at first appear. While *Macbeth* may not fit neatly into the category of a "race" play, like a specter it is nevertheless haunted by, and haunts, the rhetoric and performances of race. Unlike the over-determined role *Othello* plays in historical *and* contemporary constructions of racial difference (especially blackness), *Macbeth* often lures actors, directors, writers, and others into thinking that the "Scottish play" does not carry "the onerous burden of race" (see the Lennix essay). This lure is so powerful, as the totality of *Weyward Macbeth* demonstrates, that actors, directors, and writers often assume that they are the first to see the connections.

Quite to the contrary, *Macbeth* has long played a role in *American* constructions of race, from its appearance as the first Shakespearean play documented in the American colonies in 1699 (owned by the Virginia lawyer and plantation owner Captain Arthur Spicer [Teague 14]) to a proposed Hollywood film version with an all-black cast (as of this writing, to star Terence Howard, Sanaa Lathan, Harry J. Lennix, Blair Underwood, and Danny Glover). In the nineteenth century, *Macbeth* provided the context for the black actor Ira Aldridge's experimentation with whiteface performance (see the Lindfors essay) as well as the classist-, racist-, and nationalist-based Astor Place Riot in 1849 (see the Nathans and MacDonald essays). Although Orson Welles's 1936 Federal Theatre Project version of *Macbeth*—commonly referred to as the "Voodoo" *Macbeth*—is often discussed as the creation of Welles's singular and immense creative genius (see the Rippy essay), there was an all-black FTP production the year before in Boston (see the Simmons essay). Likewise, there have been numerous contemporary theatre companies that assume that they are unique in attempting to re-stage Welles's "Voodoo" version (see the Newstok and Sloan essays). Moreover, there have been African-American, Asian-American, Native American (Alaskan and Hawai'ian), and Latino theatre companies that turn to *Macbeth* to help stage their own unique racial, ethnic, and cultural identities (see the Lennix, Huang, Maynard-Losh, Carroll, and Esquea essays).

Macbeth's lure—that it does not come with "the onerous burden of race"—comes at a price, however. Despite the fact that many of the non-traditionally cast productions bill themselves as unique translations—the first to change Scotland to the Caribbean, Africa, the Alaskan tundra, a multiracial post-apocalyptic future, etc.—they can only do so by employing a type of historical amnesia. Unlike *Othello*, *The Tempest*, or the other supposed "race" plays, *Macbeth* does not readily announce itself as already weywardly racialized. The writers, directors, performers, and scholars of weyward *Macbeth*s are thus enabled in forgetting (or never knowing) the play's long history of literary and performance intersections with race. This is precisely the void *Weyward Macbeth* seeks to fill. This collection is designed to combat the historical amnesia that this weyward play, its performance history, and its scholarship engenders.

Signifying Something

Unlike Macbeth's fears that life is a "brief candle" and a "tale / Told by an idiot, full of sound and fury, / Signifying nothing" (5.5.22, 25–27), *Weyward Macbeth* demonstrates the eerie quality of the play's ability to reignite itself all the while signifying a great deal within racial discourses. Thus, we begin with the weyward qualities of the playtext itself. The aspect of *Macbeth* that has invited and enabled numerous adaptations and appropriations—the witches' embodiment of the metaphysical—was actually penned by a ghostly, decidedly non-Shakespearean, hand—Thomas Middleton (see the Daileader essay). Beginning with this spookily invisible yet visible presence, *Weyward Macbeth* then turns to the early American intersections with *Macbeth*. As I indicate above, the "Scottish play" that explores the essence of lineage, bloodlines, and bloodlusts haunts debates about freedom, slavery, and racial/national identity. With four essays that address antebellum moments, political figures, and cultural forces (see the Nathans, Briggs, Lindfors, and MacDonald essays), and one that addresses the early twentieth-century hauntings (see the Moschovakis essay), Section Two establishes the complex and ambiguous history of early American allusive references to *Macbeth*. As all of the essays identify, *Macbeth* does not fit seamlessly into either pro- and anti-slavery/segregation rhetoric. Perhaps because of the fact that Macbeth's characterization is ambiguous (he is both a hero *and* a rogue), the play circulates through disparate speeches, images, poetry, and performances about America's racial politics. Shakespeare's equivocal character, then, provides an uncanny medium for our nation's divided self/other racial constructions.

Section Three moves firmly into the twentieth century to focus on Federal Theatre Project(s). Beginning with an often overlooked 1935 Boston FTP production, and ending with the 2001–2005 choreo-drama inspired by the famous 1936 "Voodoo" *Macbeth*, this section demonstrates the strangely amnesiac quality to appropriations of the play. Although Orson Welles's production is frequently cited as radically unique—with an all-black cast and sold out audiences in Harlem—the repetitious nature of constantly re-doing the voodoo forces one to contemplate the strange whiting out of performance histories. The productions also bring to the fore the central role the federal government has played—and perhaps might play again—in supporting unconventional and non-traditional productions.

Addressing the early twenty-first century, Section Four provides snapshots of five distinctly racialized adaptations of *Macbeth*. These short pieces provide an insider's view to why *and* how *Macbeth* is translated to a fictitious diasporic island, ancient Japan, an Alaskan native community, a multicultural future, and a Hawai'ian island. Once again these essays reveal the beckoning and yet contentious relationship between the text and the modern racialized settings. Although the productions often depict distinct cultural and racial environments that are separated from dominant Western influences, the appropriation of Shakespeare's play nevertheless reveals the fiction of this isolation.

The final sections of the collection address different facets of *Macbeth*'s allusive force in music, film, and drama. Although we end with specific allusions in black poetry and theatre to analyze the often combative relationship black artists have with Shakespeare (see the Kolin, Erickson, and Gainey-O'Toole/Alexander essays in Section Seven), Section Five demonstrates the ways American musical renditions of *Macbeth* appropriate the tragedy within dialogues about race and performance. While non-traditional casting in opera still inspires critical controversy, Verdi's *Macbeth* provides a unique exception, with a surprisingly long history of black principals as Macbeth and Lady Macbeth. Yet one must ask how and why this exception comes about, and what racial stereotypes enable this non-traditional casting (see the Cheatham essay). Likewise in musical adaptations, dialogues about race are often coded into the analysis. For example, Duke Ellington's turn to Shakespeare in his 1957 jazz suite *Such Sweet Thunder* frequently invites the speculation that he was not actually familiar with Shakespeare's plays. Ellington's *Macbeth*-inspired piece, "Lady Mac," however, reveals his revisionary engagement with Shakespeare (see the Lanier essay). Tracking black popular music from jazz to hip hop, it is also important to analyze the role *Macbeth* plays in modern rap adaptations of Shakespeare's works (see the Barnes essay). In other words, appropriative gestures that unite and divide Shakespeare and modern racial constructions do not solely occur in the realms of theatre and literature, as musicians have grappled with issues of race through *Macbeth* from the middle of the twentieth century onward.

Adaptations of, and allusions to, *Macbeth* in contemporary cinema often reveal a great deal about the racialization of whiteness. Thus, Roman Polanski's 1971 rendition of *Macbeth* and Nina Menkes's 1996 *The Bloody Child: An Interior of Violence* exploit the weyward qualities of the playtext to comment on the vulnerabilities of whiteness (see the Royster and Lehmann essays). These films bring into shocking focus the many ways the twentieth-century world renders whiteness visible and racialized. And finally, interracially cast television shows and films that allude to *Macbeth* often perform a rhetoric of colorblindness that conflicts with how color-sited (and color-sighted) audiences remain (see the Scott-Douglass essay). It turns out that the plot of *Macbeth* provides a surprisingly adept frame to reveal these tensions precisely because of the strange relationship between the Macbeths.

As this collection is designed to combat the historical amnesia about this play's weyward history within dialogues about race, we have included an Appendix of non-traditionally cast productions. Although it is impossible to catalogue every performance (even if one focuses primarily on professional productions in the United States), over one hundred productions featured in the Appendix reveal how often producers, directors, actors, and reviewers imagine themselves working in a vacuum. Documenting the frequency of these productions and analyzing the adaptations, appropriations, and allusions, *Weyward Macbeth* positions the "Scottish Play" in the center of American racial constructions. Shakespeareans alone could not tell this eclectic story: we needed Americanists, filmmakers, musicians, musicologists, actors, directors, film theorists, and artists to tell a tale that signifies something about *Macbeth*, race, and the American imagination from as many viewpoints as possible.

"Note on Commercial Theatre"

When Langston Hughes penned "Note on Commercial Theatre" in 1939, he was protesting the absence of black narratives in American popular culture. As he saw it, white producers, artists, and performers were appropriating black cultural production, like the blues and spirituals, and putting them in the Hollywood Bowl, in symphonies, in musicals, and on Broadway. The problem with this cultural theft, Hughes intones, is that these appropriations white out the narratives, struggles, and triumphs of the "Black and beautiful" (line 15). Even though these stolen spirituals are promoted as expansive, they nevertheless include "everything but what's about me" (line 11). Coupled with this socio-artistic larceny, Hughes warns, comes the denial of the specificity and uniqueness of racially-informed cultural production. Therefore, Hughes places *Macbeth* in the center of his lament (line 9 of 20) precisely because he identifies the recently-produced "Voodoo" production (1936) as a forced marriage in which the black actors must mute their racial identities in order to fit into the dominantly-restricted Shakespearean narrative.

Looking back at the Welles-directed "Voodoo" *Macbeth* with over 70 years of hindsight, I find Hughes's poem both prescient *and* limited. It is prescient in the way he locates the tensions of being "put" into "*Macbeth* and *Carmen Jones* / And all kinds of *Swing Mikados*" (lines 9–10). The forced nature of these translations, revisions, and adaptations is never as seamless and invisible, neither as logically coherent nor stable, as the creative teams desire and envision. Disruptive junctions inevitably become highly visible precisely because Shakespeare was not writing about modern American racialized environments: he knew about witches, but not about "voodoo."

Yet Hughes's view is also limited in the way he envisions the cultural legacy of Welles's FTP production. The force of Hughes's employment of the verb "put" imagines that the black presence in the "Voodoo" *Macbeth* has less cultural resonance than the Shakespearean text itself. Yet it is not, in fact, *Macbeth* that is remembered, as the iconic image that graces the cover of this volume demonstrates: the word *Macbeth* may be central, but it is the dynamic presence of black Harlem in the 1930s that draws one in. Likewise, it is the fantasy of a black primitive culture that the "Voodoo" *Macbeth* constructs that subsumes the play. The drums, the dancing, the largesse of the cast and the audience (reported crowds of over 10,000 people at opening night!), and the performance of "voodoo" are what are remembered and replayed (through photographs, annotated scripts, production drawings, and one four-minute film clip) over the placement (the "put") in *Macbeth*. While Hughes is correct in thinking that he must, and will, play a crucial role in creating ("put"ting *on*) the stories, songs, and poems about a "Black and beautiful…me," many other black and beautiful, Asian and beautiful, Latino and beautiful, and Native American and beautiful artists have seen themselves in *Macbeth* precisely because it is weyward: that is, weird, fated, fateful, perverse, intractable, willful, erratic, unlicensed, fugitive, troublesome, and wayward. This is not to suggest that these visions are uncomplicated—that the translations are not somewhat forced (the "put")—but the fact that these visions so often occur within the framework of *Macbeth* is remarkable and necessarily challenges one's notions of what is—even what can be—"about me."

It is fitting that the collection ends with an epilogue about the current socio-political moment: the presidency of Barack Hussein Obama. While several of the essays allude to the era of Obama and what that might mean for the future of racial politics and performances, several others emphasize the importance of renewed federal funding for the arts. After all, we would not have the legacy of Welles's "Voodoo" *Macbeth* without the federally-funded Federal Theatre Project of the 1930s. Yet as Richard Burt makes clear in the epilogue, there is a distinct irony in the way *Macbeth* is recalled in these statements in both the popular and academic media. The historical amnesia that I describe above is evident in the way that the "Voodoo" *Macbeth* is re-visioned as solely progressive; its complications, its *weyward* qualities, are forgotten and white-washed in favor of simplistic imaginings of Shakespeare's universal applicability.

Thus the epilogue returns us to the beginning of this introduction: that is, to the weyward nature of historical transmission. As we hope to convey through the creation of our Obamicon-Shakespearicon image, progressive change never comes about through simplistic transferences, applications, or substitutions. Putting Shakespeare in Shepard Fairey's iconic Obama "Change" poster does not make him radical, nor does casting *Macbeth* with non-traditional casts/settings make the play radical (see Figure S.8). The very weyward qualities—in all of their myriad complexities—of the playtext, the performance history, and the critical scholarship must be brought into focus so that our desires for progressive acts are not disabled before they even begin. Thus we include some two dozen concise essays addressing everything from Frederick Douglass' allusions to the play, to hip-hop adaptations on YouTube, to Duke Ellington's revisionary musical rendition, to multiracial prison productions, so that we might learn what it means to remember one facet of our cultural legacy. A conscious remembering, revisioning, and restaging is the true first step to change and progress.

Notes

1. For an example of a book that draws upon the subversive powers of the "wayward," see Carter's *Wayward Girls & Wicked Women*.
2. For another critical collection that capitalizes on Shakespeare's ambivalent meaning of "weyward," see Callaghan, Helms, and Singh's *Weyward Sisters*.
3. For an example of the elision of the witches' chant, listen to the NPR piece, "Second Graders Take on *Macbeth*, Halloween," in which the "second grade Shakespeare scholars" at Washington DC's Maret School omit the racially inflected lines.
4. "Let not light see my black and deep desires" (1.4.51); "My hands are of your colour; but I shame / To wear a heart so white" (2.2.62–63); "black Hecate's summons" (3.2.42); "night's black agents" (3.2.54); "you secret, black, and midnight hags" (4.1.64); "black Macbeth / Will seem as pure as snow" (4.3.53–54); "black scruples" (4.3.117); "The devil damn thee black, thou cream-faced loon!" (5.3.11). At least one all-black production deliberately modified such lines: a reviewer of Peter Coe's 1972 *Black Macbeth* in London lamented that "One or two verbal alterations ('The Devil damn thee white') would have been better avoided" (Lambert 23). See Daileader's following essay for further critical reflections on these passages.

2

Weird Brothers: What Thomas Middleton's *The Witch* Can Tell Us about Race, Sex, and Gender in *Macbeth*

Celia R. Daileader

A rainy night in London, 1616. A writer in his mid-thirties sits down to file his quill pen. He has collar-length brown ringlets, a neat goatee with a few silver hairs, and a face that can turn women's *and* men's heads. He slips the pen into the inkwell and begins to write, by the flickering oil-light and an occasional flash of lightning, on the pages of a text whose title page reads *The Tragedy of Macbeth*.

He begins by drawing a large X across the first page.

This is not a scene from the sequel to *Shakespeare in Love*: this is a scene from history. The writer with the ringlets to his collar and the neat goatee is Thomas Middleton; Shakespeare has recently died. The playwright turns a page, reads, "When shall we three meet again? / In thunder, lightning or in rain," and makes a satisfied grunt (1.1.1–2). Then he alters the stage direction "*Enter three fairies*," replacing "fairies" with "witches," and inserts in the dialogue the following three lines: "You should be women / And yet your beards forbid me to interpret / That you are so" (1.3.43–45).

For the past four hundred years, readers of the play known as "William Shakespeare's *Tragedy of Macbeth*" have taken for granted that this unusually compact, bewitching tragedy emerged unaided in its current state from the creative imagination of the so-called Bard of Avon; few if anyone dreamt that the ambiguously-gendered "weird [weyward] sisters" and their cauldron were the brainchild of an equally prolific, equally popular junior contemporary and sometime collaborator. And so it goes without saying that Orson Welles, in conceiving and directing his "Voodoo" *Macbeth*, was equally in the dark about the degree to which the spookiest, "queerest," and most spectacular elements of Shakespeare's text—the very elements that made it voodoo-able—were the products of what many Shakespearean textual scholars have since considered a rogue editor, an interloper, even a nemesis.

In this essay, I outline *Macbeth*'s complicated textual history as well as what little we can surmise about early performances, in arguing that it is to Middleton's interpolations and alterations that we owe the ambivalent, though certainly unwitting, legacy of "racialized" interpretation. Namely, I theorize the manner in which elements of music and spectacle (dance, stage machinery, and "special effects") somehow call for (in the minds of a classically-trained modern dramaturgical community) an "exotic" setting and cast. A separate but related line of inquiry concerns the degree of overlap and divergence between Shakespeare's and Middleton's racial politics, legible through their blackface roles as well as their use of an often gendered figurative blackness. What does the transformation of Holinshed's and Shakespeare's prophesying fairies to Middleton's hags to Welles's Haitian witches tell us about Shakespeare, about Middleton, and about early modern racial and gender prejudice, and how do the former prefigure the modern and postmodern "non-traditional" productions analyzed elsewhere in this collection? I cannot answer all these questions definitively, but rather I wish to illuminate, if not unravel, the race/gender double-helix that haunts the play's production history.

Who the Hell Is Hecate?

That Middleton revised *Macbeth* is no news to Shakespeare scholars. Indeed, as early as 1788, it was observed that "two songs identified in stage directions in 3.5 and 4.1 appear in Middleton's *The Witch*" (Taylor 2007, 384). By the mid-1980s the evidence for Middleton's interpolations and abridgments to the tragedy in the first Folio was widely accepted, and, in many cases, bemoaned. Lay ignorance of Middleton's hand in the play, however, rages on. Indeed, critics themselves habitually attribute the play to Shakespeare—*even when they know Middleton adapted it*. By the same token, the play's very popularity in theatres is in part due to its co-author: Middleton probably cut a good "one quarter" of Shakespeare's original text, streamlining the action in a way that makes less work for directors (Taylor 2007, 397). And, most importantly for the project of this collection, Middleton provided spectacular scenes involving supernatural entities, complete with music and stage machinery. For example, the bewhiskered hags that are now standard Halloween fare were Middleton's re-imagining of the "three women fairies or nymphs" who prophesize to Macbeth in the lost original.[1] Also, Middleton inserted the character of Hecate—who presides over the witches—along with a talking cat.

Hecate and her cat seldom find their way into modern productions. Imagined descending from the ceiling of the Blackfriars Theatre, these characters and the accompanying stagecraft are now considered over-the-top. Interestingly, though, the weird sisters and their cauldron—also Middletonian interpolations—have become a theatrical staple. Inga-Stina Ewbank's introduction to new Oxford edition states that "Twentieth-century distaste for the Hecate scenes is part of a more general purism, the theatrical equivalent of the scholarly search for authenticity, for the 'original' play" (1165). Such "purism," however, takes on a different shade of meaning when we consider *Macbeth*'s amenability to exoticized settings and an interracial cast. In such cases, Shakespearean "purism" might be doffed or suspended—though at the risk of disappointing the purists in the audience.

It is not my project to recuperate those non-traditional performances that might have preserved Hecate's speech in 3.5 or other extraneous and spectacular Middletonian material. My hunch, however, is that Hecate's erasure from *Macbeth*'s modern production history is not just a measure of the distaste for generic mixture, musical interludes, and/or talking animals on stage (for heaven's sake, look at the longevity of *Cats*). For the Hecate dances are weirdly subversive—even feminist—in a way three witches (Haitian or not) hunched over a cauldron most decidedly are *not*. But to appreciate that, we have to look at Middleton's *other* witch play, his other *Macbeth*.

Fairies and Witches and Blacks—Oh My!

Evidence of Middleton's co-authorship of *Macbeth* can be found in the following verses:

> HECATE: Black spirits and white, red spirits and grey,
> Mingle, mingle, mingle, you that mingle may.
> Titty, Tiffin, keep it stiff in;
> Firedrake, Puckey, make it lucky;
> Liard, Robin, you must bob in.
> ALL: Round, around, around, about, about,
> All ill come running in, all good keep out.
> (4.1.44–50)

The song occurs twice in *The Witch*, with minor variations, neither far from the wording of their source in Reginald Scot's *The Discoverie of Witchcraft* (1584): "*White* spirits, *black* spirits, *grey* spirits and *red* spirits. / Devil-toad, devil-ram, devil-cat, and devil-dam" (*The Witch*, 1.2.3–4; my italics). A different fairy rainbow appears in Shakespeare's *Merry Wives of Windsor* (1597): Falstaff is accosted by "Fairies black, grey, green, and white" (5.5.34). How would such color-coding have been presented on stage? Clearly, colored leotards and "vizards" or masks would be crucial, but could the actors also have worn face paint? If this were the case, we are looking at blackface—and white-face, red-face, gray-face, and green-face.

In fact, *The Witch* may have more in common with *Merry Wives* than *Macbeth*—at least in tone. For the witches in *The Witch* are strangely benign. In contrast to the weird sisters, their activities have no effect on the plot. In her introduction to the Oxford edition of *The Witch*, Marion O'Connor observes, "The courtiers who consult [Hecate] are neither her victims nor her converts but rather simply her clients. That they are neither harmed nor corrupted by their encounters with witchcraft is a flat violation of the rules of demonology as articulated in Elizabethan and Jacobean treatises on the subject" (1126). At the same time, the stage business was to be sensationally grotesque, calling for the skinning of live snakes and the flourishing of a real baby slated for sacrifice. When the play was suppressed, it is no wonder Middleton and the King's Men were eager to rescue the Hecate material by transposing it to *Macbeth*.[2] Ewbank details the effects of the interpolation on the pre-existing plot structure: "What the introduction of the song and dance in 4.1 achieves...is to make the scene less focused on the moral destruction of Macbeth and to shift the emphasis on to the witches as being in command, free and unbounded" (1168). The witches represent a community of women entirely outside—even, in the stagecraft for 3.5, literally *above*—monarchical control, emphasizing more starkly Macbeth's isolation and increasing impotence. With spirits singing, "Come away, come away" (3.5.34), Hecate is hoisted into the heavens (together with her cat), boasting that she is beyond the reach of "cannons' throat"—the phallic cannon here metonymizing male political authority at its most menacing (3.5.69). Ewbank rightly calls this moment "subversive in an un-Shakespearean way in so far as it enacts an alternative not of benevolent monarchy but of positive, joyful anarchy" (1168). The subversion is even more apparent in the overlap with *Merry Wives*, where the victim of the fairies' foolery—and the women's plotting—is a bumbling, fat lecher, rather than a noble over-reacher.

Though Macbeth calls the witches "secret, black and midnight hags" (4.1.64), neither they nor "black Hecate" (3.2.42) seem to have blacked up. On the one hand, this is not surprising given the moral blackness of the human characters, which I address below. On the other hand, the Renaissance commonplace about the Devil's blackness would seem to require it. Further tantalizing hints as to early performances come from the witches' dance in the anti-masque to Ben Jonson's *Masque of Queens* (1609), which Middleton would not have witnessed in performance, but might have read. Jonson imagined the dance as a "spectacle of strangeness," involving "twelve women in the habit of witches or hags," emerging from and then disappearing into a flaming and smoking onstage hell (321). Interestingly, that number corresponds to the number of masquers in another "spectacle of strangeness"—Jonson's *Masque of Blackness* (1605), which titillated and offended viewers for presenting the Queen and her attendants in blackface. I note only briefly the curious fact that *Blackness* describes these figures as "twelve nymphs, Negroes, and the daughters of Niger" (315): that alliterative linkage not only seems meant to startle—as did, of course, the entire spectacle—but also seems meant to invite us, however briefly, to rethink our assumptions about "nymph-ness" and skin color. The choreography of the witch-dance, however, bore little resemblance to that elegant—if somewhat scandalous—first of Jonson's court entertainments: the witches in the *Masque of Queens* danced "back to back, hip to hip, their hands joined, and making their circles backward to the left hand, with strange fantastic motions of their heads and bodies"—all consistent with the preposterous reversals of the black mass (330). Jonson's text refers to Hecate but does not include her among the dramatis personae; however, Jonson does provide the hags their presiding

"Dame," who enters "naked-armed, barefooted, her frock tucked, her hair knotted and folded with vipers; in her hand a torch made of a dead man's arm, lighted, girded with a snake" (323). These Jonsonian witches, of course, are sent back to hell once they have had their fun, and—as Ewbank points out—they reappear at the end of the masque only in order to be bound and driven like so many cattle before the queens' chariots (1168).

Did Middleton transform Shakespeare's fairies to witches in order to indulge in similar pyrotechnics? Perhaps. Yet rendering them gender ambiguous also opens up possibilities for casting. Gary Taylor notes that "If the three witches have beards, they can be played by bearded adult actors," and this change would "transmit the acting burden to the men," thereby freeing "the company's boys to handle the interpolated material in 3.5 and 4.1" (2007, 389). Since most of this material involved singing, the boys would be particularly useful. Clearly, boys played the witches (or fairies) in 1606, but by 1611 "those boys would either have become adult actors (in which case they could keep the parts they already knew), or stopped acting altogether (in which case someone else would have to learn the lines, and it might as well be a man as a boy)" (Taylor 2007, 389). Above all, though, Middleton's transformation of Shakespeare's prophesizing women to spectacular supernatural beings of indeterminate gender—along with his reluctance to send them down the trapdoor into a below-stage "hell" consistent with Jonson's—reflects his attraction to their anarchic power.

If she was not darkened cosmetically, how was "black Hecate" embodied, as queen of this coven? Did she wear a snake headdress, like Jonson's presiding witch? She does handle a pair of snakes in *The Witch*. I find it hard to imagine a snake-haired Medusa figure being hoisted aloft in *Macbeth* without completing freaking out the audience—but maybe that was the point. Whatever her costume, Hecate's flight flies in the face of early modern notions of spatial, cosmological, and religious hierarchy, as does the stage direction "*a spirit like a cat descends*." Contemporary discussions of Hecate describe the goddess as tripartite due to her association with Diana and Persephone, with the moon, earth, and the underworld: some icons grant her three faces. Perhaps this, too, could have been pulled off in performance.

Much good scholarship has been generated by the few instances of blackface in Shakespeare (see Vaughan). As yet, however, no one has undertaken an exhaustive study of Middleton's black characters. He was, as Taylor duly notes, the first English author to positively represent an African king and queen, giving them lengthy and sympathetic speeches in his civic pageant, *The Triumphs of Truth* (1613) (Taylor 2005, 123–32). Similarly, *The Triumphs of Honour and Virtue* (1622) presents "a black personage representing India, called, for her odours and riches, the Queen of Merchandise, challenging the most eminent seat...upon a bed of spices" (*THV* 42–45). The ideological frames of these pro-blackness speeches seem distinctly Middletonian. "This black is but my native dye; /.../ I'm beauteous in my blackness," the Indian Queen boasts—the sartorial pun on "dye" highlighting, along with the preposition, a logic clearly distinct from the apologetic rhetoric of The Song of Songs ("I am black *but* beautiful," in most translations) (*THV* 54–58). That prepositional logic is reiterated when the character of Honour, mounted atop of an enormous globe, flatters the Lord Mayor with the words, "the more white / Sits Honour *on* thee" (emphasis added; *THV* 300–01).

Chronologically sandwiched between these two overtly racialized episodes of blackness we find the *Masque of Heroes* (1619), which contains a startlingly anti-essentialist use of blackface in the ante-masque: "The three Good Days, attired all in white garments...the three Bad Days all in black garments, *their faces black*...the Indifferent Days in garments half white, half black, *their faces seamed with that parti-color*..." (*MH* 266sd; my italics). The actors' faces, fascinatingly, are "seamed" like garments, with the radical implication that moral binaries can likewise be turned inside out. Here I am reminded of a minstrel version of *Othello* that stages miscegenation by bringing out a small child so painted (Lhamon 2003, 383).

Comparison of Middleton's racial politics in his masques and pageants to those of his contemporaries fairly consistently highlights his resistance to the usual essentialisms and a surprising degree of sympathy. For instance, Anthony Munday's 1609 show for the Ironmongers presented a "Black More"—sensationally—in the mouth of a whale. John Squire's *Triumphs of Peace* (1620) featured allegorical figures for America, Africa, and Asia, who crown Europe as "Empress of the earth"—an act, of course, endorsing their own subservience (Loomba 2007, 1714).

Masques and pageants are by definition political, so it is in these genres that one expects the most xenophobic and nationalistic representations of England's others. But, in fact, it is in Middleton's plays that dark-complected figures are portrayed in a negative light. Most apparently ideological is his explosive *A Game at Chess*, wherein the Black House of Catholic Spain represents unmitigated corruption, greed, and debauchery. Indeed, the Spanish Ambassador took issue with the play specifically for styling King Philip of Spain "*el rey de los negros*" ("King of Negros") (Coloma 866; see Taylor 2005, 132–39). This quote, however, says as much about the Spanish Ambassador's racism as it says about racial blackness in Middleton's play: in fact, the word "negro" never appears in *Game*, and, apart from one remark about a character's "complexion," blackness is entirely sartorial and/or moral. Differing drastically in tone, *More Dissemblers besides Women*—set, like *The Witch*, in Italy—incorporates a comic gypsy sequence, wherein actors in blackface (including one female character meta-theatrically "disguised" as a gypsy) boast of their trickeries in ways that recall the canting scenes in *Roaring Girl*, as well as, more pertinently, the witch scenes in *Macbeth* and *The Witch*. Indeed, as the comparison underscores, Middleton deploys racist stereotypes in much the same way he deploys demonology—for comic and theatrical effect. Moreover, these two plays reveal a departure from a geopolitical model which locates blackness outside of Europe. This is distinct from Shakespeare's highly essentialized, negative representations of blacks in his tragedies. So much for *literal* blackness in Shakespearean/Middletonian canon—what of those tropes of blackness that tempt directors to cast black actors in "white" roles (see Daileader 2000)?

THE RHETORIC OF BLACKNESS IN
SHAKESPEARE VERSUS MIDDLETON

Though relatively few productions have chosen to cast a black actor as Macbeth (or Lady Macbeth), there would be a textual justification for doing so: the word "black" occurs as often in *Macbeth* as in *Othello*, and the moniker "black Macbeth" echoes "black Othello" (2.3.28–29). Moreover, the rhetoric of blackness in *Macbeth* might have eclipsed that of *Othello* had the former not been so drastically cut. Of the eight occurrences of the term "black," Middleton might be responsible only for "you secret, black and midnight hags"; he was definitely responsible for the "black spirits" of the song inspired by Scot.[3] The other references to blackness, placed all together, seem distinctly Shakespearean:

> MACBETH: Let not light see my black and deep desires (1.4.51)
> MACBETH: Ere the bat hath flown
> His cloistered flight, ere to black Hecate's summons (3.2.41–42)
> MACBETH: Whiles night's black agents to their preys do rouse (3.2.54)
> MALCOLM: It is myself I mean, in whom I know
> All the particulars of vice so grafted
> That when they shall be opened black Macbeth
> Will seem as pure as snow (4.3.51–54)
> MALCOLM: Macduff, this noble passion,
> Child of integrity, hath from my soul
> Wiped the black scruples (4.3.115–117)
> MACBETH: The devil damn thee black, thou cream-faced loon (5.3.11)

Comparable to the last quote is Macbeth's reference to "pale-hearted fear" (4.1.101), along with Lady Macbeth's tropes of mother's milk and her scorning "To wear a heart so white" (2.2.63), a devaluation of whiteness as bloodless effeminacy or cowardice that echoes Aaron's defense of his blackness in *Titus Andronicus* and Morocco's in *The Merchant of Venice*. These moments enrich the play's color imagery and theoretically pave the way for a racialized reading—and hence, an interracial cast.

How does the rhetoric of blackness in *The Witch* compare? The "black spirits" in the two songs make up two of only five references, and all but one (the song's repetition in 5.2) occur in the same scene. Boasting of her interlude as a succubus with "the Mayor of Whelpley's son," Hecate says, "I knew him by his black cloak, lined with yellow" (1.2.31). This is probably topical, referring to the execution of Ann Turner—one of the many allusions to the Howard/Carr scandal that may have caused the play to be repressed (both she and the executioner wore black and yellow). A few lines later, Hecate describes the sacrificial baby as "black i'th' mouth" from ashes and magical herbs he has been forced to ingest (a phrase Middleton elsewhere uses to decry smoking) (39). Finally, one of the mortals seeking Hecate's magical aid praises her "black power" (125). When contrasted to the rhetoric of blackness in *Macbeth* (and, for that matter, *Titus Andronicus*, *Othello*, *Romeo and*

Juliet, and the Sonnets), what emerges is the distinctly un-Shakespearean feel of *Witch*'s negro-linguistics.

Tropes of blackness seem to work differently for Middleton than for Shakespeare, particularly with regard to gender and sexuality. Concordances to the works of both authors reveal comparable frequency of usage. However, more than a third of Middleton's references occur in *A Game at Chess* and correlate to the titles of the chessmen/characters and the moral qualities allegorized therein; Shakespeare's rhetorical blackness is more evenly spread throughout his works. Although both authors frequently invoke an allegorized "black night," Shakespeare almost always genders the figure female, as in Juliet's famous invocation of "Thou sober-suited matron all in black" (3.2.11). Second and third in the density of blackness references in Middleton we find *The Black Book*—a satire narrated by Lucifer—and his moral tract *The Wisdom of Solomon*; in each, sartorial and figurative/demonic blackness feature prominently. Only one of these thirty-five entries under "black" associates the term with a female entity (that is "black Megaera's power" in *Solomon* 6.214). Entirely absent from Middleton is the usage of "black" to denigrate a woman's appearance (often, also, implying unchastity), as in *Othello*'s quibbling about women who are "black and witty" (2.1.134) and in the protracted discussion of a "black" mistress in *Love's Labour's Lost* and the Sonnets. Along the same lines, it is Shakespeare, not Middleton, who invokes a figuratively Ethiopian/blackamoor undesirable woman (as in *Two Gentlemen of Verona* 2.6.26, *Much Ado* 5.4.38, *The Passionate Pilgrim* 16.16, and *Troilus and Cressida* 1.1.74). Three times in the Middleton canon a racial term is used figuratively to insult someone's appearance, but all three instances can be attributed to a collaborator.[4] Middleton uses racial markers literally, not figuratively; nor is he unwilling to acknowledge the attraction of literal, racial blackness, as when he cites the proverb "A black man's a pearl in a fair lady's eye" (*Your Five Gallants* 5.1.176).

By pointing out that Middleton less frequently associates blackness with women and sexuality I am not arguing, obviously, that Shakespearean blackness—figurative or racial—is always female. Such a claim would be absurd given the 1/3 ratio of female to male black speaking characters in his plays (that is counting Cleopatra, whom I take at her word when she calls herself "black")—not to mention his five references to Edward the "black Prince" in the history plays. It can be argued, however, that when Shakespeare invokes blackness there is *almost always* a feminine point of reference, be it primary or secondary. Think, for example, of Shakespeare's use of black men as objective correlatives to a woman's interior moral blackness in *Titus Andronicus* and *Othello*: for Tamora's "honour" is of "[Aaron's] body's hue" (2.3.73), and Othello and Desdemona become virtually indistinguishable—morally if not complexionally—by the end of their tragedy (see Daileader 2005). It is also very frequently the case that a man's moral blackness or physical ugliness can be blamed on a woman. Indeed, might not all or most of Macbeth's evil be attributed to his wife's sexual blackmail of him? And what does it mean that Macbeth can only be defeated by a man *not* "of woman born"? Does this not also implicate "black" Macbeth's mother in his downfall?

Finally, a close reading of *sexual* blackness in the two authors highlights Shakespeare's antifeminist bias. This is nowhere more evident than in the contrast between Shakespeare's *The Rape of Lucrece* and Middleton's *Ghost of Lucrece*. Rhetorical blackness features prominently in the former. Of these ten references, though, only three refer to the rapist—and only indirectly, through the act which is described as a "black...deed" (226), as "black payment" to—not the victim herself—but her husband (576), and as resulting from "black lust" (654). Notably, all three characterizations of the rape as "black" occur before that act takes place: the seven subsequent references to blackness target the feminine entity of night, whom Lucrece initially blames for the rape, *and* target Lucrece's own body, wherein the "blood [is] chang'd to black in every vein" (1454). In contrast, Middleton's Lucrece poem never once genders blackness or describes the victim as internally blackened by the rape: rather, she speaks of "black spirits" (44) and "black appetites" (242) that will perform the "black story" (280) of her tragedy on the "stage" (241) of her heart. Middleton's poem, moreover, inverts the Shakespearean image of the victim's blackened blood, presenting her ghost as a "bloody crystal" (12). Indeed, if anyone in this poem is black, it is Tarquin, six times called a "night owl" (206; see Daileader 2008).

What emerges, thus, from an examination of the works of both authors is evidence that *Macbeth*'s demonized—and sexualized—rhetorical darkness is a distinctly Shakespearean feature, while the play's potential for vivid theatricality, its elements of pageantry and carnival, seem owing to Middleton. From here it is a relatively short step to those "voodoo" productions and—given the misogyny implied in *Macbeth*'s rhetoric of blackness—to reviews of Angela Basset's 1998 performance of Lady Macbeth that harp on the black actress's sexuality (Brantley 1998).

Middleton's calling *The Witch* "ignorantly ill-fated" seems a likely indication that it was suppressed for alluding to the Howard-Carr scandal. Then again, the play's praise of "black power" might have made it equally unpalatable.

Conclusion

A rainy night in London, 2012. A thirty-something director with a neat black goatee sits at his computer. On his screen glows the following message:

> *Damn good question. How to make it fresh. Here's an idea. Let's set the play in Haiti, throw in some bongo drums, cast a woman of color as Lady Macbeth and/or a witch. Hell, have all three witches be black. Just no talking cats, you hear? No flying witches and no talking cats. You want color, sure: throw in some black actors. The blacker the better. But no talking cats.*

After all, this is Shakespeare.

Notes

1. The phrase is from Simon Forman's account of a Globe performance on April 20, 1611 (quoted in Ewbank 1166).
2. Middleton's epistle explains that the text was "recovered into [his] hands (though not without much difficulty)" and calls the play "this (ignorantly ill-fated) labour of mine" (Epistle 3–5).
3. "Hags" is definitely Middleton; Shakespeare might have written the rest of the line, but he did not imagine the weird sisters as witches.
4. In a scene of *A Fair Quarrel* authored by Rowley, a woman rebuffs a man with "Away, you're a blackamoor" (3.2.100). Likewise, it is Thomas Dekker who penned the line "Be he a Moor, a Tartar, though his face / Look uglier than a dead man's skull...up then he gets" (*The Patient Man and the Honest Whore*, 6.391–92) as well as "I wash a negro" (*The Roaring Girl* [2.185]).

2

Early American Intersections

Figure S.2 "A Proslavery Incantation Scene, or Shakespeare Improved." David Claypool Johnston, 1856. Courtesy, American Antiquarian Society.

3

"Blood Will Have Blood": Violence, Slavery, and *Macbeth* in the Antebellum American Imagination

Heather S. Nathans

> Pour in the streams of Freemen's blood:
> Then the charm is firm and good.
> To see the blood of Freemen spill
> To hear the widows' cries so shrill
> To know my Kansas-Nebraska bill,
> Has caus'd these woes; to me is joy,
> Here and at home in Illinois.

This doggerel graces a bitter cartoon from 1856 entitled "A Proslavery Incantation Scene, or Shakespeare Improved" (see Figure S.2).[1] The image shows the "witches" of the proslavery party clustered around a bubbling cauldron whose heat is fed by the flames of burning abolitionist newspapers. The "witches," Stephen A. Douglas, James Buchanan, and Preston Brooks (among others), declaim spells against the nation's antislavery party. Slave manacles and a whip adorn Douglas, while Brooks clutches his famous cane—the one with which he attacked and nearly killed abolitionist, Charles Sumner. An earlier image from an 1850 *Harper's Weekly* entitled "The Hurly Burly Pot" features David Wilmot (creator of the infamous Wilmot Proviso), journalist Horace Greeley, abolitionist William Lloyd Garrison, and proslavery politician John C. Calhoun cavorting around a boiling cauldron, disguised as witches (see Figure 3.1).

Each is tossing an element of discord into the pot, and each recites his own version of the "toil and trouble" charm. This *Harper's Weekly* image indicts both Southern *and* Northern politicians and activists for the controversy over slavery. The effectiveness of these cartoons relies on two things: the audience's understanding of the contemporary political debates invoked, and the audience's familiarity with Shakespeare's *Macbeth*.

The play was ubiquitous in antebellum American culture. *Macbeth* was among the most popular Shakespeare plays on the national stage, performed

Figure 3.1 "The Hurly-Burly Pot." *Harper's Weekly*, 1850. Library of Congress, Prints and Photographs Division.

by major British and major American stars, from Edmund Kean to Edwin Forrest (Grimsted 249–61). While it was frequently adapted, cut, or excerpted, it appeared in theatres from South Carolina to Kentucky, from Ohio to Massachusetts. Lines from the play were cited in schoolbooks, newspapers, and women's periodicals.[2] Thus, many nineteenth-century Americans would likely have recognized the allusions to *Macbeth* in the cartoons described above—at the very least they would have caught the parody of the lines, "Double, double, toil and trouble," which appear in both cartoons. However, familiarity does not necessarily equal relevance, and the choice of the play as a vehicle for political satire on slavery raises intriguing questions about antebellum Americans' understanding of the cultural resonance of *Macbeth*. If *Othello* had become a referendum on black and white amalgamation, if *Julius Caesar* had morphed into a lesson on republican virtue, how did audiences interpret a play in which the main character murders his rightful ruler, installs himself on the throne, and leads his country into chaos and civil war?

In this essay, I examine the multiple meanings Shakespeare's *Macbeth* had for both white and black audiences on the eve of the Civil War. I examine it in performance and also as its images migrated through American culture. For some, the play warned against the dangers of overweening ambition, for others it offered a cautionary tale against internal rebellion. Some spectators imagined Macbeth as a democratic hero, rising up against a tyrant (like the Southern Confederacy defending its states' rights against Northern oppressors), while others conjured parallels between the "un-sexed" Lady Macbeth and the female antislavery agitators. The repeated image of blood in the play awakened memories of Nat Turner, Harpers Ferry, and the ongoing violence of slavery. Perhaps more than any other Shakespearean text, *Macbeth* fit the mood of the troubled nation on the eve of war.

"Streams of Freemen's blood"

In a letter to *The Liberator* written from Macbeth's Castle on December 26, 1845, abolitionist Henry C. Wright told fellow antislavery activists that "all the scenes of that tragedy [*Macbeth*] are before me and around me." The haunted hills recalled to him "the...dagger, the witches, and all the scenes and characters." He added, "This region is rich, but most painfully interest(ed) in events of the past." For Wright, neither the past nor the future seemed as urgent as the *present* crisis facing the nation—the crisis in which men continued to see their destiny as one tied to slavery and the slave system. Wright's language—with its focus on present action, its determination to shake off the yoke of the past—echoes the rhetoric of the immediatists in the American antislavery movement. At the close of his letter, Wright implored, "May the God of nations speed the car of revolution!" (*The Liberator*, February 6, 1846).

Why did Wright choose *Macbeth* as a metaphor for his message of immediatism and rebellion? Perhaps the choice lies in the contradiction the play presents between destiny and individual agency. Wright sees in the American people the power to resist the siren song of slavery. Just as Macbeth might have resisted the witches' temptation, or what Wright terms "the hell of an unjust, hating, revengeful heart, and a polluted life," so might his fellow Americans resist the hell of the slave system (*The Liberator*, February 6, 1846). Without the willpower to break free of slavery, he fears they will be led inevitably to destruction.

The shift from gradualism to immediatism had come fifteen years before Wright's impassioned plea. In 1831, William Lloyd Garrison launched *The Liberator* in which he proclaimed: "I will be as harsh as truth, and as uncompromising as justice....I am in earnest—I will not equivocate—I will not excuse—I will not retreat a single inch—AND I WILL BE HEARD. The apathy of the people is enough to make every statue leap from its pedestal, and to hasten the resurrection of the dead" (*The Liberator*, January 1, 1831). This image of the dead haunting the living, of the past shackling the present and future, is the fear that seems to have linked *Macbeth* to slavery in the minds of many abolitionists. Indeed, the character of Banquo drifted through antislavery rhetoric for more than two decades. For example, in 1834 *The Liberator* printed a letter from an activist who conjured Banquo's ghost as a symbol of the "souls and bodies" of the "two hundred thousand native Africans annually" who had been sacrificed to slavery (*The Liberator*, March 8, 1834). The author imagined these men haunted by the ghosts of their misdeeds, crying aloud like Macbeth at the feast. But, as the writer warns the guilty nation, "murder will out...America has been in this too-long protracted tragedy, a conspicuous and guilty participator. It is high time for her to awake out of sleep and wash her bloody hands" (*The Liberator*, March 8, 1834).

Though Northern abolitionists frequently invoked *Macbeth* as an indictment against Southern slaveholders, some Southern authors tried to set Banquo's ghost on their Northern counterparts by underscoring the North's complicity in the slave system. In the throes of the Civil War, angry Georgia slaveholder Ebenezer W. Warren wrote *Nellie Norton, or Southern Slavery and the Bible*, accusing Northerners of starting the war to assuage their own guilt over slavery:

> The "evils of this horrible system" have as much real existence as the ghost of Banquo that gave such alarm to Macbeth. The latter had murdered the former, and he imagined he saw his ghost, and was greatly alarmed. You of the North first introduced slavery and the slave trade: when you found it unprofitable you sold them into Southern slavery, into his "horrible system." Do your consciences torture you for it? Do you see the ghost? Does it harass your minds and disturb your dreams? (Warren)

This passage aligns the South with the innocent observers at Macbeth's feast, reversing more than half a century of rhetorical gambits, which persistently cast the guilty South as the murderous Scottish king who plunged his nation into war. Though the North and South obviously disputed where the guilt for slavery lay, neither could deny that its bloody consequences would continue to haunt the nation for decades, if not centuries, to come.

ANTISLAVERY HERO OR CONFEDERATE CHAMPION: MACBETH AS THE MAN OF ACTION?

Eyes flash. Muscles bulge. Hands clutch a gleaming weapon. Legs brace for attack. These phrases describe three iconic images of violence in antebellum American culture rendered in paintings, pamphlets, engravings, and cartoons: slave rebel Nat Turner, freedom fighter John Brown, and theatrical luminary Edwin Forrest in the role of Macbeth. The parallels among these images are simultaneously striking yet puzzling. Could Forrest, the incarnation of nineteenth-century heroic masculinity, play a villain? Could Turner and Brown, bywords for bloody rebellion among the Anglo-American community, appear as heroes?

As the captain describes him in the play, Macbeth becomes a new example of passionate masculinity in nineteenth-century performance culture:

> For brave Macbeth—well he deserves that name!—
> Disdaining fortune, with his brandished steel,
> Which smoked with bloody execution,
> Like valour's minion
> Carved out his passage till he faced the slave,

> Which ne'er shook hands nor bade farewell to him
> Till he unseamed him from the nave to th' chops,
> And fixed his head upon our battlements.
> (1.2.16–23)

Noah Ludlow, a performer who toured the States throughout his career and who observed some of the nation's most prominent performers, wrote extensively about Forrest's effect on American audiences. In his *Dramatic Life as I Found It*, Ludlow cites a review that described Forrest's Macbeth as, "the ferocious chief of a barbarous tribe...the courtly guise, the old world conventionalisms, which 'hedge in the divinity of kings,' and the polished graces that surround the great and high-born are not held by Mr. Forrest as the imperative auxiliaries of his acting" (690–91).

This image of "brave Macbeth" resonated among both black and white communities, and within both pro- and antislavery factions. Each group took up the character, refashioning his motives and his deeds to suit its political purposes. Yet the play proved an imperfect and troublesome vehicle for these multiple agendas. For example, in his *Oratory: A Unique and Masterly Exposition of the Fundamental Principles of True Oratory*, Henry Ward Beecher, abolitionist, preacher, and orator *par excellence* abjured young speakers to infuse *action* into their words, proclaiming, "Let the man act out his theme.... Potency links itself with personality." He describes the "resolute stride of a defiant Macbeth," and "the wild despair of the blood-stained Lady Macbeth" as part of the orator's repertoire of gesture, look, and tone. Beecher's passionate exhortations for would-be speakers to embody violent emotions and to use their powers to sway audiences to frenzied response, certainly underscores the shift in American rhetoric and American politics in the antebellum period. Gone were the restrained models of classical republicanism, and in their place were heroes whose "eyes and face will tell the scathing flash of hate or pity's melting mood" (74–79).

The flaw in this comparison comes in the call to "act out" the theme. Beecher certainly does not advocate that his pupils emulate Macbeth's deeds, but rather that they expand their emotional and psychological vocabulary to encompass violence as well as pacifism. Yet while Beecher argues for more persuasive and performative rhetorical language, he falls into the very trap that Macbeth did on the eve of Duncan's murder—the gap between words and deeds. This same dilemma appears in *Macbeth*, embedded in the lines, "If it were done when 'tis done, then 'twere well / It were done quickly" (1.7.1–2), or in Lady Macbeth's rebuke to her husband, "Art thou afeard / To be the same in thine own act and valour / As thou art in desire?" (1.7.39–41). It is this "gap" that creates problems in assigning the role of Macbeth to antislavery activists such as Nat Turner and John Brown. Despite their obvious "potency" and "personality" (to use Beecher's terms), Turner and Brown were instead consigned to the fantastical and feminized roles in Shakespeare's drama. By linking Turner with Lady Macbeth and Brown with the Weird (or "weyward") Sisters, anxious observers could disassociate Shakespeare's/

Forrest's masculine incarnation of Macbeth from the abolitionist drama taking place outside the playhouse. Excising the masculine hero (or Forrest's "ferocious chief") at the center of the antislavery movement and substituting the "unnatural" figures of Lady Macbeth or the witches rendered the uprisings more comprehensible and ultimately less immediately menacing.

Though no explicit comparisons seem to have been made between Nat Turner and Macbeth at the time of his 1831 rebellion, the *manner* of his strike against his white oppressors must certainly have raised echoes of Duncan's murder in the popular imagination. Turner, guided by "voices" (like Macbeth), launched his attack in Southampton, Virginia on August 21, 1831. Horrified accounts of his uprising flooded national newspapers. Reports describe how he and his followers crept into the rooms of sleeping white men and women and stabbed, mutilated, or beheaded them. Descriptions of the violence, of "little helpless lisping infants," who "had their brains dashed out against the house side and fences" (*The Christian Register*, October 1, 1831), recall Lady Macbeth's ruthless vow:

> I have given suck, and know
> How tender 'tis to love the babe that milks me.
> I would, while it was smiling in my face,
> Have plucked my nipple from his boneless gums
> And dashed the brains out
> (1.7.54–58)

Even the account of Turner in prison echoes the images of Lady Macbeth after madness had overtaken her in the wake of the murders. Turner's "confessor" wrote, "I shall not attempt to describe the effect of his narrative as told and commented upon by himself...the calm, deliberate composure with which he spoke of his late deeds and intentions, the expression of his fiend-like face when excited by enthusiasm, still bearing the stains of the blood of helpless innocence about him" (*The Liberator*, August 8, 1861). The language recalls the sleepwalking scene in which Lady Macbeth re-lives the murders, sees the blood of Duncan, and muses, "Here's the smell of the blood still" (5.1.42). It also conjures the horror of the Doctor watching her as he cries, "My mind she has mated, and amazed my sight. / I think, but dare not speak" (5.1.68–69). Yet Turner could not be imagined (at least publicly) as a hero—his act of rebellion had been too gruesome and overt for a nation still making the rhetorical, moral, and intellectual shift toward immediatism. While many African Americans may have privately rejoiced at Turner's bloody action, his rebellion became a synonym for unlawful murder, rather than righteous rebellion. Thus, he had to be imagined more in the character of the unnatural Lady Macbeth, rather than her husband.

John Brown, another icon of abolitionist violence, also invited comparisons to *Macbeth*. On October 16, 1859, Brown launched a raid on the arsenal and armory in Harpers Ferry, West Virginia. His capture of the arsenal was to be his first step in a general uprising and war against slavery. His rebellion failed, and he was captured, tried, and executed. His name, like those of earlier rebel leaders, became entwined with the violence of slavery. Yet contemporary poets and artists did not link Brown to the character of Macbeth, but rather to that of the *witches*. As Herman Melville wrote in his 1859 poem, "The Portent":

> Hidden in the cap
> Is the anguish none can draw;
> So your future veils its face,
> Shenandoah!
> But the streaming beard is shown
> (Weird John Brown),
> The meteor of the war.

In Melville's poem, "Weird John Brown" becomes, like the Weird Sisters in *Macbeth*, the portent of doom for the nation. Melville's invocation of *Macbeth* is infinitely more subtle than many of the examples cited thus far, and yet it also suits Melville's tone of prophecy. Few could doubt that Brown's bloody uprising and his execution (which many viewed as martyrdom) would hasten the impending battle that threatened to engulf the nation. Yet, at the same time, Brown could not be fitted to a model of heroic masculinity (such as Forrest embodied in his performance of Macbeth). His detractors viewed him as a bloody madman, while his supporters explained his actions as a combination of his passionate sympathy for his black brethren and his desperate sorrow over the death of his son.

The "Whitening" of Macbeth

While abolitionists and proslavery speakers cite *Macbeth* as a metaphor for political violence, the Astor Place Riot of 1849 obviously offers the most visible incident of the play being used as a testing ground for real-life political activism. The furor erupted on May, 10, 1849 at New York's Astor Place Opera House where British star William Macready was appearing in the title role of *Macbeth*. Edwin Forrest, Macready's long-time rival, had publicly expressed his disapproval of the Opera House's choice to bring in a "foreign" star to play Macbeth. Forrest's followers took up his cause and threatened action against the house. While Macready offered to withdraw, the city's elite encouraged him to stand his ground. On the night of May 10, a crowd gathered outside the theatre to protest. The city's police force and the National Guard, warned to expect trouble, gathered to disperse the crowd. Violence erupted and by the end of the evening, more than 100 Americans lay dead or injured in what would become known as the worst theatre riot in American history.

Theatre scholars and cultural historians generally trace the riot's origins to an admixture of class tension and anti-British sentiment ("Shakespeare in American Life: The Astor Place Riot"). Race is curiously missing from the conversation, prompting the question: Is 1849 the moment when Macbeth became "white"?

The Bowery B'hoy, the figure most often associated with the rioters outside the playhouse, and with Edwin Forrest's followers, was a character deeply embedded in the complex racial politics of antebellum New York. His native habitat—Five Points—seethed with racial and ethnic unrest. For example, George Caitlin's 1827 painting (also entitled "Five Points") shows a neighborhood completely racially integrated and consumed by encounters (alternately friendly, sexual, or violent) among blacks and whites. Blacks and whites would have had opportunities to share the same music, the same popular entertainments, and the same access to information. Theatre riots were not exclusively the province of white theatre-goers, nor were New York theatre audiences exclusively white. The fact that African Americans performed and cited *Macbeth* so frequently in speeches or newspaper articles demonstrates a certain cultural familiarity with the text. Where, then, is the black American presence in the most famous riot in the play's history?

To suggest that the Astor Place Riot was deliberately engineered to claim Macbeth as a "white" character would certainly be going too far. However, this may have been the net effect of the incident, which carefully positioned the play and "ownership" of the role in the *white* American community. The riot may also be viewed as part of the ongoing struggle to preserve the Union in the face of increasing sectionalism, and thus, perhaps, as an "acting out" of the dilemma that faces Macbeth's kingdom after Duncan's murder. Should the nation accept the tainted but powerful presence of a leader who offers stability, or should it risk civil war to pursue the cause of justice?

Forrest, as I have noted, was a Democrat, and had often campaigned or spoken on behalf of the party that still supported slavery. The year before the Astor Place Riot, New York's *The Albion* noted: "The masses are with him; and if acting, as an art, is supposed to be an exponent of nature, Mr. Forrest, in thus conciliating the suffrages of the million, must have touched the chords which vibrate in the breasts of men as a body, or he could not obtain that supremacy over the feelings of his auditors he has so long and so triumphantly exercised. . . . Mr. Forrest in his acting is not merely the embodiment of a national character, but he is the beau ideal of a peculiar phase of that character,—its democratic idiosyncrasy." The paper described Forrest's almost hypnotic power over his audiences, "Witness the furor of audiences subjected to his control, the simultaneous shouts of applause which follow his great efforts, see the almost wild enthusiasm that he kindles in the breasts of his auditors, and who will deny that Mr. Forrest has got the heart, nay, the 'very heart of hearts,' of the masses" (quoted in Ludlow 690). By 1849, Forrest had become the hero of white working class audiences, and he served as a convenient rallying symbol for unscrupulous Tammany Hall leaders such as Isaiah Rynders—one of the men charged with inciting the Astor Place uprising, and a notorious proslavery, anti-abolitionist figure. Rynders claimed Forrest's Macbeth not only as a quintessentially American and working-class champion, but also as an implicitly "white" defender against the creeping tide of abolitionist sentiment he saw overtaking his city.

Rynders rose to power during the election of 1844, when he and his cronies used "intimidation and outright violence" to prevent Whig voters from casting their ballots. Many attributed James K. Polk's "razor-thin" victory to Rynders's machinations (Anbinder 141–42). He was one of the primary instigators of the Astor Place Riot, and by 1850, his confidence had increased, as had his visibility in policing "white" American culture. Rynders famously quoted entire passages from Shakespeare in his political speeches, and infamously attacked black abolitionist Frederick Douglass during an 1850 abolitionist lecture when he and a group of his followers tried to disrupt Douglass's speech, claiming that blacks were inferior to whites (Anbinder 141). By 1857, Rynders had become a U.S. Marshal, charged with returning fugitive slaves to their masters in the South. Though Forrest's public allegiance with the proslavery Democrats was never as charged as Rynders's, he still provided the medium through which Rydners and others could negotiate new, "white" images of American masculinity, and ultimately, white rebellion against antislavery forces. Unlike the mad Lear, the racially problematic Othello, or the indecisive Hamlet, Forrest's Macbeth embodied the brutal violence that Rynders and his followers understood as the new reality of antebellum American politics (see the MacDonald essay).

Likewise, the failure of two Macbeths by popular African-American actors seems to confirm this whitening process. James Hewlett had been the star of New York's African Grove Theatre in the 1820s, and had established an independent touring career after white protestors closed the playhouse. As Shane White and Marvin McAllister have noted, Hewlett's performances of traditionally white Shakespearean roles often synthesized a blend of parody of white performers with his own interpretations of the characters. White spectators frequently satirized Hewlett's performances (most notably Mordecai Noah of the *National Advocate*, the theatre's chief opponent), without seeming to understand the subtle appropriation being staged under their very noses (White 84–86).

Yet interestingly, what seems to have been an acceptable level of credulity, superstition, or ferocity in white interpretations of *Macbeth* struck audiences as inappropriate when portrayed by black performers (McAllister 70–71). While Hewlett became known primarily for his solo performances, Ira Aldridge emerged as the first African-American star to forge a successful international career as a classical performer. Aldridge received widespread acclaim in many of his roles, from Othello to Shylock, but Macbeth appears to have been one of his few failures. In 1853, *The Spirit of the Times* printed a report from an American correspondent in Paris who had seen Aldridge play the role and who described him as "very bad" (*Spirit of the Times*, March 12, 1853). Again, what made Aldridge's performance less effective than either his other Shakespearean performances or than those of his white counterparts seems unclear, but it may be linked to the aggressive white masculinity that had become so irretrievably identified with the role over the past fifty years (see the Lindfors essay).

"Unsex Me Here"—Lady Macbeth and Female Abolitionists

> A Yankee, who lately went to see *Macbeth*, gave the following as his notion of the tragedy: "[Macbeth's] lady appeared to me to possess a tarnation dictatorial temper, and to have exceedingly loose notions of hospitality, which, together with an unpleasant habit of talking to herself, and walking about *en chemise*, must make her a decidedly unpleasant companion."
>
> <div align="right">(<i>Godey's Lady's Book</i>, October 1866)</div>

Thus far I have suggested the ways in which the characters from *Macbeth* were applied to male figures in the ongoing debates over race and slavery, noting that commentators appropriated both male and female characters indiscriminately in their arguments. Yet this series of appropriations elides the role of women in the struggle, and it leaves questions about how or whether Lady Macbeth figured in the characterizations of female abolitionists. The passage from *Godey's Lady's Book* cited above conjures a decidedly negative image of Lady Macbeth—a character often impugned for her unwomanly behavior. The description also echoes stereotypes of female behavior attributed to the women involved in the abolitionist movement, who were often ridiculed for having "unsexed" themselves in the antislavery struggle, or perhaps even worse, who were imagined as consorting with devils or witches in an unholy alliance against the slave power.

For example, in 1852, New Orleans's *Daily Picayune* likened Harriet Beecher Stowe, author of *Uncle Tom's Cabin,* to Lady Macbeth, accusing her of "unsexing" herself. In 1853, the *Southern Ladies Book* took the accusations and comparisons a step further, claiming that Stowe: "has unsexed herself...as a mother she has left obloquy to her children." The journal described her treachery to the United States as "black and bitter," and the product of, "the cold venom of a nature naturally and irremediably vile" (quoted in Roppolo 353–54). Nor was Stowe the only female antislavery activist likened to Shakespeare's infamous queen. Abolitionist and co-founder of the Boston Female Anti-Slavery Society, Maria Weston Chapman was often slighted as the "Lady Macbeth" of the movement (Clinton 59).

Lady Macbeth, however, was not figured as overly emotional or irrational—qualities that might render her too "feminine." Rather, she was often assumed to be too logical and direct in her ambition-fueled actions. A review of Charlotte Cushman's performance as Lady Macbeth suggests the extent to which the character was linked with unfeminine activities in the public imagination by the middle of the century. *The Albion* described Cushman's 1849 interpretation of the role as "entirely new" and an "absolutely thrilling exposition of the text." As *The Albion* observed, "There are no half measures with her—no hesitating on the means to be used to acquire her object." It was likely this kind of *logical* ruthlessness that gave the mid-nineteenth-century Lady Macbeth her resonance among those who opposed women's involvement in the nation's contentious debates over race and slavery.

(Exeunt, Fighting. Alarums)

How should contemporary scholars of American theatre, history, and Shakespearean literature gauge the pre-Civil War obsession with *Macbeth*? Allusions to the text abound throughout the nation's discourse on race, violence, and war, and it is tempting to see even more hidden references in every description of bloodied hands, guilty consciences, and treachery. Perhaps the nation was indeed like Macbeth seeing portents everywhere, as well as signs of guilt, and omens of disaster. Perhaps the modern scholar, knowing the turmoil that was about to engulf the nation cannot help but think, like Siward before the final battle scene in *Macbeth*:

> The time approaches
> That will with due decision make us know
> What we shall say we have, and what we owe.
> Thoughts speculative their unsure hopes relate,
> But certain issue strokes must arbitrate;
> Towards which, advance the war.
> (5.4.16–21)

Notes

1. While some sources place the date of David Claypoole Johnston's cartoon between 1861 and 1864, the American Antiquarian Society and the Library Company of Philadelphia suggest 1856 as the correct date.
2. The *Macbeth* performed on nineteenth-century American stages had been liberally adapted from Shakespeare's original, and often incorporated changes from seventeenth- and eighteenth-century British adaptations. The large number of adaptations makes it challenging to isolate which version of the play would have been seen by a particular audience. It is worth noting that minor characters were often condensed into one and that Macbeth's death was frequently moved onstage for better dramatic effect.

4

THE EXORCISM OF MACBETH: FREDERICK DOUGLASS'S APPROPRIATION OF SHAKESPEARE

John C. Briggs

> Oh, for a man, I say again, who will boldly climb high enough to hang our banner on the outer wall.
>
> (July 5, 1875, Series 2, 4.419)[1]

Speaking in 1875 to rouse the flagging supporters of Reconstruction, Frederick Douglass characteristically made use of his reading of Shakespeare to dramatize his call to protect and defend the Constitution and its new reconstructive amendments. In the passage cited above, Douglass's point of contact with Shakespeare is a section in *Macbeth* introducing the climactic battle between Macbeth's besieged forces and the army made up of the assassinated king's son Malcolm and his allies. Macbeth, the murderer of Duncan, is about to meet his match, but before he dies Macbeth issues a defiant challenge. He threatens the attackers, rouses his beleaguered followers, and displays to himself his own furious opposition, declaring, "I have almost forgot the taste of fears" (5.5.9). "Hang out our banners on the outward walls," he shouts. "The cry is still 'They come.' Our castle's strength / Will laugh a siege to scorn" (5.5.1–3). No matter how dire the news, Macbeth finds new freedom at each blow by breaking free of his obsession with occult influences. The heroic carelessness he shows in the play's last act asserts the persistence of his humanity, however distorted it has become through his submission to the witches' prophetic temptations: "Lay on, Macduff, / And damned be him that first cries 'Hold, enough!'" (5.10.33–34).

Variations on the phrase "banners on the outward wall" appear several times in Douglass's works. It is his most characteristic phrasal link to "the Scottish play." Yet anyone with the slightest familiarity with *Macbeth* is compelled to ask why Douglass used and repeated a battle cry from an infamous tyrant to rally others around the Fourteenth Amendment and freemen's (now new citizens') rights. Douglass's incorporation of Shakespearean language and tone in speeches and writings throughout his career is assumed to have supported his efforts to persuade audiences who were widely and often deeply familiar with Shakespearean language—audiences skeptical of, or likely to be gratified by, an ex-slave's facility with the man Douglass called "the great poet." There are allusions to at least fifteen plays in his collected works, often for purposes of immediate illustration.[2] But why does Douglass hearken back to *Macbeth*, and through this particular moment in the play, when his purposes as a reformer seem antithetical?

To resolve this interpretive challenge, we first need to appreciate the fact that Douglass often uses phrases from Shakespeare in ways that do not resonate with their original contexts (Sturgess). In fact, the "banner" phrase might not have come directly from Shakespeare: John Greenleaf Whittier, whom Douglass named "the slave's poet" in his autobiographical writings (*Life and Times* 99), embedded the Bard's line in his 1846 poem dedicated to the abolitionist Daniel Neall (Whittier 300). When he draws language directly from Shakespeare, Douglass frequently does so without apparent regard for context. In an historical lecture in 1869, he refers to the Dutch Protestant fighters against the Spanish as "tempest-tossed" (Series 2, 4.190). The phrase most likely originates from *Macbeth*, this time in the witches' curse on an unfortunate sailor in the first act (1.3.23–25). The phrase seems only a shiny coin—an interesting isolated metaphor—in Douglass's transaction with his audience's political and literary imagination. Likewise, in a later address in which he objects to efforts to move Washington, DC to a new location, he converts another phrase from *Macbeth* into eloquent ammunition, apparently without care for its origin: "Every stone in the massive marble and granite walls [of Washington, DC] cries out, *trumpet tongued* against the expense and folly of removal" (Series 2, 4.473, *Macbeth* 1.7.19).

The more one glosses such passages, however, the more one notices that Douglass is often doing something more. He must have known the context of Whittier's "banner" line as well as its context in Shakespeare. The abolitionist Neall was known for his heroic stands and "deeds of love." Whittier calls him a "[l]over of peace, yet ever foremost when / The need for battling Freedom called for men / To plant the banner on the outer wall." Neall combined opposites that moralized Macbeth's fury and naturalized his "milk of human kindness" (1.5.15). He was "[g]entle and kindly, ever at distress / Melted to more than woman's tenderness, / Yet firm and steadfast, at his duty's post / Fronting the violence of a maddened host" (300). Douglass, by contrast, quotes Shakespeare's line with unalloyed *aggressiveness*. Whittier praises an embodiment of chivalrous hyperbole: a stolid defender and a more-than-female model of kindness. Douglass looks for a bold man who will climb high on the walls to outface the enemy, even when confronting his doom. If Douglass's allusion is paying homage to Whittier and Neall, he is also distilling the lines' martial meaning from his direct reading of *Macbeth*.

When we look at the more plainly direct appropriations, we see that Douglass draws frequently upon the context and dramatic resources of the

Shakespearean source to help make his point, or to modify it. He uses Shakespeare to deepen, expand upon, and direct his thought as well as his audience's response. Thus the "tempest tossed" Dutchmen who resist religious oppression in the passage above are, in Douglass's overall description, battling against forces not unlike the witches' powers: an empire that Protestants associated with superstitious practices and an intolerance of liberty. In saying the rebels are "tempest tossed" in their struggle for freedom of religion, Douglass aligns them with the witch-tormented, innocent sailor in *Macbeth*. The witches—who in *Macbeth* have no power to kill souls directly—subject him to arbitrary storms in hopes of taking away his will to live. Facing a seemingly overwhelming foe, the Dutchmen resemble that sailor in the face of tyrannical forces that threaten their will to resist. The test that Douglass is highlighting is one of maintaining one's sanity and moral purpose under circumstances that forcefully push toward a loss of self-control.

Douglass's reference to the "trumpet tongued" stones of the capital's buildings, speaking as if they were prophets protesting the contemplated crime against them, comes to life in the phrase's resonant context in *Macbeth*. There Macbeth describes Duncan's virtues as things that speak to him and to others in defense of themselves. As he contemplates his intention to destroy his innocent king, they declare their resistance. His guilt-ridden anticipation of the crime, which includes his persistent awareness of angelic voices that oppose him, is an expression of horror at the depth of his incipient transgression:

> ...this Duncan
> Hath borne his faculties so meek, hath been
> So clear in his great office, that his virtues
> Will plead like angels, trumpet-tongued against
> The deep damnation of his taking-off
> (1.7.16–20)

In the case of Douglass's Dutchmen and the stones of Washington, DC, the struggle is not only between life and death; it is between the blameless desire to live and an almost unearthly purpose to erase the hope that liberty can be established to last. The struggle to keep that hope alive, as well as the moral compass it sustains, is internal, psychological, a struggle of the spirit. In his own way, Macbeth shares in both sides of that struggle.

Speaking in 1863, Douglass quotes from *Henry VI, Part 2* to explain the connection between moral certainty and free strength:

> Thrice is he armed that hath his quarrel just;
> And he but naked, though locked up in steel,
> Whose conscience with injustice is corrupted.
> (3.2.233–35)

In the same speech, he associates the perverted conscience with malevolent uncanny influences—a giving in to "ghastly powers of error" that degrade the progress toward liberty for all. Such powers can and must be defeated, psychologically as well as physically, only with what Douglass calls a deep love of truth "in her simple beauty and excellence" (Series 2, 3.552–53).

For Douglass, the measure of that love is to be understood as it is spoken, written, and enacted in what he considered to be the great cause of liberty and the progress of man. It is the office of the truth-teller to identify the inadequately acknowledged fact that "[t]here are two principles in this country—Slavery and Liberty" (Series 2, 3.99), and that liberty must be won in the internal as well as external struggles of free human beings. The opposite of slavery is liberty, not equality (Series 2, 2.458). The great struggle for liberation is for a liberty that manifests a shared human nature: "One touch of nature makes the whole world kin," he says quoting *Troilus and Cressida* (3.3.169). Drawing from *Hamlet* in 1871, he links the freemen's newly won rights to liberties secured in the name of a common humanity—the "loyal blood"—of whites and blacks, through which "all of us of this color" are "to be or not to be" (series 2, 4.286). The love of truth is dedicated to liberty of the soul and intellect as well as the body. It transcends "the grim and ghastly powers of error" despite "pride of race" and "prejudice of color" (Series 2, 3.553). "[T]he cause of freedom...makes friends of all its friends" (Series 2, 3.188).

It is for this remarkable confluence of reasons that Macbeth is for Douglass an exemplar of the haunted man trying to break the manacles of his diabolical enchantment, his psychological and sleepless slavery. Unlike most other abolitionist references to the drama, which focus on Macbeth the murderer and on Banquo's ghost giving ghoulish testimony of crimes, Douglass approaches the play with the assumption that the chief action is not only psychological and moral, but also within the psyche and spirit of the protagonist.[3] In the last act of his play, Macbeth, for all his crimes and perversity, reaches for liberty when he breaks from his submission to the witches, from the hopes and fears they toyed with and helped make monstrous in him, to tell Macduff he knows his own bloody transgressions and is willing to face death.

Hanging his banner from the outer wall, Macbeth freely faces death. In a famous passage in *My Bondage and My Freedom*, Douglass finds his freedom in a remarkably similar moment. When he fights with his diabolical master Covey, Douglass explains: "I had reached the point, at which I was not afraid to die. This spirit made me a freeman in fact, while I remained a slave in form" (*Oxford* 199). Covey's refusal to admit that a sixteen-year-old slave had beaten him revealed the tyrant's weakness, which became apparent when Douglass recognized that his master's obsessive surveillance and duplicity were used to deny the idea that freedom existed.

What is that freedom a freedom *from*? For Douglass the answer seems to be a sinister enchantment, which he remembers from his childhood into the years after the Civil War. As a little boy who was denied the chance to visit his mother's deathbed, he was—as he described the moment in *My Bondage and My Freedom*—under the influence of "[t]he ghastly form of slavery" that "rises between mother and child, even at the bed of death" (*Oxford* 200). A half century later, he speaks of slavery as a ghoulish presence after its death: "Slavery is indeed gone; but its long, black shadow yet falls broad and large over the face of the whole country" (Series 2, 5.13). The autobiography Douglass became tired of telling, but kept re-telling through three iterations, documents this haunting. What forgetfulness can obliterate the memories of atrocities absorbed into the bones: the violation of women for the sake of selling their offspring, who resemble the violating masters who sell them? How can one declare independence from a legacy of hopes raised and dashed, of sanity undermined, by the manipulations of men bent on destroying hope itself?

> It is only when we contemplate the slave as a moral and intellectual being, that we can adequately comprehend the unparalleled enormity of slavery, and the intense criminality of the slaveholder. I have said that the slave is a man. "What a piece of work is man! How noble in reason! How infinite in faculties!" (Series 2, 1.2.258, *Hamlet* 2.2.293–95)

Here again Douglass identifies himself and his country with the mournful, *haunted* Hamlet in his effort to rid himself of the dreadful legacy of murder and adultery presented by his dead father's ghost. The enchanted prince, paralyzed by indecision and repeatedly tempted by the prospect of ending it recklessly, moves between manic elation and despair. Douglass, upon news that post-war civil rights legislation had been nullified, observes that "[m]y outlook has been sicklied o'er with the pale cast of thought." The "ghost of a by-gone, dead and buried institution" fills, he says, "the very air with a malignant prejudice of race." It acts like the psychologically contagious poison Claudius used to slay his brother: "It has poisoned the fountains of justice, and defiled the altars of religion. It acts upon the body politic as the leprous distilment acted upon the blood and body of the murdered King of Denmark." The murder and the victim have fused. The resultant haunting perpetuates fears that poison the prospects of freedom. And yet, Douglass argues, the specter of slavery can be, must yet be, "conquered and driven from the breasts of the American people" (Series 2, 5.410, *Hamlet* 1.5.64).

Here the foe, unnamed in its uncanny influence, is all the more ubiquitous and powerful in its being without a label that confines it to a rational category, and yet Douglass speaks to apprehend it. When he refers to "the tempest in my mind," he alludes to the greater influence of the slave's mental agony than his physical pain, which Lear on the stormy heath declares he "will endure" (Series 2, 3.11, *King Lear* 3.4.13). Whether it is the spirit of oppression or the inheritance of the history of slavery, Douglass speaks of it in his appropriation of Hamlet as though it were a curse. It almost inhabits the innermost recesses of the spirit, like poison of the nerves. Like the ghost in *Hamlet*, whose mystery Hamlet must resolve and confront by counterplot and resistance, the legacy of slavery is, for Douglass, something against which he and his audience have the ability to contend.

"The Scottish play" had an enormous following in antebellum America. It continues, of course, to haunt the stage. Its power is indicated by the fact that many actors still do not refer to it by name, but the blood-steeped, witch-goaded protagonist is hardly remembered for his heroic demise. Douglass almost never refers to the play, preferring to identify himself more directly with Othello, Portia in *The Merchant of Venice*, and Hamlet, whom he alludes to more frequently and in greater detail than any other Shakespearean character. It is difficult for modern audiences to imagine how *Macbeth* might be a source of inspiration, not only for Douglass but also (for somewhat different reasons) for Lincoln.[4] These very facts lead us to other parts of Shakespeare's plays that offer ways of understanding the context and meaning of Douglass's appropriation of *Macbeth*. The more we look, the more Douglass's use of Shakespeare sheds light upon his allusions to Macbeth's discovery and vehement affirmation of his own freedom.[5]

The most interesting evidence of that connection lies in Douglass's use of Shakespeare to elucidate his discussions of the Constitution, which, early in his abolitionist career, he came to see as the legal basis—indeed the legal foundation—for the abolitionist cause. Differing from the disunionists led by William Lloyd Garrison, who argued that the Constitution accepted and affirmed the legitimacy of the institution of slavery, Douglass uses Shakespeare to describe the Constitution and the Founders' intentions as bulwarks against it. This doubly difficult argument—complex and against the grain of much abolitionist thought—manifests itself most interestingly in appropriations of Shakespeare. Echoing and reversing Hamlet's fantasy about the many shapes of clouds as whales, camels, and weasels, Douglass confronts the idea that a fatally-conceived Constitution somehow haunts those who seek the abolition of slavery:

> What is the constitution? It is no vague, indefinite, floating, unsubstantial something, called, according to any man's fancy, now a weasel and now a whale. But it is something substantial. (Series 2, 3.346, *Hamlet* 3.2.345ff)

The Constitution is a most substantial thing—in fact a very good thing, not an illusion or a phantom law. It stands against the haunting power of slavery. In his famous "Fourth of July" address, Douglass makes a somewhat similar argument using lines from Macbeth's mouth, again from the last defiant scene of the protagonist's resistance. If it is true that the Constitution protects the right "to hold and to hunt slaves," Douglass argues, then the Fathers of the Republic are "the veriest imposters that ever practiced on mankind." They are like those who abandoned Macbeth in his final, heroic moment of need, whom he condemned as persons who "palter with us in a double sense / That keep the word of promise to our ear / And break it to our hope" (Series 1, 2.384, *Macbeth* 5.10.20–22). The fact is, "[i]t is a slander upon [the Founders'] memory" to think so. Paralleling the argument Lincoln brought forward in the 1850s, Douglass argues that the words "slaves" and "slavery" do not appear in the document, and that the Founders intended it not to be corrupted by language institutionalizing slavery in the new American republic. He is adamant on this point: "I...deny that the constitution guarantees the right to hold property in men" (Series 2, 3.346). For him the unjust Fugitive Slave Law is a modern invention; it is not in the Constitution.

Douglass frames this argument not so much against fellow abolitionists as against the South's attempt to evade the Founders' intentions, and the North's failure to understand the full implications of the Constitution it fought to protect (see Series 2, 4.64). The detailed shape of this central argument is strikingly elucidated in one of Douglass's speeches in the midst of the Civil War, in an argument related to his various allusions in other speeches to the "bond" in the *Merchant of Venice*. It is instructive to read a naked version of the argument without Shakespearean allusion, and then a version of it clothed in Portia's clarifying interpretation of the bond of flesh. In both, Douglass presents the Constitution as something solid standing against fatal distractions to its slavery-resisting purpose:

> I hold that the Federal Government was never, in its essence, anything but an anti-slavery Government. Abolish slavery tomorrow, and not a sentence or syllable of the Constitution need be altered. It was purposely so framed as to give no claim, no sanction to the claim of property in man. If in its origin slavery had any relation to the Government, it was only as the scaffolding to the magnificent structure, to be removed as soon as the building was completed. (Series 2, 3.596)

> I admit nothing in favor of slavery when liberty is at stake.... [I]f you want to prove a bad thing, if you want to accomplish a bad and violent purpose, you must show it is so named in the bond. This is a sound legal rule. Shakespeare noted it as an existing rule of law in his *Merchant of Venice*: "a pound of flesh, but not one drop of blood." The law [the Constitution] was made for the protection of labour; not for the destruction of liberty; and it is to be presumed on the side of the oppressed. (Series 2, 3.359, *The Merchant of Venice* 4.1.302ff)

What Douglass means is elucidated further in other speeches, in several other allusions to Portia's judgment on the bond of flesh. Slavery may have "sternly demanded its pound of flesh, no matter how much blood was shed in the taking of it," but the Constitution did not (Series 2, 5.120). For good reason, slavery is not "nominated in the bond" (Series 2, 3.152, *The Merchant of Venice* 4.1.254). Like the drop of blood that Portia's judgment found missing in Shakespeare's play, slavery cannot be claimed to be a portion of the Constitution.

It would be wrong, therefore, to give the impression that Douglass's affinity for Macbeth—most notably, for Macbeth in his final hours when he discovers his freedom facing death—is the full account of the great abolitionist's appropriation of Shakespeare. It is arguably his most charged connection to the man he called "the great poet," and the energy it manifests is evident throughout his writings and speeches. But Shakespeare helps Douglass place Macbeth within the poet's broadly ameliorative understanding of the world, an understanding Douglass directed not only toward his former owner, but also toward Lincoln's role in the Civil War, and the duty of the living to acknowledge "the ten thousand benefits we have received from those who have gone before us" (Series 2, 2.201). For Douglass, Shakespeare's works were a repository of the idea that human beings share a common nature and dignity ("What a piece of work is man!").

> According to the great poet, he who knew more of the human soul than any one who went before him, or any who have come after him: "Man is an animal looking before and after." He alone, in this respect, is capable of giving and receiving...[a process that is] one of the grandest perfections of mankind. (Series 2, 5.201, *Hamlet* 4.4.[9.27])

Liberty tests that human nature by freeing it to show its true character in the struggle for freedom and the "giving and receiving" that Douglass considered to be the essence of education and service. To be educated, in Douglass's terms, is to be able to receive learning in settings where it is not easily given: "from tongues in trees, books in running brooks, sermons in stones, and lessons of education in a portrait" (Series 2, 4.358, *As You Like It* 2.1.16–17). In its own right, book-learning opens the understanding when nothing else can. Thus Shakespeare's birthplace is inaccessible to the visitor who "knows nothing" of the plays (Series 2, 5.284). Shakespeare is more than a field guide. He demands a sympathetic engagement that draws audiences beyond themselves. He creates characters "clothed with flesh and blood," and "warm with the common sympathies of the race,...not angels nor demons—but much of both in their tendencies and possibilities. They are our brothers and sisters, thinking, living, acting very much as we easily fancy we might have done in their places" (Series 2, 3.463). Macbeth is one of these.

Does his affinity for Shakespeare mean that Douglass uncritically mixed his aesthetic education with his politics, unaware of how far apart they were in his own thought and action? Did he indulge his Shakespearean sympathies only in spaces protected from the hard tasks of liberation, war, and moral and political reconstruction? In fact, his reading of Shakespeare offered him a way to project his view of the world into minds and events with striking philosophical depth. In his postwar speech on human progress, he uses Shakespeare to explain this process with a theory of oratorical/poetical leadership:

> [H]e who speaks to the feelings, who enters the soul's deepest meditations, holding the mirror up to nature, revealing the profoundest mysteries of the human heart to the eye and ear by action and by utterance, will never want for an audience. (Series 2, 3.462, *Hamlet* 3.2.20)

And yet there is another side to truth-telling in this Shakespearean frame:

> A man can have a permanent hold upon his fellows—by means of a falsehood....With a clear perception of things as they are, must stand the faithful rendering of things as they seem. (Series 2, 3.462)

The great orator must not lie, but the great orator must conceive of his duty as more than merely speaking truth. He must hold the mirror up to nature by faithfully rendering the visible, seeming world. When he enters "the soul's deepest meditations," he is revealing mysteries in his faithful account of appearances, which takes them in without presuming to decode or agree with them, or reduce them to something not fully human. In this Shakespearean vision, he sees the human potentialities for good and evil more fully, without repudiating his moral compass.

In 1876, in the spirit of this great poet/orator, Douglass wrote a profoundly Shakespearean memorial for Lincoln by holding up the mirror to nature. In it he observed how Lincoln's complex moral and political accomplishment created the institutional grounds for humane mercy, judgment, and the exorcism of Macbeth:

> Viewed from the genuine abolition ground, Mr. Lincoln seemed tardy, cold, dull, and indifferent: but measuring him by the sentiment of his country, a sentiment he was bound as a statesman to consult, he was swift, zealous, radical, and determined. Though Mr. Lincoln shared the prejudices of his white fellow-countrymen against the negro, it is hardly necessary to say that in his heart of hearts he loathed and hated slavery. He was willing while the South was loyal it should have its pound of flesh because he thought it was so nominated in the bond, but farther than this no earthly power could make him go. (Series 2, 4.436)

Notes

1. See Series 2, 5.121 and elsewhere. All Douglass citations, unless otherwise indicated, are from the Yale edition of *The Frederick Douglass Papers*, ed. Blassingame et al.
2. *2 Henry VI, Richard III, 1 Henry IV, 2 Henry IV, Romeo and Juliet, As You Like It, Julius Caesar, The Merchant of Venice, Hamlet, Othello, Macbeth, King Lear, Troilus and Cressida, The Tempest,* and *Henry VIII*.
3. For evidence of an appeal to psychology in the midst of moral condemnation, see Horace Mann's attack on Daniel Webster in September 16, 1851: "It remains to be seen, whether the political Macbeth [Webster] shall succeed to the Banquo [Northern refusal to accept fugitive slave laws] he spirited away, though the 'weird' brethren of the slave-mart...still tempt him onward by their incantations" (547). Mann apparently thought Webster's Macbeth had it in him to become a virtuous Banquo again.
4. For Lincoln, Macbeth seems to be not only a tyrant murderer but also strangely sympathetic and thus a haunting example of fate-distracted ambition in a man of conscience. He is a profound warning to unreflective democratic ambition, both on the part of popular, dictatorial leaders and a populace prone to mob rule. See chapter 2 of my *Lincoln's Speeches Reconsidered*.
5. See L. Levine for an understanding of the popular setting of Shakespearean performance in nineteenth-century America—a setting that fostered expectations of mimetic, creative, and satirical uses of Shakespeare that were not entirely inconsistent with Douglass's. See D. Anderson for an understanding of Douglass's mastery of mimicry, including both biting satire and emulation.

5

IRA ALDRIDGE AS MACBETH

Bernth Lindfors

Macbeth has often been examined as a drama about equivocation. In the world of the play, nothing is as it appears to be. Ambiguities abound, and it is impossible to tell where the truth lies. Beliefs are susceptible to contradiction, strong assumptions vulnerable to surprise. Ira Aldridge, one of America's first professional black actors, engaged in a form of histrionic equivocation. By playing Macbeth and other white roles within the artifice of theatre, Aldridge sought to establish a truth about racial equality (Marshall). Despite the fact that Macbeth was neither Aldridge's most famous role nor one of his most frequently performed roles, the reception of his unscripted actions in *Macbeth* reveals the import of this role in his approach to Shakespeare, race, and performance.

ALDRIDGE'S EARLY SELF CONSTRUCTION

Ira Aldridge is said to have made his first appearance on stage at a private theatre in which all of his amateur fellow-performers were of his own complexion (*Memoir* 10). This was probably New York's short-lived African Theatre, which staged at least fifteen plays, pantomimes, and musicals between 1821 and 1823 before its owner went bankrupt (G. Thompson 3–23, 227). The known repertory of the African Theatre during this period consisted of three Shakespearean plays—*King Richard III*, *Macbeth*, and *Othello*—as well as a variety of British and American entertainments, several of which dealt with social and political issues involving African, Indian, and European characters (G. Thompson 27–30; see McAllister). Aldridge made his debut in Richard Brinsley Sheridan's popular melodrama *Pizarro* as Rolla, a Peruvian commander who heroically leads his troops against the army of a malevolent Spanish conquistador. He next performed as Romeo in a production of Shakespeare's *Romeo and Juliet*, but there is no hard evidence to confirm that either of these plays was put on in New York. They may have been staged in the winter of 1822 in Albany, where the African Theatre troupe retreated to escape an outbreak of yellow fever as well as further harassment by white ruffians who had interrupted a performance, assaulted the actors, and vandalized the playhouse the preceding summer (G. Thompson 15). So Aldridge appears to have had his start in a provincial setting as a performer of anticolonial racial melodrama and Shakespearean tragedy. This was to prove a portent of his subsequent career in the British Isles.

In the 1820s, there were no black actors employed at established theatres in New York or any other American city. The same was true of theatres in Britain, but, armed with a letter of introduction from Henry Wallack, a visiting British actor at New York's Chatham Garden Theatre who had employed Aldridge as a costume carrier in the fall of 1824, Aldridge crossed the Atlantic and, at age seventeen, secured his first professional engagement at the Royalty Theatre in London's East End on May 11, 1825. He opened as Othello, a role he had "studied, rehearsed and played with approbation, within forty-eight hours" (*Theatrical Times*, April 15, 1848), and his success was such that he was retained through the spring and summer to perform in a succession of melodramas featuring virtuous, vengeful, or villainous slaves (see Lindfors 2006). The *Public Ledger and Daily Advertiser*, the only London paper to comment on offerings at the Royalty, praised Aldridge as "a wonderful performer for a Gentleman of Colour" (May 24, 1825).

This small bit of favorable publicity must have caught the attention of managers at the Royal Coburg Theatre in Lambeth, a rural industrial region south of the Thames, for they recruited Aldridge in the fall to play Oroonoko, a noble Angolan prince; Sambo, a cheerful, loyal West Indian servant; Zarambo, a rebellious slave; and Christophe, a Haitian despot. Except for Sambo, these were further melodramatic roles that called upon Aldridge to display various shades of emotional intensity—love, rebellion, tyranny (see Lindfors 2007). Not everyone was pleased with the novelty of such roles being entrusted to "a *Black Actor*, a right earnest *African Tragedian*," as the *Sunday Monitor* derisively put it, thinking his appearance would be a "monstrous exhibition" (October 9, 1825). However, this paper changed its tune completely after Aldridge performed: "His performance was one of much talent...[and] throughout was marked by feeling, devoid of the least extravagance, a quick perception, and to which may be added a degree of dignity" (October 16, 1825). Aldridge turned out to be a better actor than most theatregoers anticipated.

After seven months in London, Aldridge embarked on his first provincial tour, stopping wherever he could find employment in England, Wales, Scotland, and Ireland. In those days, theatres mounted at least two plays each night, usually a tragedy or melodrama followed by a comedy or farce. To complement his heavy roles, Aldridge quickly added Mungo in Isaac Bickerstaff's *The Padlock* to his repertoire. As this happy-go-lucky servant who gets drunk and cheekily confronts his master, he displayed considerable talent as a comic actor. Mungo soon became a reliable staple for him, a role he performed as frequently as Othello. He developed the habit of appearing in both plays on opening nights in order to impress new audiences with his versatility.

This was an effective strategy, for he often took spectators completely by surprise. Many provincial theatregoers, like those in London, assumed that the representation of Othello by a black actor would be a disaster. They came prepared to laugh at the antics of an incompetent thespian, so they were astonished when they found that Aldridge handled the part with remarkable skill. Their appreciation grew when they saw him reappear as the jovial, bumbling, dialect-speaking, drunken servant in *The Padlock* because this was the kind of black man some of them had expected to see play Othello. But now, having witnessed the dignity and passion of his performance as the jealous Moor, they knew he was acting the fool, and they applauded him even more enthusiastically.

Aldridge appears to have deliberately set audiences up for this kind of surprise. He introduced himself not as Mr. Aldridge but as Mr. Keene, a mischievous pseudonym conjuring up a shadowy image of Edmund Kean, who was then regarded as one of the greatest Shakespearean actors in Britain. Also, after the London *Times* mockingly referred to him as an "African Roscius," a name alluding to the eminent Roman actor Quintus Roscius Gallus, Aldridge adopted this ironically honorific title as his theatrical trademark, brandishing it in bold print on playbills advertising his performances (October 11, 1825). Provincial audiences were thus led to believe that he was an ambitious black overreacher destined to fail. A reviewer in the *Hull Packet and Humber Mercury* reported that "We attended the theatre impressed with an idea that the playing the part of *Othello* by a native African—or at least by one born of African parents, and bearing a complexion of the purest and deepest tint—would turn out to be what is sometimes called in theatrical *parlance 'a gag'*" (February 2, 1832). However, audiences quickly learned that the joke, such as it was, was on them. And in the process, many of them learned a valuable lesson; to those who were open to new ideas, Aldridge's cultured refinement and sense of humor came as a welcome revelation. People suddenly were made aware of their own prejudices, their own absurd assumptions about Africans as uncivilized brutes. One commentator in the *Manchester Gazette* said, "He performs in a manner which practically contradicts the argument of the advocates of slavery, that the sable races are deficient in intellect" (February 24, 1832). Another observed that "the mind displayed in his performance was of a character completely to put to shame those calumniators who have stated that the intellects of the negro race, are inferior to those of their white brethren" (*Hull Packet*, February 2, 1832). This was a political point that Aldridge was eager to make, so it is not surprising that he continued for the rest of his career to present himself as an African. He even invented a fable about his past, claiming he was the son of a Christian Fulani prince from Senegal (*Memoir* 8–9). He was engaging in identity politics, initially to encourage attendance at his performances but also ultimately to argue for the abolition of slavery. After all, here was a spectacle that had never been seen before: a "savage" doing Shakespeare. The draw was irresistible.

However, to keep himself employed for more than a few nights in each new venue, he knew he needed to expand his repertoire. There were only a limited number of starring black roles available, so he began experimenting with tawny and swarthy characters: Moorish princes, a Turkish emperor, a Hindustani warrior, a Carib Indian, an Eskimo chief, a mulatto servant, an Afghan king, an Arab robber, and a Venetian Jew. From there, it was only a short step to white Shakespearean roles, first Richard III, then Macbeth, and even on a few occasions, in response to audience demand or managerial insistence, as Iago. When he succeeded in these, he was emboldened in the years that followed to try almost any kind of character, regardless of race.

The Difference of Macbeth

Aldridge's first recorded appearance as Macbeth occurred in Paisley, Scotland on June 25, 1830 as part of a Farewell Benefit for the African Roscius. Earlier in the year, Aldridge began performing scenes from *Richard III* in whiteface. When this proved a success, he added Macbeth (see Lindfors 1999; see Figure 5.1).

A review in a Glasgow literary journal stated that:

> The African assumed *Macbeth*, but we cannot speak so highly of his assumption of this character, as we did of his *Othello*. His *Macbeth*, as a whole, was more a chaste and beautiful piece of acting, than a powerful portraiture of the passions essential to the proper delineation of such a character. (*The Thistle*, July 3, 1830)

In contrast to this tepid reception, many later found Aldridge's Macbeth to be both emotionally and physically charged, and, thus, a rival to performances by the most famous white actors of the time. In Tipperary it was said that:

> in the "dagger scene"—that awful SCENA where the imagined, but impalpable instrument of the coming murder, appears in air before the frenzied eye, he more than realized our anticipations. The crazed brain, where the instrument of destruction is painted, directing the willing hand to clutch it, which grasps, and grasps, and nothing finds, was bodied forth with a life and energy of which McCready [sic] or Keane [sic] might well be proud. (*Tipperary Free Press*, October 1, 1836)

In Devonport, where he played only the final scene,

> it was a curious and not uninstructive thing to find a black man playing the European character of Macbeth—enunciating the exquisite poetry of our great bard with a pathos and cultivated judgment.... [I]t was the strongest denial of the monopoly of intellectual power by any one portion of the human race.... [T]he performer had studied the character, with exceeding care, and had employed the highest charms of his art in its delineation. (*Devonport Independent*, December 26, 1846)

Figure 5.1 Ira Aldridge as Macbeth. Courtesy of McCormick Library of Special Collections, Northwestern University.

A few months later, Aldridge appeared in a full production of the play in Ipswich: "The chief points in Macbeth's character as drawn by the pen of Shakespeare, were all truthfully embodied. Every nice distinctive feature was wrought out, and the representation stood before us grand, coherent, perfect" (*Suffolk Chronicle*, February 6, 1847). Encouraged by such responses, Aldridge went on to offer *Macbeth* in its entirety to audiences in Leeds, Plymouth, and Edinburgh. He never did the full play in London.

When Aldridge tried his luck on the Continent—after twenty-seven years on tour in the British Isles—the reception of Macbeth was often in strong contrast to that of Othello. At the heart of the schism was a debate about whether Aldridge's "lurid" style was the product of Nature or of Art (*Deutsche Theatre-Zeitung*, June 4, 1853). Was it authentically African or theatrically British? One answer given was that it was natural because Africans were by nature theatrical. Another was that it was a learned art, proving that Africans could be educated and civilized if given an opportunity to develop their intellect and talents. A third opinion was that it was a perfect blend of native wildness and European restraint and polish.

This debate came to a head when Aldridge was invited to perform in Berlin in January 1853. After his first appearance, a fierce argument broke out in the press over whether his acting was good or bad. At issue was Aldridge's interpretation of Othello, which some attacked as utterly repulsive and artistically degenerate and others applauded as ineluctably African, providing a savage truth to nature.[1] When Aldridge took the stage a week and a half later as Macbeth, the terms of the discussion changed but the opinions of the critics remained polarized. Since this was not a black role, the focus was on his artistry, not on his aboriginal Africanness. The *Neue Preussische Zeitung* hailed Aldridge's performance as radiating the deepest understanding of Shakespeare (January 15, 1853). *Die Zeit* praised his majesty and grace as sublime touches that rendered his interpretation of this role as perhaps even more perfect than his enactment of Othello (January 15, 1853). The *National-Zeitung* still favored his Othello but admired his Macbeth for the wonderful surprise and beautiful, profound pleasure his acting offered to the audience (January 15, 1853). However, *Königlich Privilegirte Berlinische Zeitung* felt that Aldridge was still learning the role and played it too slowly and in a lukewarm fashion (January 15, 1853). And the *Berlinische Nachrichten* complained that his Macbeth was merely a tamed, colorless Othello who had little in common with Shakespeare's character (January 15, 1853). These last two papers, both of which had condemned Aldridge's acting in *Othello* as too passionate and unbridled, were now criticizing his Macbeth as not passionate enough. The *Deutsche Museum*, on the other hand, continued to maintain that Aldridge was overacting; having played Othello like a drunken sailor, he was compounding the error by playing Macbeth like a boisterous drum-major (January 27, 1853).

This debate, which appears to have been a contest between younger critics who liked Aldridge's unconventional approaches to Shakespearean roles and older, more conservative critics who did not, attracted considerable attention outside Berlin, giving Aldridge a great deal of free publicity. In other cities with strong dramatic traditions, such as Vienna and Prague, the same kind of controversy resurfaced, but the thunderous applause and numerous curtain calls he received at the end of each show suggest that he had the majority of the public on his side.

Not long after Aldridge's two-week run in Berlin, His Majesty Frederick William IV, King of Prussia, who had seen him act both in Berlin and in a special performance in Potsdam, awarded him the Prussian Gold Medal of the First Class for Art and Science, a harbinger of the many decorations and honors he was later to be given in Europe. From then on, he was in demand everywhere. Theatres not only in Vienna and Prague but also in Bratislava, Pesth (Budapest), Posen (Poznan), Königsberg (Kaliningrad), Krakow, and Zurich wanted him, so he began touring more widely, sometimes stopping in smaller cities and towns along the way.

For most German theatregoers, the English language remained an obstacle to full appreciation of the plays that Aldridge's British troupe presented. Aldridge himself employed such eloquent gestures and facial expressions that audiences had no trouble understanding the meaning he was attempting to convey, but there were frequent complaints about his supporting players whose professional competence was considered inferior to his own. There was also some criticism of the severe cuts that had to be made in the Shakespearean plays because Aldridge's company was too small to represent all the characters involved. By the end of 1853, Aldridge had taken the step of sending his entire British troupe home, and for the next fifteen months he performed exclusively with local companies. He continued to do all his Shakespearean roles in English while his fellow actors spoke in German, Polish, Hungarian, or Dutch. Audiences still regarded this as strange and sometimes absurd but the strength of Aldridge's acting always won them over. In the years that followed, he returned to the Continent frequently, performing in Germany, France, Switzerland, the Baltic countries, and as far east as Kiev, St. Petersburg, Moscow, the Russian provinces, and Constantinople. He died on tour in Poland in August 1867.

It may be well to take a look at responses to his performance of Macbeth in some of these countries in order to see how he interpreted the role. Press reports in major cities on the Continent were often much fuller than those in the British provinces, so they give us a more detailed account of what Aldridge did on stage and how audiences responded to him. When he was traveling with his British troupe, one paper noted that they performed the initial encounter with the witches and then only 1.5–2.4 and 5.5–5.7 (*Leipzig Tageblatt und Anzeiger*, November 29, 1852). Nine characters were missing in this condensation, and a critic in the *Berliner Feuerspritz* remarked that "under the present circumstances we received a half dozen admittedly excellent genre pieces rather than a great historical painting" (January 17, 1853). Yet audiences liked what they saw, and most commentators were impressed with Aldridge's subtle representation of Macbeth that contrasted sharply with his impassioned rendering of Othello.

The scenes that were most consistently singled out for praise were his interactions with Lady Macbeth, his vision of the dagger, and his address to the bloody ghost of Banquo, but there were two scenes in particular that were mentioned in nearly every review of his performance: his murder of Duncan, and his duel with Macduff. In both Aldridge went well beyond the text to invent new stage action. Here is the way the murder scene was described by a critic in Vienna:

> This was the moment that we consider the most perfect of the role. Quietly he strides to the door where Duncan is sleeping—slowly he reaches out his hand to open it, hesitantly his foot follows—a thunderclap, one last warning from the above resounds through the quiet nocturnal chamber—the murderer trembles—flinches, looks warily about—he lays his hand on the door—and creeps into the chamber to murder the royal guest in his sleep! These last minutes before the deed had a powerfully gripping effect due to Aldridge's dumb show. (*Weiner Zeitung*, February 24, 1853)

And in Berlin:

> The struggle within the hero's soul before the deed...filled the spectators with fear and pity and arrested their very breathing. He did not plunge or storm into the bloody work of the night. He crept or prowled in the dark to the king's chamber, which he wanted to change into a murderous den; even at the threshold he stops repeatedly, falters. It was as if his guardian angel wanted to drag him back by the hair.... What deep natural truth lay in this vacillation and anxiety between the decision and deed!...It was as if the colored hero were to translate Hamlet's famous monologue "To be or not to be" into "To murder or not to murder!" with his gaze, facial expressions and gestures. (*Neue Preussische Zeitung*, January 15, 1853)

When Aldridge appeared as Macbeth in Russia more than ten years later, he had added further refinements to this memorable scene. An actor who performed with him in 1865 in Kazan recalled that:

> Aldridge walked toward the door for at least five minutes, the public was afraid to take a breath. Here he is walking up the steps, on one of them he slips and falls! The gripping scream of deathly terror, a new terror for the public. Macbeth rises with some difficulty; already his face and shoulders are hidden by the door, and his leg is incapable of stepping over the entrance. It trembles and when it disappears, the observer in his mind follows Macbeth into the bedroom of Duncan. It has occurred. Slowly, deathly pale, shivering, Macbeth returns and it is impossible to describe his scream in that place—when he is scared of the touch of his wife's hand. (Vekhter 2)

The battle scene with Macduff at the end of the play was also enhanced by unscripted actions. Continental audiences had never seen such a vigorous, prolonged, artistically controlled gladiatorial combat. Experts declared it a "masterpiece in the art of fencing.... The fighters fenced so that the sparks came down like a rain of flames and they used the most exceptional subterfuge. The gradual dissipation of strength after being injured, the act of dying itself, was superbly acted by Mr. Aldridge" (*Allgemeine Theatre-Chronik*, December 3, 1852). In this duel "his sinews gradually began to lose strength, finally, like a wounded boar, he drags his sword after him, reaches for his dagger, and falls—a truly tragic hero" (*Neue Preussische Zeitung*, January 15, 1853). "This is not the dueling of play-actors, but of heroes who truly burn with rage and vengeance, and with what a high conception of the nature of Shakespeare's hero this Macbeth continues to fight the air with the hilt of his broken sword, even as death is already burrowing into his heart and he no longer belongs to the realm of the living.... These were unforgettable, terrible moments" (*Die Zeit*, January 15, 1853). The duel "presents an image which is as shattering psychologically as it is physically" (*Deutsche Theatre-Zeitung*, January 22, 1853). To these commentators it did not matter that Shakespeare had intended for Macbeth to die offstage; as one of them said, an actor "able to execute the portrayal of the fall with as much mastery as Aldridge surely has the right to amend the poet" (*Breslauer Zeitung*, February 6, 1853).

A Swiss critic contrasted Aldridge's Macbeth with what theatregoers usually saw when German-speaking actors played the role:

> We are accustomed to regarding Macbeth as a Germanic character. His indecision makes him into a brooder. His evil desires and his moral or cowardly reservations take on the character of syllogisms and do battle in his head rather than in his heart.... Ira Aldridge, in contrast, develops an almost idyllically soft side of his spirit... [taking] this softness to the point of heartfelt pity for Duncan.... [A]ll the horrors of the bloody deed descend upon his agitated imagination, and the viewer begins to doubt whether he will perform it, whether the power of conscience will make him shy away.... This Macbeth is nothing but passion and feeling. At first I asked myself: is he perhaps an African Macbeth? But then I... found the answer: No, that is no German, but no African either; this Macbeth is a Gaelic hero through and through. Truly! Shakespeare imagined the Scottish king so soft, so wild, so ruled by violent imagination. (*Solothurner-Blatt*, June 24, 1854)

This remark is borne out in the two scenes just described. Aldridge, the hesitant murderer, displays empathy for his victim and remains for a moment irresolute, yet in his battle with Macduff, he struggles to the end, knowing he is destined to lose his life. He is both a man of conscience and a man of action, a thinker and imaginer as well as a bold doer, a reluctant overreacher who, having stepped in blood so far, chooses to go down fighting.

To convey this interpretation of a profoundly conflicted individual to audiences unaccustomed to hearing English spoken on the stage, Aldridge drew upon his consummate skills as a mime to represent Macbeth's fluctuating moods and feelings. Such acting required no translation, for people could see the mind's construction in his face and witness his thoughts crowned with acts. It was a moving picture, enabling spectators to understand and appreciate a complex Shakespearean character in a new light.

What made the performance even more remarkable was that audiences were well aware that the actor behind the whiteface makeup was black, not white. Most of them had already seen him as Othello and Mungo, opposing roles he played convincingly without makeup to comment contrapuntally and ironically on accepted notions of racial difference. But now, "with his dusky epidermis coated first with bismuth and afterwards toned up with carmine and burnt umber, to depict not only Macbeth's nationality, but the effect of his vigorous and open-air life," he was disguising his race to take on a totally new identity as a conscience-stricken Scottish regicide (*Otago Witness*, October 25, 1884; see Boyd). Like Macbeth himself, he was dressing in borrowed robes, and although observers knew he was not at all the man he seemed, they accepted and applauded him as a highly gifted usurper of the role.

NOTE

1. See respectively: *Königlich Privilegirte Berlinische Zeitung*, January 5, 1853, and *Neue Preussische Zeitung*, January 5, 1853.

6

MINSTREL SHOW *MACBETH*

Joyce Green MacDonald

At first glance, *Macbeth* does not seem to be a very likely subject for minstrel show parody. After all, it does not even have any black characters in it, unlike *Othello*, which has elicited a wide range of blackface performances—both parodic and straight—into the twenty-first century. In February 2008, white British tenor Ian Storey, for example, commented in the *Los Angeles Times* on the various kinds of costume and makeup issues that have accompanied his performances of the title role in Verdi's *Otello*: "It's a really sensitive subject.... How dark should you go?" (Ng).[1] But as Patricia Parker persuasively argued about *Hamlet*, a play does not have to include black characters in order to be deeply informed by Renaissance moral vocabularies of fair and dark (Parker). More to the point of my discussion here, Shakespeare's proneness to racial appropriation operates regardless of the presence or absence of black characters. What matters more to this appropriation than a storyline involving black characters is the pre-existence of a broadly racialized explanatory narrative of social relations, which is then mapped over the cultural currency of the Shakespearean text. Given *Macbeth*'s popularity on the antebellum stage, it was perhaps a logical site for blackface reaction. The examples I discuss here are not so much about putting *Macbeth* fully into blackface—I was able to discover the existence of only one such work, irresistibly titled *Bad Breath, or the Crane of Chowder*—as they are about *Macbeth*'s (and more broadly, Shakespeare's) utility to the racial unconscious of the antebellum stage. Going dark, I argue, was often about other things than race, and helped bring other things to light.

For example, *Christy's Nigga Songster*, a collection of songs performed by the famous blackface troupe the Christy Minstrels, contained a three-song sequence called "Nigga's Description of Shakespeare." Along with lyrics inspired by *Romeo and Juliet* and *Hamlet*—and set to popular music of the day—*Christy's* also contains "Nigga's Description of Macbeth":

> Oh! If 'twere done when 'tis done,
> Though the deed's a little wrong,
> 'Twere well it were done quickly,
> But the double trust is strong;

> Oh! Can this be a dagger
> Attendant on my will,
> A real Arkansas snagger,
> Which I have not, but see still?
> (*Christy's Nigga Songster* 213)[2]

This riff on *Macbeth* barely seems racially coded at all when compared to the *Hamlet*-inspired lyric, which was set to the tune of the most popular minstrel of all:

> Oh! 'tis consummation
> Devoutly to be wished
> To end your heart-ache by a sleep,
> When likely to be dish'd.
> Shuffle off your mortal coil,
> Do just so,
> Wheel about and turn about,
> And jump Jim Crow.
> (*Christy's Nigga Songster* 214)

The hallucinatory image of Hamlet jumping Jim Crow suggests how blackness was appropriated to the cause of nationalist self-assertion, as a disorderly black body performatively erases the Dane's melancholy and substitutes for it the jolly, inassimilable verve of the mid-nineteenth century popular stage. In his analysis of pre-Civil War blackface minstrelsy, W. T. Lhamon argues that—at least before the crisis of the 1860s—blackface minstrelsy harnessed its racial caricatures to the service of a burgeoning working-class consciousness (Lhamon 1998, 1–55). This consciousness asserted itself against elites' claims of social dominance as strongly as it would later attempt to advance it through racial representation (Lott). This white obsession with fantasies of racial identity was mapped over and helped to focus the simultaneous drive to develop a distinctively American national culture, independent from the shaping authority of British institutions. One place where these two cultural movements—toward racialization and away from Britain—converged was Shakespeare.

Look again, for example, at the *Christy's Macbeth* lyrics. The "Arkansas snagger" in line seven apparently refers to a tool used for clearing brush, an implement associated with farmers and woodsmen much more than Shakespeare's vaguely aristocratic and archaic "dagger" could have been. The "Nigga" in the title may be associating himself with all laborers at the same time as he parades in blackface, which, as Lhamon reminds us, was a kind of performance enjoyed by both black and white audiences at the same time in the same (racially-separated) playhouses, at least in the years before the Civil War. The mix of high and low, Shakespeare and American slang, exactly illustrates the appropriation of whites' perception of black style to the causes of a wider class and national self-assertion. Blacking up allowed whites to voice class bravado—and class resentment—at the same time that it allowed them to resist the implications of the unsettling possibility that culturally dispossessed white workers may have more in common with truly marginalized blacks than they were ready to admit.

This class-marked, nativist conviction swirled around the performance of Shakespeare's *Macbeth* in 1849, when a riot against the English tragedian William Macready's appearance in the title role rocked New York City. Driven from the stage by an organized mob under the leadership of Tammany Hall head Isaiah Rynders, Macready decided to cancel the rest of his American tour before being persuaded to go on by a group of prominent Manhattanites, including Washington Irving, the publisher Evert Duyckinck, and a young Herman Melville (Berthold 429–30). All were associated with the Young America movement, inspired by the European republicans of 1848 to advocate the growth of a stronger national—instead of merely sectarian and regional—identity for the United States. But the segment of the white urban working class mobilized by Rynders—a class in the process of defining itself against the tensions generated by immigration, the explosion of mercantile capital, and the economic competition of Manhattan's proud community of free blacks (Kaplan)—seized on the occasion of Macready's *Macbeth* to assert its vitality as political actors in the nation's new public sphere. Rynders' Native American Party treated the Young America petition as a provocation, and responded with an even bigger, and advertised, riot: "All public men" were invited to the "English Aristocratic Opera House" (actually the Astor Place Theatre on East 8th Street) to express their displeasure at this foreign imposition on American taste (Berthold 430).

The political rejection of Macready and imported high taste that came to a head in the Astor Place Riot had already been performatively embodied by the American tragedian Edwin Forrest's shadowing of the Macready tour. Forrest followed Macready along many stops of his U. S. tour, usually playing the same roles—including Macbeth—to much partisan acclaim (Leonard 52–56).[3] He was the undisputed favorite of the American audience, an eminence he enjoyed since his debut in the title role of the Indian king in John Stone's 1829 *Metamora, or the Last of the Wampanoags*, which he continued to play for thirty years or more. His status as the first American acting icon was thus strongly rooted in his powerful portrayal of the emergence of a distinctively American identity through the instrumentality of racial suppression—in *Metamora*, suppression of the New England Indians.[4] While some New Yorkers despised what Walt Whitman called Forrest's "loud mouthed ranting style" (quoted in Rebhorn 455), many more thrilled to it. The Forrest-Macready rivalry, with all it implied about nationalism, high culture, class identity, and the role of race in conceiving them all, was a powerful subtext in the *Macbeth* riot of May 10, 1849. As many as 15,000 people, apparently mostly men, gathered in response to Isaiah Rynders' call, showering the theatre with thrown rocks and bricks and its upper-class patrons with shouted invective. The crowd outside the theatre was only dispersed when the National Guard—called out for the first time in response to an urban disturbance—fired into it, killing at least twenty-two people.

As satisfying as it might be for Shakespeareans to imagine Shakespeare at the center of a public riot, it is regrettably probably more accurate to say that *Macbeth* was only the occasion, rather than the cause. The Astor Place Riot appropriated Shakespeare as the marker of high culture to larger conflicts arising from the struggle to define the new polity being forged during the Jacksonian era (Wilentz). Whereas the nativist strain of the emerging Democratic Party would certainly come to advance an implicitly anti-black agenda as the nation moved closer to civil war, blackface seldom seems to appear for its own sake in the mid nineteenth-century *Macbeth* parodies. What is more common is a kind of commonsense reclamation of Shakespeare from the perceived forces of European cultural oppression in which ruffling the dignity of the Shakespearean text becomes a way of declaring aesthetic independence from its power. In the preface to the British *Macbeth Travestie* of 1813, for example, the anonymous author excuses the liberties he takes with the play by saying he is only doing what others have already done with other Shakespearean plays, and anyway, even "though abounding in more fancy than any other play by Shakespeare, [*Macbeth*] is not for force of sentiments, beauty of imagery, and solemnity of conduct, equal to the Prince of Denmark" (*Accepted Addresses* 80). Mocking *Macbeth* is not as serious an affront as mocking *Hamlet*, because it is not a play we have to take as seriously (although, of course, *Hamlet* became the subject of *Hamlet Travestie*, also in 1813); it simply "admits of more burlesque situations" (78).

A second, American, *Macbeth Travestie* (1847) retains this emphasis on the eminent mockability of *Macbeth*. The lead actor, William Mitchell, was commended for his unique style, "half tragic, half comic, half Macready, and half funny Mitchell. The vision of the wooden dagger, and the ludicrous horror with which Macbeth looks upon his bloody hands after the murder of Duncan, were points that (to use a technical phrase) told amazingly, and proved the manager to be also the true artist" (Northall preface). Just as the idea of tormented Hamlet jumping Jim Crow strikes modern readers as bizarre, so does the image of an actor spoofing the horror of Shakespeare's guilt-stricken Macbeth as the image of the bloody dagger overwhelms his conscious mind. The Northall-Mitchell burlesque, however, is more fully informed by the kind of social anxiety and even outright fear that led to the Astor Place Riot than is the first *Travestie*. Mitchell was an Englishman who immigrated to America and, by the end of the 1830s, established his own Olympic Theatre in Manhattan. There, he produced a series of shows dedicated to burlesquing the seriousness of high-culture imports from Europe, for the Olympic's largely male working-class audiences. He not only cross-dressed on stage; he frequently appeared in blackface in such shows as *Lucy Did Sham a Moor* and *Lucy Did Lam a Moor*, which were takeoffs on Sir Walter Scott's 1819 novel *The Bride of Lammermoor* (Rourke 101–03; to "lam" someone was to beat him up).

Although he did not apparently black up in his *Macbeth*, Mitchell's *Macbeth Travestie* did include more explicit acknowledgement of the growing racial consciousness specific to Jacksonian New York. In Shakespeare, one of the witches recounts an unpleasant encounter with a woman:

> A sailor's wife had chestnuts in her lap,
> And munched, and munched, and munched. 'Give me,' quoth I.
> 'Aroint thee, witch,' the rump-fed runnion cries.
> Her husband's to Aleppo gone, master o'th' *Tiger*.
> But in a sieve I'll thither sail,
> And like a rat without a tail
> I'll do, I'll do, and I'll do.
>
> (1.3.3–9)

Mitchell's *Macbeth Travestie* renders these lines as:

> A Loafer's wife had peanuts in her lap.
> And cracked—and cracked—and cracked!
> Give me, quoth I. Oh, get out, now—she nuttily did snigger—...
> Her husband is a-fishing gone with a great fat nigger;
> And in a boat I'll to him roll.
> Without a cent to pay the toll.
> I'll go!—I'll go!—I'll go!
>
> (Northall 7)

Mitchell exchanges chestnuts for peanuts and the idea of a world-ranging voyage to Aleppo for a strictly local encounter. Exotic Aleppo is excised in favor of the here and now, which includes "a great fat nigger." The term "nigger" shows up again in Mitchell's second act, corresponding to Shakespeare's scene of Macbeth's sighting of Banquo's ghost at the conciliatory banquet:

> What man dare, that I dare, e'en though it were to tussle
> With Shakespeare in a song, quite a la Russell.
> To dance a Pas de Deux in public square.
> With dancing dogs, or rugged Russian bear.
> Or wrestle for a cent with some Herculean nigger—
> Anything, but look upon that horrid figure.
> (Northall 24; *Macbeth* 3.4.98–102)

Mitchell perhaps exchanges Shakespeare's "Hyrcan tiger" for his own "Herculean nigger," again editing out the original invocations of the exotic and the foreign—"Hyrcan tiger," "arm'd Rhinoceros"—in favor of an urban carnivalesque that mixes high and low ("To dance a pas de deux in a public square"). The exchange works to knock Shakespeare off the cultural pedestal that nativist philosophy suspected that he occupied, reclaim him from what the working-class ethos of the Olympic Theatre identified as the unnecessarily alien and elite, and reproduce him firmly within the increasingly racially-conscious terms of the theatre's core audience. Such travesties as Mitchell's performed a more aggressive inquisition into the value of the Shakespearean text and Shakespearean performance in American urban life than Forrest's parallel Macbeth. Mitchell's *Travestie* is so eager to reorient the play to the perceived tastes of his audience that it uses the racial slur even when it does not quite logically fit. "The devil damn thee black, thou pale faced figure," his Macbeth shouts at an entering guard; "Who put that chalk upon your face?" The officer tries to tell him that Macduff's men are arriving: "There's fifty—" but Macbeth interrupts him: "Geese, nigger?" (33). If the entering officer was "pale faced," how could he be a "nigger" at the same time? With the superfluous term, Mitchell runs roughshod over the original passage's indication of the servant's extreme fear of Macbeth's imminent downfall (indicated by his pale face), choosing instead to harness and racialize the free-floating anxieties of his audience of male urban workingmen.

We might think of the gestures toward race in Mitchell's *Travestie* as small evidences of blackface's transatlantic stature in the antebellum era. As far as I have been able to tell, Mitchell did not discover blackface until he came to America, while British actors like Charles Mathews (1776–1835) found it here and took it back to British stages. These class-marked racial impersonations circulated in the Atlantic basin and would come to inform the cultures of the modern industrialized West. Just as the figure of the disappearing Indian in plays like Forrest's *Metamora* infused the brutal facts of removal and territorial expansion with a kind of nostalgic sublimity, these burlesque applications of blackface to Shakespeare would come to soothe American whites'—perhaps especially white males'—doubts about their ability to compete for social place and social presence in the decades before the war. The difference in tone, from invocations of the noble savage in the Indian plays to that of social-climbing black jokesters and buffoons in the blacked-up Shakespeares, marks the difference between a culture that was in the process of confidently staging a negotiation of imperialist aims and one that was not yet able to assume as confident a posture toward managing the black Americans in its midst. The joke was between the white (frequently male) spectator and the blacked-up white man on stage, soliciting racial solidarity around their own precarious senses of class position and against the black interlopers who seemed to be newly invading urban space. Wendy Jean Katz cites an 1841 letter to the editor of the *Cincinnati Daily Enquirer*: "White men...are naturally indignant when they see a set of idle blacks dressed up like ladies and gentlemen, strutting about our streets and flinging the 'rights of petition' and 'discussion' in our faces." The letter was signed, "A Workey" (66).

One blackface Shakespeare that clearly illustrates this point of racial and class solidarity against black *arrivistes* and their preposterous claims to civil rights was the anonymous *The Hop of Fashion; or The Bon Ton Soiree* (1856). The show combines two pet peeves of the working-class audience at houses like the Olympic: black social mobility outside stations of servitude to whites, and perceived upper-class hostility to the ambitions of the white working class. Race here is mobilized to do the work of both class and capital. Eighty years before the publication of *The Hop of Fashion*, Bostonians disguised themselves as Indians to protest the tax on imported tea despite the fact that there had been very few actual Indians in New England since the end of King Philip's War 100 years earlier. By the appearance of this play, almost 100 years after the Declaration of Independence, it was blacks and not Indians who became instrumental to the emergence of white male American citizens. Despite the fact that the great majority of blacks—at least outside the great cities of the north—were living in slavery, the caricature of a pretentious black dandy was a lightning rod for the very real fear generated by the rapidly changing conditions of life in the Jacksonian metropolis.

As the play opens, Peter Slim, who has just won the lottery, is making plans to hold a masquerade ball "at my mansion dis evening," and reads the invitation aloud: "Your company am most 'specfully s'licited to 'tend de fust grand annual sore-eye dis ebenin' at de house ob your ole friend and whitewasher—Captain Slim" (56). "Whitewasher" is obviously a malapropism for the formulaic "well-wisher," but more significantly, the term also winks at the blackface convention in play onstage. Captain Slim is both a "black" man aspiring to whiten himself socially through throwing a fancy dress ball in the new McMansion that he acquired through blind luck and not any particular merit, and a white actor aiming an inside joke directly at the whites in the audience: aggressive black social climbers grabbing all the goods of a materialistic society is an obvious fiction, but a comforting and serviceable fiction nonetheless.

The guests at Captain Slim's party include a couple of New York stereotypes: Mose, "one of the b'hoys" (a mid-century slang term like "workey," whose spelling was probably meant to approximate the sound of an Irish accent); and Lize, "one of the Sykesy crowd" (presumably a prostitute like Nancy, in love with the vicious thug Bill Sykes in Dickens' *Oliver Twist*). They also include three Shakespearean characters out of context and out of their more familiar cultural register, Richard III, Macbeth, and Lady Macbeth. The *Hop*'s Macbeth echoes—and renders nearly unintelligible—the lines from the banquet scene as he approaches to seek admittance to the ball from Captain Slim's manservant Anthony:

> Can such things be, an' o'ercome us without our special wonder; and now I do behold you keep the natural ruby of your cheek; while mine is blanched with fear. Approach thou like the rugged Russian bear, Hurcan tiger, or the armed rhinocros [*sic*]—take any other shape but dat.... If trembling Janipathy protest me de baby of de girl. (*The Hop of Fashion* 58)

As more and more guests attempt to enter—and as the Shakespearean characters get mixed up with the contemporary stereotypes—Mose gets angry at an anonymous Irishman daring to dance with Lize, who feels faint. Anthony rushes in with a cup of coffee for her, but in the hubbub it gets spilled on her dress, and the fancy dress ball turns into a melee: "A general run ensues. All kinds are trying to pursue the nigger waiter." The play ends in "a general row," with "Lady Macbeth flying to and fro all over the stage" (64). Again, *Macbeth* is not so much woven into the fabric of a new play as, due to its familiarity, it is cherry-picked for moments that can be memorably turned to contemporary purpose. Interestingly, the Shakespearean characters are associated with Anthony by their common recourse to minstrel show black dialect. Although Mose does let loose with some "dis" and "dat," this may be part of his stage Irish identity rather than "black" dialect; much humor is made of the fact that Anthony cannot even understand the speech of the play's other Irish character. The play's Richard III minstrelizes his opening soliloquy: "Now is de winter ob our discontent made glorious summer by de son ob York, and all de clouds dat lowered upon our house am in de deep bosom ob de ocean buried" (62).

Elsewhere, I have written about just how much racial outrage the idea of black people performing Shakespeare induced in New York's white observers in the 1820s (MacDonald 1994). *The Hop of Fashion* solicits another emotion with its conjunction of Shakespeare and racial representation: mockery. Here, in a contest for social agency, blackface is the weapon turned against Shakespeare (both the popular *Macbeth* and *Richard III*, the play that members of New York's African Grove Theatre were arrested for performing) and the city's growing diversity of races, classes, ethnicities, and subcultures. Both Shakespeare and a changing New York are tokens of the degree to which the white native spectator, for whose enjoyment *The Hop of Fashion* was designed, felt alienated from a city whose competing new cultures are leaving him behind. Indeed, Mose and the nameless Irishman allude to the social ill-ease of the play's target audience, despite being stage caricatures of just what members of that audience were so worried about. After brushing off Lady Macbeth's alarm at their rough behavior—"Oh, go way—you're a woman," Mose tells her—the Irishman asks for a light for his cigar. "Are you a know noffin?" Mose demands first. "Yes, I am. Every inch of me," the Irishman proudly answers. Mose gladly offers him the light.

Manhattan in the 1850s was a hotbed of the nativist Know-Nothing Party, formed around conservative ideals of national unity, cultural homogeneity, and an orderly social hierarchy of rights and responsibilities (B. Levine). Now, obviously, the Irishman's response to Mose has a double meaning; he's a know-nothing, all right, because he's an argumentative lout. But just as the play allows the rules of the game behind blackface to show earlier in its use of "whitewashed," it is also tips its hand on class position. The spectators for this particular play are both white and socially beleaguered, and the play is ready to admit it. Paradoxically, poor Anthony—completely unable to control admission to the glittering fancy dress ball at Captain Slim's or to stop the mayhem that breaks out there—may speak as much for that white Know-Nothing voter as against the interests of the city's free black "Ethiopian mobility," whom the play's action burlesques. "Mobility" was a seventeenth-century word, which was subsequently shortened to "mob." The notion of an "Ethiopian mobility" projected a particular racial identity onto antebellum social ferment, naming it a phenomenon that was strictly the fault of black people (Lhamon 2003, 14–19). The blackface disguise similarly estranged a white spectator from the sneaking suspicion of his own social impotence as embodied in frantic Anthony, relying on the spectator's white-skin privilege—the one source of distinction and superiority that remained sure in a shifting cultural landscape—to do so. Such were the conundrums of the constructions and imposition of racial identity.

The progress of minstrel show *Macbeth* demonstrates how Shakespeare—particularly this popular play—was appropriated to the social struggles of antebellum New York. Beloved enough to command an audience at three different playhouses during the week of the Astor Place Riot, this play about violently rejecting traditional hierarchies of duty and obligation thus coincidentally figured in the ideological struggles accompanying the birth of the Democratic Party in the city, as the old Whigs were splintered by more assertively nativist elements. *Macbeth* helped the rioters mobilized by Rynders to organize a distinctively American response to perceived British cultural bullying: "Working men, shall Americans or British rule in this city," demanded the flyer that his shock troops posted (Berthold 430). They seized Shakespeare and claimed him for the likes of the "loud mouthed" Forrest, violently recontextualizing the play and its author within an aggressive American cultural nationalism. This new American nationalism drew on the infinitely flexible resources of blackface for its self-expression. Blackface was instrumental in performing the dislocation and alienation of the whites in 1850s New York through these skewed *Macbeth*s, a dislocation whose depth was directly indexed to its choice of the most high culture of authors for its parodic expression. At the moment when the implications of Shakespearean value for the progress of the republic were being contested in the streets, the choice of blackface signaled the emergence of an equal and opposite and deeply American popular response. To black up the bard was to deconstruct class and cultural authority by asserting white racial authority over blacks and their representation. In a changing world, it was the one kind of power that remained constant.

Notes

1. The article further notes that
    The first time he performed the opera, he wore a wig with dreadlocks down to his waist.... In another production, Storey had to wear nothing but a thong in the climactic scene when he strangles his wife. That required him to endure daily applications of full-body makeup, plus a fixative spray so that the makeup wouldn't rub off. (Ng)
2. I thank the Center for Popular Music at Middle Tennessee State University in Murfreesboro for providing me copies of this material.
3. Both actors even played Macbeth in Philadelphia on the same night in November 1848; three actors—Forrest, Macready, and American Thomas Hamblin—were playing the role on the same night in New York the first time Macready was driven from the stage.
4. *Metamora* was only the most popular of the "disappearing Indian" plays staged in the early republic. For a fuller account of these fascinating dramatizations of the connection between race and nationalism, see M. Anderson.

7

READING *MACBETH* IN TEXTS BY AND ABOUT AFRICAN AMERICANS, 1903–1944: RACE AND THE PROBLEMATICS OF ALLUSIVE IDENTIFICATION

Nick Moschovakis

> Whoever thinks that Shakespeare's theatre has a moral effect, and that the sight of Macbeth irresistibly repels one from the evil of ambition, is in error.
> —Friedrich Nietzsche, 1880 (140)

> [I]s self-obliteration the highest end to which Negro blood dare aspire?
> —W. E. B. Du Bois, 1897 (184)

Until the end of the eighteenth century, most critical and dramatic interpreters of *Macbeth* agreed that the play had been written to expose and condemn—in Nietzsche's words—"the evil of ambition." But in the nineteenth and early twentieth centuries some began to question this long-settled assumption. Gradually it became acceptable to understand the tragedy less as an object lesson in crime and punishment and more as a glimpse of fatal struggles seen through the turbulent minds of modern subjects. Such subjects might not regard the desires of Macbeth and Lady Macbeth as absolutely wrong, or the claims of Duncan, Malcolm, and Banquo as absolutely right (see Moschovakis 13–19). By 1877 the critic Edward Dowden would remark in his popular primer on Shakespeare that "[g]ood and evil" were not quite "clearly severed from one another...in *Macbeth*" (134). A door was being opened to later twentieth-century critics and directors who would find *Macbeth*'s vision of strife and suffering to be morally problematic—not morally dualistic.[1]

This trend toward a more problematic interpretation of moral issues in *Macbeth* coincided with modern culture's broader shift toward the premise that morality, in practice, is often anamorphic or equivocal. By the later nineteenth century, both "old regimes" and revolutionary regimes had shown how a society's laws and institutions could cloak or excuse terrible violations of its established precepts. "Good" and "evil" were historically intertwined and confused. An egregious example in America was the enslavement of Africans by professed Christians. Another was the continued oppression of African Americans in the ostensibly democratic United States after the Civil War.

Writers of the late nineteenth and early twentieth centuries tended to address such ethical paradoxes in two ways. One was dualistic: expose the paradoxes, decry them, and call for the righting of manifest wrongs. The other was more modern and problematic: see moral conflict and ambiguity as an effect of competing social interests (a view that could be grounded in a Nietzschean or a Marxist ethic, with its dialectical account of how ideologies arise and compete). To the problematic modern mind "good" and "evil" were relative terms, with different senses for groups in and out of power.

Accordingly, writers who quoted and alluded to *Macbeth* in the late nineteenth and early twentieth centuries tended to do so in one of two distinct ways, one implying a dualistic reading of the play, the other a problematic reading. On the one hand, the play was presented as a universal moral touchstone, praising good and damning evil. On the other hand, it was used as a rhetorical tinderbox to ignite the courage of the weak—to inflame their hopes and their courage to oppose the strong. The contemporary emergence of morally problematic *Macbeth*s alongside morally dualistic ones, both in criticism and on the stage, meant that modern allusions to the play did not always require their readers or hearers to presume that Macbeth's ambition was simply "evil." An allusion might implicitly identify a person with Macbeth, or a cause with Macbeth's cause, in a way that expressed ambivalence about the character, or, most problematically, solidarity with him.

This essay explores how African Americans, and others writing about race, took both dualistic and problematic views of *Macbeth* when quoting and alluding to the play between the Civil War and the end of World War II. Many allusions dualistically framed such outrages as slavery, lynching, and Jim Crow as the deeds of evil latter-day Macbeths. Others implicitly identified Shakespeare's regicidal thane with the very men and women who fought for African-American rights and freedoms—implying a more problematic reading of *Macbeth* through a more complex valuation of Macbeth's motives. Such problematic allusions suggest how advocates for black rights might have aspired to resemble Macbeth (or, to a degree, Lady Macbeth) in several ways: in being transgressively ambitious, boldly insurgent, and at the last heroically unyielding.

RACE AND THE DUALISTIC MACBETH: THE RECRIMINATING SPECTRE

One of the most prominent *Macbeth* allusions in African-American writing occurs near the opening of *The Souls of Black Folk* by W. E. B. Du Bois (1903). In an unattributed quotation from the banquet scene, Du Bois makes

Banquo's risen ghost embody the legacy of slavery and the lack of true progress for blacks since Emancipation:

> Years have passed away since then,—ten, twenty, forty; forty years of national life, forty years of renewal and development, and yet the swarthy spectre sits in its accustomed seat at the Nation's feast. In vain do we cry to this our vastest social problem:—
> "Take any shape but that, and my firm nerves
> Shall never tremble!"
> The nation has not yet found peace from its sins; the freedman has not yet found in freedom his promised land. (Du Bois 1903, 10; *Macbeth* 3.4.101–02)

Du Bois's "spectre" of Banquo is a "swarthy" ghost, haunting and mutely indicting the white masters who preside over "the Nation's feast."[2]

That Du Bois should choose to quote *Macbeth* is not very striking in itself. Such quoting is common in rhetorical compositions of the nineteenth and earlier twentieth centuries, and Du Bois was, among other things, an accomplished orator.[3] Yet the way Du Bois inserts this quotation into a meditation on the history of American slavery is significant. It comes out of a specific, longstanding pattern of quotation and allusion based in morally dualistic readings of *Macbeth* by antebellum abolitionists. So William Wilberforce, in 1799, writes about how a certain book "pleads, trumpet-tongued, against" slavery (Wilberforce 340, *Macbeth* 1.7.19). Again, Lydia Maria Child, writing in 1859, suggests that John Brown's rebellion ought to warn those who preside over the "peculiar institution" of slavery to heed an ominous passage from the same soliloquy: "this even-handed justice / Commends the ingredients of our poisoned chalice / To our own lips" (Child 110, *Macbeth* 1.7.10–12). Both abolitionists' allusions to *Macbeth* work by identifying slaves implicitly with Duncan and their masters with the murderous Macbeth.

Du Bois's allusion, thus, draws on an old tradition of applying the lessons of the dualistic *Macbeth* to the sin of racism. Moreover, his particular use of Banquo's ghost as a figure for the wrongs perpetrated by slavery—against slaves, freed slaves, and descendants of slaves—also has many nineteenth-century antecedents. Here is one, from an 1876 article in *The Republic* discussing the lasting pain and unfinished business of the Civil War:

> Those few statesmen left who passed through the ordeal of the battle for the Union naturally recognize the terrible legacy it has left the people.... The war has left records, traces, scars, and wounds that, like Banquo's ghost, will not down at the mere bidding. It is in vain to cry, like lady Macbeth, "Out damned spot," the stain is on the palm. ("The Bloody Shirt" 90)

Similarly, in 1882, George Washington Williams alludes to the murders of Duncan and Banquo in his *History of the Negro Race in America From 1619 to 1880*. Williams uses Macbeth's bad conscience as an example to explain the psychological motives that made white Virginians enact anti-miscegenation laws:

> In order to oppress the weak, and justify the unchristian distinction between God's creatures, the persons who would bolster themselves into respectability must have the aid of law.... Macbeth was conquered by the remembrance of the wrong he had done the virtuous Duncan and the unoffending Banquo, long before he was slain by Macduff. A guilty conscience always needs a multitude of subterfuges to guard against dreaded contingencies. (121)

In this recurrent trope of nineteenth-century *Macbeth* allusion, slaveowners and racist whites were dualistically identified with the evil Macbeth, American blacks with "the virtuous Duncan and the unoffending Banquo." Du Bois deliberately adapted that trope in 1903 when he represented blacks—or, more generally, the race problem—as a spectral intruder at America's banquet.[4]

The allusive identification of Banquo's ghost with the scars left behind by slavery proved compelling, echoing into the mid-twentieth century. One example is in Melvin Tolson's long poem "Dark Symphony." A prizewinning lyric at the American Negro Exposition in 1939 (Dove 1999, xiv), "Dark Symphony" is impressive both in its rousing dualism—here inflected by a Marxist-Leninist spirit of prophecy—and in its prosodic and semantic sophistication. It begins by invoking memories of patient suffering in the past: the slaves were "Loin-girt with faith that worms / Equate the wrong / And dust is purged to create brotherhood" (Tolson 37). The speaker then makes a direct *Macbeth* allusion (perhaps cued by the word "dust" as a reminder of the "Tomorrow" soliloquy, 5.5.18–27). In that direct allusion the anticipated redemption of blacks is contrasted with the just judgment that awaits their oppressors:

> No Banquo's ghost can rise
>   Against us now,
> Aver we hobnailed Man beneath the brute,
> Squeezed down the thorns of greed
>   On Labor's brow,
> Garroted lands and carted off the loot.
>   (37)

Tolson's black speaker attests to a clearness of conscience that whites will soon envy, as the coming Communist revolution unexpectedly realizes the eschatological promises of Christianity. The white Macbeths will then confront slavery's accusatory ghost.

Another poem by Tolson, "An Ex-Judge at the Bar," also identifies white society's rulers with Macbeth—but does so from the rulers' point of view. A 32-line dramatic monologue printed in the same volume as "Dark Symphony" (*Rendezvous with America*), "An Ex-Judge at the Bar" punningly refers both to the "bar" of justice and to a "bar" where Southern white men drink. It is spoken by the ex-judge of the title, who is white, to the "Bartender."

Tolson's white ex-judge explains how once, in another "bar," he faced "The Goddess Justice" when she appeared suddenly and confronted him "in naked scorn" (19).

> I listened, Bartender, with my heart and head,
> As the Goddess Justice unbandaged her eyes and said:
> "To make the world safe for Democracy,
> You lost a leg in Flanders fields—*oui, oui?*
> To gain the judge's seat, you twined the noose
> That swung the Negro higher than a goose."
> Bartender, who has dotted every *i*?
> Crossed every *t*? Put legs on every *y*?

The ex-judge then answered his accuser with words quoted from *Macbeth*:

> Therefore, I challenged her: "Lay on, Macduff,
> And damned be him who first cries, 'Hold, enough!'"
>   (19–20, *Macbeth* 5.10.33–34)

After this, the poem concludes enigmatically, with the white ex-judge ordering three drinks, "One for the Negro...one for you and me" (20).

Has the encounter of Tolson's racist ex-judge with Justice led the ex-judge to see equal treatment as a moral necessity? Or has it merely put him on the defensive and led him to offer "the Negro" a concession—a drink—in hopes of blurring blacks' perceptions of their situation? In either reading, Tolson compares the white-run justice system of the 1940s to the Macbeth of act 5, while presuming that the reader will bring to the poem a dualistic view of the play. The ex-judge, who once condoned Jim Crow and lynchings, now feels culpable and embattled, with his hypocrisies exposed, his heroism unmasked, and his mojo spent.

Race and the Problematic Macbeth: "Lay on, Macduff"

In contrast to allusions that presumed a morally dualistic *Macbeth*, other allusions presumed a morally problematic *Macbeth*. African Americans could be identified as Macbeths (or Lady Macbeths), forming resistant thoughts and sometimes boldly acting against the system that kept them down—an equation with moral implications rather too complex to be reduced to the simple terms "good" and "evil." Readers were invited to envision the race problem as an *agon*, not so much between absolute right and wrong, as between the particular, opposed wills of the oppressed and their oppressors.

As earlier essays in this collection have shown, some nineteenth-century allusions to *Macbeth* had already made this problematic identification between those who opposed or resisted white supremacy and Shakespeare's ambitious couple. In a revolutionary age the murder of Duncan might be freely interpreted as a deathblow against tyranny.[5] Some allusions to *Macbeth* in the 1910s, 1920s, and 1930s repeated this identification of self-assertive African Americans with Macbeth and his wife.

In 1917, for example, the reformer Ida B. Wells-Barnett used the very lines from *Macbeth* that Tolson had appropriated to express the desperate ferocity of a doomed Southern white supremacist—"Lay on, Macduff, / And damned be him that first cries, 'Hold, enough!'" (5.10.33–34)—to express the resolve that African Americans ought to show in fighting racism. A violent action by black soldiers incensed at their treatment in the U.S. Army had brought death to twenty-one whites and six blacks in Houston, Texas. "The army secretly court-martialed the soldiers involved, hanged nineteen of them without benefit of appeal, and sentenced many more to long prison terms" (Norrell 79). Wells-Barnett declared at a public gathering:

> I would consider it an honor to spend whatever years are necessary in prison as the one member of the race who protested, rather than to be with all the 11,999,999 Negroes who didn't have to go to prison because they kept their mouths shut.... Lay on, Macduff, and damn'd be him that first cries "Hold enough!" (quoted in Schechter 157)

Although she clearly held "the government" to be in the wrong, Wells-Barnett invoked "honor" and racial solidarity—not absolute moral or Christian virtue—as the principles requiring her and other blacks to speak out.

Rather than conjure with Banquo's aggrieved but silent ghost, Wells-Barnett assumed the more defiant posture of Macbeth's final hour, when he resolves to "try the last" and so to reclaim through self-assertion the honor he has lost (5.10.32). Her allusion intimated that Wells-Barnett was more a problematic than a dualistic reader of *Macbeth*. Honor is, arguably, what drives Macbeth to kill Duncan. Duncan has slighted Macbeth in bestowing the "honour" of the Scottish succession on his son Malcolm—while denying it to Macbeth, whose warcraft saved his regime (1.4.39). Just so, "honor" for Wells-Barnett meant fulfilling one's duty to one's kindred who are denied due respect and recognition. In choosing lines from Shakespeare to end her speech, Wells-Barnett might well have given thought to this analogy between Macbeth's deed and the soldiers' insurrection.

Noting Wells-Barnett's clear allusive identification of herself with Macbeth in 1917, we may find similar identifications in several well-known African-American lyrics from the following decade. One is Claude McKay's sonnet "If We Must Die." As James Weldon Johnson wrote in *The Book of American Negro Poetry*, McKay—"preëminently the poet of rebellion" (166)—composed the poem in response to the "terrifying summer of 1919, when race riots occurred in quick succession in a dozen cities" (167). McKay's speaker cries out in defiance, his words recalling (among other possible models) the same speech of Macbeth in act 5:

> If we must die—let it not be like hogs
> Hunted and penned in an inglorious spot,
> While round us bark the mad and hungry dogs...
> If we must die—oh, let us nobly die...
>   ...then even the monsters we defy
> Shall be constrained to honor us though dead!
> ...
> Like men we'll face the murderous, cowardly pack,
> Pressed to the wall, dying, but fighting back!
>   (McKay 168–69)

These lines echo Macbeth's defiant words, "At least we'll die with harness on our back" and "They have tied me to a stake. I cannot fly, / But bear-like I must fight the course" (5.5.50, 5.7.1–2).[6]

Comparable to McKay's "If We Must Die," yet more uneasy about its speaker's problematic likeness to Macbeth, is Countee Cullen's sonnet "Mood," from his 1929 collection *The Black Christ and Other Poems*. Cullen's speaker reacts to the agony of daily subjugation (if not to outright violence) with words that suggest sympathy for the resentful, yet ambivalent, Macbeth of act 2:

> I think an impulse stronger than my mind
> May some day grasp a knife, unloose a vial,
> Or with a little leaden ball unbind
> The cords that tie me to the rank and file....
> The meek are promised much in a book I know
> But one grows weary turning cheek to blow.
>   (Cullen 187)

Lyrics like these do not quote or explicitly invoke *Macbeth*. But they must have made many readers in the 1920s recollect Macbeth as the most famous literary prototype for their speakers' self-portrayals: a proud man baited by dogs and fighting to the death, a restive soldier tempted against his own rational "mind" to "grasp a knife."

The most sustained allusions to Macbeth as a figure for African-American self-empowerment and self-advancement at this time were not in speeches or sonnets but in drama and fiction. I will cite three examples that are linked by their shared focus on the Haitian revolution of the 1790s and early 1800s: Leslie Pinckney Hill's closet play *Toussaint L'Ouverture: A Dramatic History* (1928), Langston Hughes's theatrical play *Emperor of Haiti* (1936), and Arna Bontemps's novel *Drums at Dusk* (1939).

Since it was common for critiques of racism to allude to *Macbeth*, we should not be surprised to find three works about Haiti abounding in such allusions. Yet these three texts make Shakespeare serve different purposes. Two, Hill's *Toussaint L'Ouverture* and Bontemps's *Drums at Dusk*, suggest a dualistic reading of *Macbeth*: they allusively link the words and actions of Shakespeare's Macbeth to repugnant actions or characters. But *Drums at Dusk* also conveys a strong sense that rebellion is morally problematic. Finally, Hughes allusively identifies his protagonist in *Emperor of Haiti* with Macbeth—while also radically problematizing any attempt to distinguish morally between this Macbeth and his enemies.

Hill's *Toussaint L'Ouverture* hints at its allusive intentions early, in an exchange that glances at Macbeth's striking comparison between dogs and men (3.1.94ff):

> BOUKMANT: ... these pale gad-flies
> So sting and torture us that we have lost
> The attributes of men.
> BIASSOU: Say rather we
> Have lost the better qualities of dogs
> Without a dog's excuse.
>   (21–22)

Soon thereafter the entrance of "two old black conjure women... followed by a tattered company" announces a "Voodoo gathering" that unmistakably recalls *Macbeth* (27).

Like Shakespeare's witches, Hill's Voodoo women chant in tetrameter couplets (27–30). And like the witches in *Macbeth,* they meet the protagonist only on their second appearance, when a witch greets Toussaint with these words:

> We are sisters whom the Fates
> Choose to be their intimates,
> Blood relations of disaster,
> Whom the nether king, our master,
> Grants the canny power to see
> What the future is to be.
>   (67)

As if to confirm the link with *Macbeth*, the Second Old Woman's speech will include a "double"/"trouble" rhyme (67). Yet Toussaint contemptuously rejects the witches' magic.

Hill uses these allusions not to compare Toussaint to Macbeth, but to draw a strict contrast between them. Toussaint's moral "exaltation" (an adjective applied to Toussaint in Hill's preface, 10) is the enlightened opposite of Macbeth's superstitious awe. Hill reaffirms this point in a later scene, when, in a desperate situation after a betrayal by his associates, Toussaint encounters the women again; this time his speech rhymes "won" with "done" (125). Once more, Toussaint will reject the charms the Voodoo women offer him. These allusions support Hill's explicit vision of Toussaint as a pure and noble man, albeit one with a tragic fate (see L. Hill's preface, 7–10). Hill is relying on us to keep in our minds the "evil" Macbeth of the dualistic tradition as a foil for his superior Toussaint.

In *Drums at Dusk*, Arna Bontemps also alludes to *Macbeth* dualistically, but he does so more in the manner of nineteenth-century abolitionists: he identifies the worst people in Haiti's slave-owning aristocracy with the tyrant Macbeth. The novel's main character is a white youth, Diron Desautels, who belongs to the French abolitionist group *Les Amis des Noirs* and joins the revolution of Haitian blacks in the 1790s (5). The story follows his fortunes through the early stages of the revolution, before Toussaint's ascendancy.

Bontemps depicts the slaveowners' sadism and the decadence of the French ruling class through a melodramatically dualistic lens. When fighting breaks out, late in the novel, the French aristocracy has shown its corruption in many ways, but most fully in the acts of the repulsive Count De Sacy, who enjoys raping and otherwise humiliating slaves. As De Sacy attends a dinner party a drum signal is heard outside. A black man dressed as a waiter—actually a "member of the rebel band that haunted the hills"—springs "to the top of the table with a single leap, his boots crashing glasses, scattering plates and bowls" (121). As he runs out he shouts, "*Vive la liberté!*" The white guests then reenact Macbeth's traditional response to the ghost's appearance in the banquet scene (3.4.48ff):

> Men guests leaped to their feet. The women screamed. Chairs toppled backwards.... De Sacy, as pale as a dead man, bit his lips, slobbered at the corners of hit mouth, swore bitterly in a soft, unnatural voice. (122)

Like Du Bois and others before him, Bontemps identifies the presence of the black man, who brings white crimes home, with the revenant who haunts Macbeth at his table. And De Sacy's story will continue to follow the pattern of the dualistic *Macbeth*; for soon the Count will suffer a grotesque death at the hands of the blacks whom he has tormented.

Drums at Dusk is a thoughtful narrative that does not simplify the moral issues raised by revolutionary violence. To this extent its perspective is morally problematic. When the enlightened white, Diron, hears the "Voodoo drums" that herald the impending revolt, he thinks of their "strangeness, that suggestion of ghostliness, weird masks in the haunted darkness, witchery" (38). And when he sees the rebels at one of their meetings, he considers them a crude, "fairly repulsive bunch," like "the rabble who stormed the Bastille...mostly a tough and sinister lot" (54–55). Bontemps seems to share Desautels's reservations about voodoo and rebels. The problem with even the most justifiable revolutions, Bontemps implies, is that they can succeed only by releasing irrational and dangerous energies.[7]

Hughes's *Emperor of Haiti* contrasts with the work of Hill and Bontemps by focusing on Jean-Jacques Dessalines, Toussaint's historical successor. That choice allows Hughes to present a far more problematic view of the Haitian revolution, while avoiding any perceived insult to the memory of

Toussaint. The play is skeptical about all leaders. It portrays Dessalines as a vain man, one who is easily undone after he acquires power. In Hughes's twist on allusions to the banquet scene, the Emperor Dessalines is disturbed by a sound of voodoo drums—which he says he wants to banish from his court and replace with European music—and is then upset by news "that the peasants have revolted." His reaction allusively identifies him with Macbeth: "DESSALINES leaps up so quickly he overturns his wine, it runs like blood across the banquet cloth" (319).

Hughes makes "the peasants" play Banquo's part in *Emperor of Haiti*. But, rather than equate blacks with Banquo and whites with Macbeth, Hughes instead sees Macbeth in Dessalines, the peasants' black leader—and their new master. The struggle for racial justice, Hughes suggests, can only be a travesty without social justice. And social justice remains distant at the end of *Emperor of Haiti*. The men who plot to displace Dessalines are, if anything, worse than he. Before they kill him, one explains how to pass off their deed as a heroic action: "Listen carefully, men! A great honor's befallen us.... We're chosen to be the liberators of Haiti... to free our country from a power-loving tax-hungry tyrant" (328). In fact the men are plotting, after they kill Dessalines, to found a so-called "Republic" of which they will be the rulers (329).

The version of *Macbeth* that appears through Hughes's allusions in *Emperor of Haiti* is deeply problematic. Both Malcolm and Macduff are low self-seekers. The play's political vision is cynical; no prospect for a triumph of the "good" is seen. However, the point of exposing leadership's failures, for Hughes, was to raise the political awareness of his audience. Similarly, other early twentieth-century interpretations of *Macbeth* adopted a problematic view of the play in hopes of spurring playgoers to imagine another kind of political world (Moschovakis 26). In this way Hughes's treatment of *Macbeth* can partly be compared to the "Voodoo" *Macbeth* of Orson Welles, produced as *Emperor of Haiti* was about to be written (1936).[8] Welles's *Macbeth* reflected the same African American fascination with Haiti that had appeared in earlier texts, such as Hill's *Toussaint*—yet it was more like the contemporary works of Hughes and Bontemps in its problematizing and (in some ways) cautionary focus on the inherent savagery of political violence (see the Rippy essay).

Conclusion

In this essay's second epigraph, W.E.B. Du Bois asks ironically whether African Americans can "aspire" to any higher end than disclaiming their race and affirming the values of whites. His defense of ambition and self-definition in the face of prejudice can seem to recall Nietzsche's praise of the ambitions that drive Macbeth and Lady Macbeth. Still, Nietzsche makes it clear that the passions of Shakespeare's couple are not just great but self-destructive. What end awaited those African Americans who heeded the call of this passage by Du Bois, or of the speech Wells-Barnett quoted from *Macbeth*—or who emulated the rebels in Bontemps's *Drums at Dusk* and in Hughes's *Emperor of Haiti*? None would wish Macbeth's tragic end on themselves. But many who fought for power could expect to meet something like it.

Early twentieth-century African-American allusions to *Macbeth* voice the moral dualism of those who are assured their cause is just. Yet they also betray a problematic identification with characters who, resenting the barriers to their advancement, embrace deadly violence. One possibility appears not to have appeared to the authors examined here: that African Americans need not identify only with the murdered (Banquo and Duncan) or with the murderers (Macbeth and Lady Macbeth), but they might instead identify with the legitimate redeemer, Malcolm, moral hero of the dualistic *Macbeth*. The disinherited might return, as heirs ascendant, to claim a kingdom rightfully their own. Now that political reality has shown something verging on this to be symbolically possible, might we expect our authors to allude to a *Macbeth* that is—once again—assumed to end in a triumph of the "good"? Might we recover our ability to read into *Macbeth* the morally dualistic, affirmative resolution that the play had for its audiences a century ago?

Notes

1. On the sense of *dualistic* and *problematic*, and on the development of this polarity in *Macbeth* performance and criticism, see Moschovakis (1–5).
2. For remarks on this passage that are compatible with mine, see Balfour 39 and 150 n.9.
3. An edition of Bartlett's *Familiar Quotations* from the same year as *The Souls of Black Folk* had eleven pages for *Macbeth*—more than for *King Lear* or *Othello*, though fewer than for *Hamlet*. See Bartlett 115–26.
4. See also the Nathans essay for two earlier, antebellum examples of this trope from the 1830s and 1840s.
5. See the Nathans essay. Somewhat similarly, Macbeth's belated attempt to wrench free of the witches' influence might seem an apt symbol for a nation straining to shed the burden of a terrible past; see the Briggs essay.
6. For the first line's partial echo of a 1911 essay by Du Bois, see Turner (65). For its near-quotation of another Shakespearean line (from *Measure for Measure*)—in a speech that, however, runs against the feeling of McKay's poem—see Maxwell's note in McKay (333).
7. For a different argument, see M. Thompson.
8. In a letter to Hughes dated November 8, 1936, Arna Bontemps remarked on reading a draft of *Emperor of Haiti*: "Most of the play, of course, presents Negroes in a mood unfamiliar to usual stage productions, but I imagine the production of such opuses as *Macbeth* has partly cleared the way" (Bontemps and Hughes 25).

3
Federal Theatre Project(s)

Figure S.3 Romare Bearden, poster for New Federal Theatre production of *Macbeth*, 1977. Art © Romare Bearden Foundation / Licensed by VAGA, New York, NY.

8

BEFORE WELLES:
A 1935 BOSTON PRODUCTION

Lisa N. Simmons

I am a filmmaker working on a documentary about the Negro Theatre Project in 1930s Boston. This project is very near and dear to my heart, primarily because my family was extensively involved in it, starting from my grandfather Lorenzo Quarles (who was the stage manager) to my aunts and uncles (who were actors and directors), including Frank Silvera and Luoneal Mason. As a contemporary producer who has produced films *and* film festivals for filmmakers of color, I am also passionately invested in recovering the histories of earlier institutions that supported black artists. Since this Boston troupe mounted an all-black classical *Macbeth* in 1935, a half-year before Welles's "Voodoo" *Macbeth*, I believe it is crucial to document this little-heralded antecedent to the far more famous Harlem production.

Boston has a rich yet often overlooked African-American cultural history. Black theatre has existed in Boston since at least the mid-nineteenth century, when William Wells Brown read his antislavery plays to interracial audiences. Blacks began performing pageants and staging plays in and around Boston in the late nineteenth century. Pauline Hopkins, a black female playwright who came to Boston as an infant, had her play *Peculiar Sam, or the Underground Railroad* performed numerous times at Orchard Garden throughout 1879 (L. Brown 111). With the establishment of The Women's Service Club (later to become the home base for the 1930s "Boston Players") and The League of Women for Community Service during World War I, cultural events enjoyed more regular sponsorships (see Fleming and Roses).

By the mid-1920s, Texas native Maud Cuney-Hare founded the first black arts center in Boston, the Allied Arts Center, including the Allied Arts Players, a theatre company. The Allied Arts Players gave talented black actors a beginning in theatre by staging plays in the Fine Arts Theatre and the Allied Arts Centre. A graduate of the New England Conservatory of Music and noted musicologist, Cuney-Hare is primarily remembered for her important study *Negro Musicians and Their Music*, one of only three books about black music by an African-American author published prior to 1960. The Allied Arts Centre's purpose was

> to discover musical, literary, and dramatic talent and to encourage the same; to arouse interest in the artistic capabilities of the colored child; to call attention to his or her aspiration and later to seek an open door of opportunity that colored youth may fittingly contribute to the new and tremendously important wave of art development in America. Through Art to cultivate friendliness with all racial groups and to become one of the noteworthy streams in the making of an ideal New England and American spirit. (Allied Arts Players)

Plays that were mounted included William Edgar Easton's historical drama based on the Haitian revolution, *Dessalines*, in 1930. Cuney-Hare's own *Antar of Araby* was produced and directed by Ralf Coleman in 1929, and was apparently "influenced by Shakespeare's diction and themes" (Koolish 95).

Cuney-Hare's Allied Arts Players were, in effect, the launching pad for Coleman's important career. Coleman was born in Newark in 1898 and was adopted along with his brother by a Baptist minister who had no love lost for the theatre (Hill and Hatch 332). To the boys, however, theatre was divine, and they found their way to the stage; Coleman would later recollect how his high school teacher, Professor Gregory, forced memorization of Shakespearean scenes, which eventually "lit the candle" of his devotion to drama (Singer). In 1921 he was the narrator of a pageant in Boston's Symphony Hall, and soon became affiliated with Ford Hall Forum, which "had a program director [who] was intensely interested in Negro Theatre in those days" (Singer). By 1927, he began working with Cuney-Hare and the Allied Arts Players. According to Coleman:

> [Cuney-Hare] had her studio at Gainsboro and Huntington Avenue and we had all the arts represented. We had dance, Mildred Davenport and a sculptress from Framingham, we had artists, Allan Crite, costume designers, scenery painters, all those things. We did studio plays in a little studio, one act plays, black plays, some written by Black authors and at the end of the year, we did a big production [at the Fine Arts Theatre]. (Coleman)

Thus, Cuney-Hare recruited widely, seeking to showcase the depth of local black talent.

By 1930 however, because of a disagreement with one of the actors (namely, Lucian Ayers, who would later be the lead in *Macbeth*), the Allied Art Players split, and a new group was formed with Coleman at the head: The Boston Players. This was a more professional theatre troupe, modeled after the famous Harlem stock company, the Lafayette Theatre. The Boston Players performed in and around Boston and were eventually signed to Broadway after a broker, Margaret Hughes, saw Coleman's production of *Scarlet Sister Mary*. Hughes asked the company if they would like to do a new play by Paul Green, *Potter's Field* (eventually renamed *Roll, Sweet*

Chariot). It was produced on Broadway in 1934. Like Cuney-Hare, Coleman had a passion for showing that his actors could perform more than just the "negro folk" plays that were so popular in the 1930s. He produced works by major contemporary white dramatists such as George Bernard Shaw and Eugene O'Neill.

Coleman's work with Green's play positioned him well to take advantage of a unique opportunity. President Roosevelt's Works Progress Administration (WPA) put Americans to work through various infrastructure programs. On a suggestion from Eleanor Roosevelt, the WPA also employed artists through a number of federal projects, such as the Federal Writers Project, the Federal Music Project, the Federal Art Project, and the Federal Theatre Project (FTP). Eleanor Roosevelt believed actors to be as important to the moral fabric of America as civil workers. Thus, serious dramas as well as plays about the daily life of average Americans were part of the ethos of the New Deal.

Unknown to many today, the Negro Theatre Unit as part of the FTP promised to be the beginning of a great era for black theatre in America. By 1935, federally supported black theater companies were established in major cities across the country (Fraden), and many of the Boston Players were placed on the payroll of the Negro Federal Theatre of Massachusetts. Between 1935 and 1939, plays were performed in and around the New England area: in Boston, Salem, Rhode Island, and Springfield. In a 1971 speech, Coleman recalled that

> [t]his was not welfare, but pay for services rendered, and for the first time and the last time in the history of this country the Government subsidized the theatre. Federal theatres were established in all the major cities across the country and I like to think that Boston was outstanding. (Coleman)

Negro folk plays were staged alongside dramas such as *The Trial of Dr. Beck*, *Tambourines to Glory*, *The Emperor Jones*, *Return to Death*, *Brother Mose*, among others.

Coleman directed an all-black *Macbeth* in the fall of 1935. On September 22, *The New York Times* reported that Coleman's

> Negro Shakespeare troupe, which had been ordered to the Saugus [Massachusetts] Town Hall by the ERA [Emergency Relief Administration], hurried through "Macbeth" in record-breaking time and was wildly cheered when it returned after a breath-taking ride from Saugus, doffed Shakespearean costumes and donned waterfront motley [to perform *Stevedore*,] a Negro waterfront strike play. ("'Stevedore' Hits a Snag in Boston")

The play was subsequently performed on October 18 at the High School of Practical Arts, Roxbury School Center. A program from that performance indicates that Lucian Ayers played Macbeth; Luoneal Mason, Banquo; Frank Silvera, Malcolm; Lorenzo Quarles, Donaldbain; and Ruth Johnson, Lady Macbeth. While I presume there must have been additional performances of *Macbeth* in the interim four weeks, I have yet to uncover further documentation confirming specific dates and venues. As the *Boston Chronicle* reported on October 19, 1935:

> Shakespeare's *Macbeth* was presented by an all negro cast last evening... Negro members of the Boston Civic Theatre appeared in the tragedy under the direction of Ralf Coleman. A musical program was provided by an all-Negro orchestra under the direction of Theodore Bailey. The entertainment was given in cooperation with the Boston School Centers as part of the program of extended use of public schools. Sponsorship was by the Emergency Relief Administration with local sponsorship by the Roxbury Joint Planning Committee of the City-Wide Emergency Committee. ("All Negro Cast Presents Shakespeare")

The Boston *Macbeth* emerged during the interesting early stages of the transition from the ERA to the FTP, as Congress finally approved funding in October 1935. From the records I have reviewed, it appears that this presentation of *Macbeth* was staged largely as Shakespeare wrote it, with the "all negro" cast playing all of the characters. It was not set in any exoticized "black" locale; it was simply to be Scotland. Unlike Welles's later "Voodoo" *Macbeth*, Coleman reproduced *Macbeth* in its original version because he strongly believed in presenting the talent and range of his black actors.

I am left to speculate whether Welles somehow caught wind of Coleman's earlier production. Welles always maintained that his wife Virginia (during a conversation with her friend Francis Carpenter) came up with the idea of performing *Macbeth* as the first play for the New York Negro Theatre Unit. Might Welles have been partially inspired by this 1935 version in Boston some six months before his production? If so, this would prove an important complication of the Welles legend. What has come to be known as a white boy genius directing the most well-known all-black Shakespeare play in American history might need to be contextualized instead by the possibility that Welles himself was following a black director's precedent in selecting *Macbeth*. Coleman's silent ghost of a production would thus "push" Welles's *Macbeth* from its "stool" (3.4.81).

9

BLACK CAST CONJURES WHITE GENIUS: UNRAVELING THE MYSTIQUE OF ORSON WELLES'S "VOODOO" *MACBETH*

Marguerite Rippy

Orson Welles directed an influential all-black adaptation of *Macbeth* for the Negro Theatre Unit of the Works Progress Administration in 1936, in which the play's setting was famously transposed to the nineteenth-century Caribbean, deliberately evoking historical resonances between Macbeth and the Haitian Emperor Henri Christophe. As with any Welles endeavor, numerous fascinating anecdotes surround the production, many of which remain unverifiable and impossible to separate from Welles's unusual talent as a *provocateur*. Most notoriously, Welles claimed after the fact to have once filled in for the ailing lead actor when the show was on tour in Indianapolis; years later, he recounted with pride how he performed the role so convincingly in blackface that no one noticed that it was a white actor, much less Orson Welles, performing as Macbeth. Such an anecdote is, by necessity, deviously unverifiable, for it turns on the supposed fluency of Welles's performative passing: if anyone had actually recognized Welles (no reviewers noted him), his "blackness" would not have been a success. Stories such as these capture the mix of clever self-promotion and audacity that would come to characterize the career of the then twenty-one-year-old artist. But they also begin to convey a more troubling blend of racially progressive politics and racially insensitive opportunism on Welles's part that are characteristic of the *Macbeth* production as a whole. The following essay complicates the "Voodoo" *Macbeth* mystique by detailing some of the ambivalently progressive and conventional aspects that it embodies.

Shakespeare's *Macbeth* captures an English fascination with Scotland, perceived in early modern England as a frightening land of the uncivilized. Welles's adaptation transposed this fascination into a postcolonial context by moving the play to Christophe's nineteenth-century court, transforming the witches into voodoo priestesses with Hecate as their (male) witch doctor.

The production was popular, as Carl Van Vechten explains to Langston Hughes after seeing it for the third time:

> Again crowds, again cheers, again all sorts of excitement! I've found out at last what Harlem really likes. Have you ever heard of any other playwright who could create standing room at every performance at the Lafayette for five weeks? (137)

Promotional material from the Federal Theatre boasted that 150,000 people saw the play in its New York run alone. *Macbeth* provided work and wages for a large segment of the black theatrical community and helped promote prominent African-American actors like Edna Thomas, Jack Carter, and Maurice Ellis. The production also showcased new performers, including a group of drummers from Sierre Leone who stole the show as the witches, receiving critical praise both in New York and on the road tour. It is this populist legacy that is often celebrated by the actors who worked with Welles.[1]

However, the concept of the Negro Theatre Unit within the Federal Theatre Project was controversial for African Americans at the time. The appointment of John Houseman to lead the unit made it vulnerable to charges of racism from the outset. While Rose McLendon, the "Negro first lady of the dramatic stage" (Effinger), was originally appointed to share responsibility with Houseman, she became too ill to remain actively involved, leaving Houseman alone as the white producer. Houseman sought to alleviate the situation by dividing the work of the Negro Theatre Unit into two categories: "work by, for and with black actors [the direction of which was soon handed over to Countee Cullen and Zora Neale Hurston, among others], and classical work performed by black actors but staged and designed by white artists" (Callow 1997, 36). In the second category, Welles's *Macbeth* mingled Haitian voodoo imagery with Shakespeare. In doing so, Welles complicatedly evoked not only a Caribbean analogue but also the recent preoccupation with Haiti by African-American intellectuals, artists, and revolutionaries of the 1920s and 1930s (K. Thompson).

The production has been criticized in retrospect for reenacting white colonial fantasies of race. Indeed, Welles incorporated modernist fantasies of the primitive into his "Voodoo" *Macbeth*, even as his production encouraged multiracial social agitation.[2] Ultimately, Welles stopped short of a blackface parody, although his production did invoke white fantasies of blackness as the dark id to a white ego. His interpretation of blackness drew upon the 1920s tradition of modernist aesthetics, depicting blackness in heavily primitive terms as something powerful, compelling, and supernatural.

The racial images in Welles's 1936 "Voodoo" *Macbeth* reflect a legacy of turn-of-the-century racism in performance. Blackness and the exotic flowed in two currents in the early twentieth century. The first George Frederickson terms "romantic racialism," inherited from nineteenth-century sentimentalism. Romantic racialism gives us the stereotypical images of the suffering mulatta, the exotic "black buck," and the endearing Uncle Tom. Its affectionate condescension laid the groundwork for modernist uses of blackness as metaphor and symbol. Modernist primitivism retained the romantic associations of blackness with dark passions and the supernatural or, as Marianna Torgovnick calls them, "our id forces" (8). For decades, blackness as a symbol of lost passion and simplicity made primitivism seem romantic, even nostalgic to white audiences. But as Torgovnick notes, by the 1930s primitivism was also criticized by fascists as a symptom of debased, promiscuous folk culture (9). "Voodoo" *Macbeth*'s association with its Harlem locale further tapped into the legacy of romantic racialism through its presentation of an "urban primitive." Peter Stanfield defines the urban primitive as a "lowbrow populist rebuke to the dehumanizing subjugation of modernity" that evokes the agrarian roots of black American culture against the modern cityscape (90). This formulation would explain the longing of many urban critics for a southern, rural African-American vernacular to replace Shakespearean iambic pentameter. In other words, the fantasy of urban primitivism would be complete if the overt evidence of white elitism were removed from the production.

The second performance tradition that Welles inherits comes in the form of expressionism, specifically the auditory notion of the black primitive drumbeat as an emblem of a primitive id that arises within the civilized ego. Welles's "Voodoo" *Macbeth* directly invoked the use of the incessant tom-tom beat found in Eugene O'Neill's *The Emperor Jones* (1920). Whereas O'Neill used a single, increasingly rapid drumbeat to represent Brutus Jones's interior struggle of conscience and guilt, Welles employed a group of supposedly "voodoo" drummers (they were actually African, not Caribbean) to represent Macbeth's similar struggle of conscience.[3] The collective performance structures of the Federal Theatre created an important distinction between Welles's adaptation and O'Neill's play, however, in that Welles's production featured a black cast at a time when being a working actor of any race was a rarity. The Harlem production provided steady employment and good publicity for approximately 125 players as well as musicians, including a broad mix of newcomers and respected veterans; even the child actors got substantial press coverage. This collective impulse contrasts *The Emperor Jones*, which highlights a single black actor and relegates the other black parts to literal shadow roles. Yet this very collectivism also served to further Welles's individual reputation, since the de-emphasis of actors-as-stars left the press seeking a focus for their publicity. As a result Welles was often positioned in reviews as the boy-genius, the auteur, the (self-professed) expert on Shakespeare who at times must act as the colonial governor paternally guiding his inexperienced cast.

Of course the funding of the Federal Theatre enabled a large cast and high production standards, since as reviewer William McDermott pointed out, "the WPA has no need of economy in the employment of labor" (10). Charles Collins of the *Chicago Tribune* noted more colorfully, "every stage director who finds WPA funds at his disposal exclaims 'Whoopee' and breaks loose on an orgy of experimentalism" (21). In this case Welles's experimentation with primitivism (conveyed through what was essentially leftist political theatre) positioned him at the intersection between the popular and the elite, the location of which Werner Sollors terms "populist modernism." Populist modernism asserts a collectivist impulse that paradoxically integrates modernist fascination with "high" art and Western elitism (8). Kobena Mercer calls populist modernism "a kind of surreptitious return of the binaristic oppositions associated with an essentialist concept of ethnicity arising in the very discourse that contests and critiques it" (245). As Mercer suggests, Welles re-articulated whites' fascination with the primitive even as he critiqued the politics of racism. In fact, a major difference between Welles and many of his contemporaries is that his interest in primitivism coincided with a deliberate effort to hire actors and actresses of color to perform his scripts, and the 1936 *Macbeth* was perhaps his most overt endeavor in this regard.

The "voodoo" witches and drummers were the most important actors that Welles employed in terms of creating a distinctive show. These characters would not have been central figures in a traditional production of *Macbeth*, but they formed the backbone of Welles's production.[4] The audio cue of the primitive drum captured the literal and figurative pulse of the "primitive" within *Macbeth*, and represented both an "authentic" Haitian feel and a white fantasy of the darkness of voodoo. Welles literally placed the drummers—and thus the witches and Hecate—center stage. These roles were consistently the focus of positive reviews, in New York as well as on tour. Although there is no existing recording of the full performance, the script describes the setting as follows:

> Before the curtain rises, drums and voodoo chanting in the darkness and the scene becomes gradually visible...Voodoo ceremony in progress. Half-circle of white-clad women celebrants—in the center, somewhat raised from the stage level, the three witches. To the left, the dark figure of Hecate. Distant low thunder and lightning (1.1)[5]

Hecate (played by Eric Burroughs) is given lines from both the witches and Macbeth himself, and even participates in Banquo's murder, making him a central character. Hecate and the three witches are described as "birds of prey" that stalk Macbeth as their victim (1.2). Because Hecate presides over all the events as a master-magician, Macbeth's fate becomes the product of his dark magic. Burroughs, one of the few classically trained actors in the production, received strong reviews, and was instrumental to the production's success. Collins remarked, "Hecate is here a male slave-driver of witches and a fiendish director of Macbeth's destiny" (21). Welles divides the text into two acts in the draft script (which would become three acts in the performance script), and each act is framed by Hecate's curses. At the end of the first act, Hecate promises supernatural tortures for Macbeth will follow:

> I will drain him dry as hay;
> Sleep shall neither night nor day

Hang upon his pent house lid:
(Drums stop)
He shall live a man forbid!
(A thump of a drum on the last syllable of "forbid")
(1.2)

Hecate often closes scenes, immersing the crowd in darkness and silence. For instance, he closes the first and last scenes by crying, "Peace," at which the "drums, army, music, voodoo voices, all are instantly silent" (2.6). Then he initiates his "spell," first on Macbeth (1.1), and later on Malcolm (2.6), by crying, "The charm's wound up!" Like a ringmaster, Hecate controls the tempo of the production through chants and drums.

In terms of its use of audio cues, the "Voodoo" *Macbeth* represents a strong connection to Welles's radio dramas, as it marked scene shifts with aural cues beyond the more conventional visual cues. It also channeled white fantasies of the exotic through the filter of claims of authenticity. The promise of an authentic Haitian voodoo experience lies at the center of several critical reactions to the show, and resonates in the memories of those involved with the production. The drumbeat simultaneously conjured modernist expressionism and a fantasy of direct experience with Haitian existence, all conveyed through the figures of the drummers. An advertisement for the Dallas tour features a graphic illustration of only the drummers (no Macbeths in sight), whereas the accompanying article describes the production as an "'empire' with kaleidoscopic variations growing out of primitive imaginations" (Rosenfield 10).

Welles's use of the drums to mark scene divisions, to bridge events, and to control the audience's perception of the play itself extended beyond O'Neill's decorative drumbeat, in no small part because of the creativity of the drummers themselves. Richard France remarks that Welles's drums created transitional markers that were "the equivalent of a film dissolve" (1974, 68). Houseman praised the drummers for creating a "supernatural atmosphere [that] added to the excitement that was beginning to form around our production of *Macbeth*" (97). And Arthur Knight articulates the centrality of the drums to the production by saying, "No one who ever saw his *Macbeth* will forget the rhythmic pounding of jungle drums as an underscore to the mounting tragedy" (quoted in France 1974, 68).

The reaction of reviewers to the voodoo drummers, however, shows how deeply the primitivist images resonated for audiences, in many ways fulfilling the fears of those who thought the production would highlight stereotypes of black culture rather than political realities or Shakespearean themes. From the outset of the production run, the drummers stole the show. Reviewers present at the opening on April 14, 1936 were fascinated by the casting and by the drummers in particular. Failing to appreciate the psychological complexity of Welles's overall adaptation, they tended to see the show as a musical. Solidifying the fears of the Black Communist Party, the word "amusing" keeps popping up in their reviews—a word not usually applied to Shakespeare's *Macbeth* (Callow 1995, 240).

Racist reactions to the all-black cast circulated around two areas: imperfect articulation of Shakespeare's language and titillation over physical allure. The bodies of the black male players in particular were prominently featured in reviews. The physical presence of the drummers created a tension between what these viewers regarded as the incompatible formality of English verse and the rhythmic drumming. John Mason Brown of the *New York Post* complained that the witches were "mumbo-jumbo agents of a fearful witch doctor," and Macbeth himself was "a sort of Brutus Jones Macbeth"; he lamented that Welles did not adapt Shakespeare's verse into dialect to make the language match "the locale" (quoted in Callow 1995, 240). A very similar review in *Variety* referred to Carter's portrayal as "a bit closer to the Emperor Jones than to Macbeth," and suggested the cast was "mouthing antiquated Elizabethan language which, quite obviously, they don't even understand" ("Macbeth" 61). The production of an African-American *Macbeth* raised the question of whether the adaption of the text to African-American dialect could have created the Tambo and Bones *Macbeth* that African-American groups feared. Welles's description of why he chose to cast and adapt his production as he did provides no reassurance that he possessed a more progressive vision of his actors' abilities. Reminiscing about the production in February of 1942, Welles recalled:

> The negro with his virgin mind, pure without any special intellectual intoxication, understands the essence of the Greek theater better, which demands simplicity and the absence of artificial interpretations. I staged *Macbeth* by Shakespeare in Harlem, New York, by a cast of Negros [*sic*] and never had I the trouble to read or to correct speech and intonations. They discovered all themselves, with a prodigious intuition that they have in a high degree for the tragic theater. And this in the face of their never having read or heard speak about Shakespeare! (Welles 1942, 5)

Brooks Atkinson, the most influential theatre reviewer of the day, reacted with a fascination for the eroticized physicality that characterized the adaptation, praising the "sensuous, black-blooded vitality" of the witches and admiring Jack Carter as a "fine figure of a negro in tight-fitting trousers that do justice to his anatomy" (25). One cannot help but suspect the tom-toms were playing a bit too loudly for Atkinson, whose review reveals more about his own desires than about the performance. Similarly, a review by Len Shaw in the *Detroit Free Press* praised the Witch Doctor Abdul as "a torso twisting witch doctor." In Cleveland, William McDermott described the witches as "shrieking sable slatterns of horrific men," and said of the show, "It is primitive, it is crude, and it has force and a kind of curious beauty. So far as Shakespearean production goes it is a stunt, but a successful stunt" (10).

The critical fascination for the play tended to emphasize the "Voodoo" *Macbeth* as a sensory experience. Robert Tucker of *The Indianapolis Star* praised the production's "savagery and voodooism and the manner in which the uncanny atmosphere is created and carried on" (11). The artistry of the drummers was continually ciphered through layers of white fantasy. Because we lack a recording of their performance—they are not included in the extant four-minute clip of the end of the production from the WPA documentary *We Work Again* (1937)—the drummers exist in critical discourse mainly as embodiments of these fantasies of the Haitian primitive. The connotation of the supernatural drumbeat spread outside the performance itself, and Welles, Houseman, and National FTP Director Hallie Flanagan all repeat a story in their separate memoirs that reflects the fascination with the drummers' purported witchcraft. According to the popular story, Abdul placed a lethal curse on the reviewer Percy Hammond after he wrote a bad review of the play. Houseman melodramatically recalled Abdul as knowing "no language at all except magic" (97). Tellingly, however, in his version of the Hammond curse, Houseman remembers Abdul approaching him with the review in hand and asking if it was the work of a "bad man" (97). It seems the purportedly pre-lingual primitive had been reading his own reviews with a critical eye. Flanagan's version is less dramatic, pointing out that the drummers were, in fact, African, but still suggesting that their curse "took the life of a dramatic critic who had underestimated either the production or the occult powers of the performers" (74).

Welles's "Voodoo" *Macbeth* reflects modernist conceits of "black" primitivism, commingling African, Afro-Caribbean, and African-American cultural referents to produce a fantasy of black culture. While his later projects became more self-conscious about deconstructing such fantasies, "Voodoo" *Macbeth* succeeded precisely because it re-articulated the primitivist aesthetic. Using Shakespeare to legitimize primitivism, Welles created a blend of high art and popular culture that drew in crowds and critics alike. His canny ability to channel racial associations from his surrounding culture bolstered his reputation as a genius, helping to catapult him from Federal Theatre Project director to co-creator and owner of Mercury Productions, which in turn would ultimately lead to his RKO film contract and his ensuing cinematic career.

The "Voodoo" *Macbeth* continues to evoke ambivalence, as it did even in its initial performances. It should be justly celebrated for showcasing a black-cast *Macbeth* on a national tour when segregation was still dominant. Yet we also need to scrutinize the production's more disturbing aspects, which too easily trade in stereotypical fantasies of the primitive. Charles Collins, reviewer for the *Chicago Daily Tribune*, captured the conflicts inherent in the performance when he reacted to the 1936 touring version:

> I wish that [Welles] had handled the Haitian aspects of his production without going on a mumbo-jumbo rampage. He might have evoked some authentic voodoo effects in the witchcraft scenes instead of fraudulent fantastics which suggest a Congo village of an exposition carnival street.... The acting of Eric Burroughs, however, prevents one from hoping that a part of the witches' curse fall upon Mr. Welles, the hardy improver of Shakespeare. (21)

Notes

1. For instance, the actor who played Hecate, Eric Burroughs, depicts Welles as a sensitive young man of genius; his memory of the production has recently been articulated by his son Norris in a graphic novel (Burroughs).
2. The role of modernist primitivism in Welles's work is explored more fully in my study *Orson Welles and the Unfinished RKO Projects: A Postmodern Perspective*; see also Halpern.
3. There is a tantalizing circularity in the relationship between *The Emperor Jones* and the "Voodoo" *Macbeth*, since O'Neill's work was itself influenced by Shakespeare's *Macbeth* in its ghostly visions and themes of ambition and revenge (Berlin 29). Fellow contributor Amy Scott-Douglass detailed further suggestive links in her paper, "Interracial Couples in *Macbeth* Spinoffs," presented at "Shakespeare in Color: A Symposium on African American Performances and Appropriations," Rhodes College, Memphis, January 25, 2008.
4. For more on the reasons for Hecate's diminished role in performances, see the Daileader essay.
5. Since the draft script of Welles's *Macbeth* contains page numbers rather than scene numbers, all references refer to act and page numbers (Welles 1936). This draft script contains only two act divisions rather than the three acts of the final production, but is cited here in order to include handwritten comments.

10

AFTER WELLES: RE-DO VOODOO *MACBETHS*

Scott L. Newstok

I'LL DO, AND I'LL DO, AND I'LL DO (1.3.9)

The *Macbeth* that Orson Welles directed in 1936 has come to attain a complicatedly iconic status for many subsequent African-American productions of the play and even for black theatre more generally. Indeed, the photo of the Welles premiere that graces our cover was included in one of the playbills of the New Lafayette Theatre, a company that from 1967 to 1972 was exclusively devoted to "the development and production of original contemporary plays about the twentieth-century African-American experience," with no Shakespearean adaptations whatsoever (Orman).[1] Woodie King, Jr., founder of the New Federal Theatre (named after the WPA's Federal Theatre), likewise cites the image as a record of a transformative moment: "Those early photographs of people crowding outside of the theatre: They are smiling, and they are happy, and they are dressed up to go to the theatre in Harlem" (2003, 106). The notoriety of the FTP production has helped to establish *Macbeth* as arguably the most popular Shakespearean play for contemporary black repertory, as the Appendix to this collection documents its frequent appearance at black theatre companies as well as at Historically Black Colleges and Universities (HBCU) drama programs. Performing in this specific play often stands metonymically for advances in non-traditional casting; for instance, when one reviewer notes that "Black actors have long felt that they are not given a fair whack at the major roles of English drama," it is telling that he envisions that progress would entail audiences who "should be able to swallow a black Macbeth in Shakespeare" (Kingston).

Indeed, it seems that nearly every black *Macbeth* production is expected to situate itself in some relation, howsoever tenuous, to that of Welles. Sometimes the Negro Theatre Unit's version is explicitly acknowledged as an historical breakthrough, serving as an archival site for dramaturgical reconstruction, as with the 1977 New Federal Theatre "revival," discussed later in this chapter. At other points, it gets cited in programs as a theatrical precedent, as in the 2002 Colorado Shakespeare Festival, which includes reflections on it in its Dramaturgical Notes; or the 1990s Haworth Shakespeare Festival, which went so far as to produce "educational materials for students tracing the history of [Welles's] production," despite their contemporary African, not nineteenth-century Haitian, setting. When the FTP production is deliberately disavowed as a model (e.g., "We talked about Welles's version, but we're not that specific about where this takes place" [Cuthbert]), the figure of Welles still seems to linger in insistent queries from audiences and reporters, as evidenced in a 2008 interview for a production at the HBCU Dillard University: "So is it modeled after Orson Welles's infamous 'Voodoo Macbeth'?" (Cuthbert). Even productions that fail to come to fruition can nevertheless reaffirm the extraordinarily lasting, generative power of the FTP adaptation. For instance, Mabou Mines founder Lee Breuer contemplated a revival of Welles's production in the 1990s; after his production failed to materialize, his planned Macbeth, Carl Lumbly, sought out the Berkeley Rep to mount a production with two black actors as the lead couple (Winn).

The insistent, one might say obsessive, return to Welles's production—and usually Welles's production alone—too often occludes the dozens of other intervening black productions of the play, producing the curious result wherein a company can overlook some half century of history and mistakenly assert that "there have been few, if any, professional African-American productions of *Macbeth* since the Mercury Theatre off-shoots by the Federal Theatre Project in the 1930s" (Bourne 13). It is difficult to think of another such Shakespearean production that has attained such a hyper-canonical status within the history of African-American theatre[2] or another production to which so many resources have been repeatedly devoted to "reviving" its entire artistic vision, thus making it more akin to a restaged ballet than a conventional Shakespearean staging. It is as if we have somehow inverted Mary McCarthy's 1944 survey of American theatre, which noted with some chagrin that

> it is significant that our white culture has had to draw so heavily on the Negro (witness the *Macbeth*, Jack Carter as Mephistopheles in Orson Welles's *Dr. Faustus*, and now Paul Robeson and the case of *Carmen Jones*) for the revivification of its classics. (66)

The reversal signifies that black culture draws heavily on the figure of white Welles for the revivification of an all-black classical production.

The "re-do" of my title deliberately rhymes with the "voodoo" tag that came to be associated with the 1936 production, and thereby echoes Cole Porter's chorus from "You Do Something to Me" (1929): "Do do that voodoo that you do so well" (117). We do well to remember this Porter for his "verbally acrobatic" translation of the primitivist "doo" to a pragmatic "do" (Bernstein 353)—coyly domesticating the exotic while still maintaining the *frisson* of its ("dark") foreign origin, much like Welles's own production. Porter's "do" here turns on the copulative overtones of the verb, which Shakespeare had earlier troped in Aaron's barbed revision of "done" from generic action ("what hast thou done?") and reputation ("Thou has undone our mother") to (racially charged) sexual act: ("I have done thy mother") (*Titus Andronicus* 4.2.73, 75–76). "Re-do" also takes us back to *Macbeth* itself, a play that not only "obsesses on the word *do*" (Rayner 59), but that furthermore enacts a constant sonic *re-doing*.[3] Welles's own production notably adds a further repetition: using Hecate's "The charm's wound up" at the opening as well as close of the play (and in his 1948 film as well). "Re-doing" thus is meant to evoke the complicated reiterative dynamics that radiate outward from the play to Welles's adaptation, to later attempts to un-do and then re-do aspects of Welles. Not surprisingly, re-doing, as an attempt at repetition, cannot quite fully recuperate whatever was (imagined to be) present in the original; any such repetition gets fraught with the difference that lapsed time must inevitably carry with it.

All You Have Done Hath Been but for a Wayward Son (3.5.10–11)

If we are to follow Lisa N. Simmons's speculations from the preceding essay, Welles himself ought to be considered the first "re-doer" of a black FTP *Macbeth* because Ralf Coleman had already directed a classical adaptation of the play when Welles and Houseman began their conversations in 1936 about what to produce for the newly formed Negro Theatre Unit in Harlem. But far beyond this inviting supposition, Welles went on to re-do *Macbeth* more than once. While no subsequent production of his revived the transposition to Haiti, at least two retained traces of the Negro Theatre Unit production that should at the least complicate the superficial impression, conveyed by some scholars, that there was no continuity between his early "black" and his later "white" productions. This perception is countered by Welles's first post-FTP revision, a 1937 radio "arrangement" in which Welles played Macbeth opposite Edna Thomas's Lady Macbeth, reprising the role she had performed onstage. As Richard France notes, this was Welles's "first directorial credit on radio...a half-hour abridgment of *Macbeth*...for the CBS Columbia Workshop, February 28, 1937" (1977, 72). There is no extant recording of this first of three different audio versions Welles made, so we cannot verify whether he maintained the drums and other sound effects from the FTP version; nor can we determine to what extent Thomas's "own way of reciting Shakespeare" would have been recognized over the airwaves as belonging to a black actress.[4] We can see, however, that Welles has at last displaced Jack Carter as Macbeth, in an interracial couple that, were it subject to the 1930 Hays Code on screen, would technically fall under the forbidden category of "Miscegenation (sex relationships between the white and black races)."

The all-black *Macbeth* thus provided Welles with his first major directing opportunity on stage, and Edna Thomas helped bridge him to his next medium, radio. Welles's first Shakespearean film was likewise *Macbeth*, albeit with a more attenuated connection to the FTP production, as no actor carried over from the original Harlem version in this all-white cast (unless, of course, we count Welles himself, from his alleged performance in blackface in Indianapolis). One recent online reviewer claims "Welles had toyed with the idea of a filmed adaptation of the so-called 'Voodoo *Macbeth*' " before "surrendering to the racial realities of the 1940s and realizing that no studio would bankroll it" (Pierce). There is no evidence to substantiate such a claim anywhere in Welles's biographies or in the Welles Papers at the Lilly Library (the reviewer might be confusing *Macbeth* with Welles's failed attempts to film a radically African-centric *Heart of Darkness*), but the apocryphally-envisioned black-cast *Macbeth* movie captures, for me, a genuine intuition: it seems entirely reasonable to presume that when Welles decided to film *Macbeth* he ought to have returned to his wildly successful 1936 adaptation.

Yet Welles often seemed at pains to deny even the merest continuity of similar production decisions. He jokingly replied to Peter Bogdanovich's query of whether there was "any remnant of your all-black stage production of *Macbeth* left in the movie?" with the disarmingly specific yet trivial detail: "Yes. When Macbeth goes up to Duncan's bedchamber, for me—I don't know why—he simply has to move stage right to left. That's all [*laughs*]" (Welles and Bogdanovich 207). Furthermore, Welles seemed surprised when an audience member at a 1977 Boston lecture asserted: "You took the set design from your Harlem theatre production of the voodoo *Macbeth*, when you made *Macbeth* into a movie," to which Welles retorted: "How do you know that?" before proceeding to defend this decision: "The basic set had the same plan which I had used in the black *Macbeth*, which I had done in Harlem, in the theatre, some years before. It wasn't the same set, but it had the same basic plan, because we were in a great rush" (French). Expediency is the justification, but numerous other connections between the 1948 film and the 1936 stage versions indicate that there were a remarkable series of overlaps, as Michael Anderegg details:

> the way the witches become a pervasive presence and voice also stems from that production, as does their final appearance at the end, chiming in on Macduff's "untimely ripp'd" and speaking the final line, "peace, the charm's wound up." Many other details could be adduced to demonstrate that, once again, Welles recycled and refurbished already existing material like the careful, thrifty craftsman that he in so many ways was. (87)

In fact, when Welles wrote the screenplay for his film, it appears that "he adapted his *Voodoo Macbeth* script, and not the original text" (Fernández-Vara 103, citing Jovicevich 233).

Welles's repetition compulsion is not unique to this play; Anderegg goes so far as to say that such a compulsion "characterizes all his work" (20). Yet what is notable here is that Welles seems to displace quite carefully the racially-inflected origins of his first professional *Macbeth*, despite the complicated filiations noticeable all the way down to the voodoo-doll-like creature that the sisters pull from a cauldron in the first scene of the movie. And this from an extraordinarily self-conscious artist who took care to remind Jean Cocteau, when they saw each other at the screening of *Macbeth* at the 1948 Venice Film Festival, that they had met earlier in Harlem; as Cocteau recalled shortly thereafter, "Oddly enough, I did not connect the young man of the Negro *Macbeth* with the famous director who was going to show me another *Macbeth*... *It was he who reminded me*" (emphasis added; Cocteau 28). It is not that Welles was unwilling to recall his Harlem production; he features it prominently, for instance, in Episode 2 of his 1955 television series *Orson Welles's Sketchbook*, in which he draws a cartoonish portrait of drummer Asadata Dafora Horton while describing the production. But there is a tendency to isolate the exoticism of the production, and not to admit some of the more charged intersections between the supposedly "classical" 1948 film and the supposedly "non-traditional" 1936 adaptation. In a peculiar way, Welles went on to argue that his 1936 *Macbeth* aimed *not* to refer to color. Note Welles's insistence, in a 1974 interview with Richard Marienstras, that he did not change *Macbeth* at all for the FTP version:

> RM: *In 1936 you did a black* Macbeth, *a voodoo* Macbeth.
> OW: It was just a way of communicating with the audience; I didn't change the play's intentions, as I saw them.... I used Haiti, voo-doo, etc., to foreground what I thought was represented in the play.
> RM: *Still, it was a way for you to make it accessible to a certain public.*
> OW: But the intention belonged to the play.
> RM: *All the same, you did change the text a little, didn't you?*
> OW: Absolutely not.
>
> (Marienstras 150)

In Every Point Twice Done, and Then Done Double (1.6.15)

Given all the "re-voo-doings" that Welles undertook, with all of their subtle displacements and disavowals of race as a formative influence, it should not be surprising that those black theatre companies that have attempted to engage with Welles's precedent should find themselves contorted into sometimes contradictory positions of drawing upon familiarity with the 1936 production at the same time they distinguish themselves from it. In fact, the very first non-Welles revival—the 1937 Los Angeles FTP production—bizarrely captures the very paradoxes such an endeavor must entail. The FTP had initially intended for their productions to be portable, not only going on tour (as Welles did with his New York Unit) but also being transferable to an entirely different unit as a kind of theatrical franchise. Max Pollock's director's report for *Macbeth* reveals how such transmission can go awry:

> Although we were under the impression that we were to receive from New York Mr. Orson Welle's [*sic*] original script, it was not forthcoming for some unknown reason. We decided to make our own treatment.... When we were through with our own script we received the New York text. We, of course, preferred our own, with one exception. We used Mr. Well's [*sic*] character of Hecate as he had

it. While this deprived us of a complete originality we felt that Mr. Welles's treatment of the witchcraft in "Macbeth" was so well suited to the Negro conception of Macbeth, and since we had to pay a royalty to Mr. Welles in any event, and time for production was pressing, we permitted his characterization to remain in our production. Had we had sufficient time we would have deleted that, too, in order to call the production completely our own. (8)

While they did hire Tommy Anderson—the stage manager from the New York Unit who had also filled in for an erratic Jack Carter on Broadway as the lead—Pollock and his crew seemed to resist the very notion that they ought to be obliged to reproduce Welles's vision at all, for the sake of their own artistic autonomy ("completely our own"), yet for financial (and I suspect symbolic) reasons decide to include *something* from the belated script.

Yet the notes on their playbill elide most of their autonomy, making them appear to hew far more closely to the New York Unit than they in fact did:

PASSION, savage lust for POWER, and the MURDEROUS FURY of King Macbeth in an AFRICAN KINGDOM—jungle drums and rhythms, VOODOO rites—this is Shakespeare's great play in dynamic, innovative splendor. The cast of 75 people is directed by Max Pollock in an adaptation based on the history-making New York Harlem production by Orson Welles. (see Figure S.9)

"Based on" seems something of a stretch if that merely entailed the modification of Hecate's role, but this is a kind of tribute to the inordinate authority Welles's production had already attained: even when you do not really re-do voodoo, you are bound to re-do it.

Conversely, even a deliberate attempt to re-do can come undone. The New Federal Theatre, which was formed in 1970 by Woodie King, Jr., made a singular exception to its decades-long mission of "presenting plays by minorities and women" to mount *Macbeth* in 1977 in a production self-consciously conceived as a "revival," following the apparent rediscovery of Welles's promptbook in 1974. Given that King was among the vehement opponents of Joseph Papp's plan, shortly thereafter, for a black-Hispanic repertory Shakespeare company (King likened it to "putting a plantation overseer in charge of black theatre. Papp's company is a perpetuation of the white power structure in America, a white concept in black face" [1979, 5]), how could King countenance the problematic precedent of Welles? Contrary to what one might expect, King seems to have sought out Welles, going so far as inviting him (via his lawyer) to join the production. While Welles did not join the New Federal in New York, he did grant King "special permission" to mount the play, which King took care to note in interviews (Lewis). Barring Welles's actual presence, King used every other available means to establish and authenticate continuities between the version he was producing and Welles's "original": newspaper advertisements sold the play as "Shakespeare's *Macbeth* Adapted by Orson Welles." The director, Edmund Cambridge, was announced as having been "involved with the original show as an apprentice. Thomas Anderson, assistant director, and Leonard DePaur, musical director of the original, were brought in as consultants" (Seligsohn); artist Romare Bearden (see Figure S.3) and Perry Watkins (John Houseman's assistant) were also brought in for talkbacks. Esther Rolle, more familiar as *Good Times*'s Florida than as Lady Macbeth, would point out in interviews that she had danced with FTP percussionist Horton's company; Lex Monson, who played Macbeth, noted in his biographical blurb that in addition to Shakespearean roles, he had originated the role of Queequeg on Broadway in Welles's adaptation of *Moby Dick*.

But perhaps by insisting so much on the proximity to Welles, the places where the production felt distant from its source were further accentuated. As Bernice Kliman notes,

> casting a woman as Hecate (Louise Stubbs), instead of a man as in 1936, visually diminished the mass of this figure—though with high cork shoes and a full white afro hairstyle she compensated creditably. The orchestra and chorus were gone and along with them most of the special music, diminishing the power found in the original. (113)

In fact, production notes in the Lex Monson collection at the Schomburg Center reveal deeper tensions. Most telling, I think, and perhaps an apt subtitle to this entire volume, is a handwritten note, reading:

> N. B. DISTURBS ME: SCOTLAND + VOODOO DRUMS—I FIND THIS INCONGRUOUS.

Apparently some of the drumming was eliminated for the production, but the complicated incongruity of the original and its revival also manifested itself in decisions about and responses to dialect, reenacting some of the same ambivalent reactions to Welles's decision not to "train" his performers in Shakespearean delivery. In the Monson file is another handwritten note, reading:

> Dialect: Everyone try for British accent but it's not upsetting if Southern or American accent comes through. As a matter of fact try to combine the Southern with the British to come up with a dialect unique. Must be used by everyone.

Such a consistently hybridized dialect apparently never was fully established; as one reviewer scoffed:

> Various Caribbean dialects are represented; British English creeps in, as does American. The murderers speak Cockney, no less. Some characters use the various dialects interchangeably from one sentence to the next. The problem goes deeper than surface inattention to language variance. Language is culture. It grows out of a specific life-experience and represents the characteristic thinking, values, and expressions of its speakers. To superimpose the King's English disregards the integrity of the culture being represented. I don't know

what the effect was for audiences in 1936, but for an aware audience in 1977, it is insulting. (Ellen Foreman, quoted in V. Clark 40)

Even when theatre companies identify and attempt to break from the problematic aspects of Welles's original production, audiences often locate the tension from the original.

Reviews of the 1977 revival frequently relate the beguilingly frustrated desire to recapture a single production from decades earlier. Robert Hapgood, the *Shakespeare Quarterly* critic who reviewed the NFT production, found himself dissatisfied with what he saw on stage, and sought instead to reconstruct a vision on the page:

> The afternoon following the performance, I went to the New York Public Library Theatre Collection at Lincoln Center, where I reenacted the original Welles production in my own mind. Spread before me were the Library's copy of the original script and publicity photographs, Richard France's article about the production in *Yale Theater* (1974), John Houseman's entertaining account of the back-stage drama of the show in *Run-Through* (1972), and the Library's scrapbook of contemporary reviews. With their aid one could imagine Welles's boldest effects (all omitted from the revival) [Hecate's bullwhip, a giant mask of Banquo, the eighteen-foot drop from Macbeth's battlements]...Sitting in the Library, I came to realize that I was finding exactly the kind of theatrical excitement that I had missed in the theatre the night before. (Hapgood 230)

Scholars are often complicit in the same troubled re-doing in which so many performers have engaged; the Library of Congress curator for the Federal Theatre Project materials, Walter Zvonchenko, confirmed that the *Macbeth*-related materials were by far the most frequently requested in a collection that includes some 70 playscripts and over 13,000 items.

What's Done Cannot Be Undone (5.1.57–58)

Welles places you in a color-bind: if you decide to perform *Macbeth* with an all-black cast, you are already following in grooves he etched long ago, which can at times feel like ruts. If, on the one hand, you attempt to reconstruct the 1936 production, you run up against impossible standards of "fidelity" all-too familiar for anyone adapting Shakespeare, but further exacerbated by the detailed yet diffuse and often contradictory archive of "what the Welles production really was like." If, on the other hand, you deny the production's influence when it clearly has established a precedent, you are liable to become irritated by its gravitational force. I sense this to be the case with Alfred Preisser, who co-founded the Classical Theatre of Harlem (CTH) in 1999, which inaugurated its first season with *Macbeth* and has since remounted this play more often than any other (five out of its first ten seasons have included *Macbeth*). Preisser stated in his correspondence with me that, while he was familiar with the Welles *Macbeth*, his "production was not influenced by that interpretation... I've never really thought about my position as a director in relationship to Welles's work." Yet Preisser admits some frustration in the reception of his all-black *Macbeth*:

> Because of the music and dance, many people perceived the production as a "voodoo Macbeth." This says more about many people's continued limited perceptions when they see African Americans onstage than it did about this particular production. In all of our dramaturgy, we never discussed the West Indies, we never touched voodoo, etc. People, including critics, would sometimes identify the costume as "African," I think because black people were wearing them, not because of what the costume actually was.

I found myself conflicted in my correspondence with Preisser, who was gracious in replying to my queries with reflective candor. As was frequently the case when researching this essay, I felt as if I were too narrowly seeking to confirm that a particular director had engaged with the FTP production.

Yet I think the question remains a fair one, given that this was a predominantly black company, in Harlem, launching their existence with *Macbeth*. I am not convinced that you can trumpet in a press statement "the first professional production of *Macbeth* in Harlem since the historic 1936 production directed by Orson Welles" without then exploring the dynamics of this redoing more systematically. Preisser occupies a fascinating space, in a theatre company he is careful to describe as *Harlem* rather than *black*, working with a play that he wants to stand autonomously but one that inevitably gets recognized as inviting parallels to Welles. Not surprisingly, "The parallel between Welles's *Macbeth* in Harlem and CTH's does not sit well in many quarters" (Rux 84). But to his credit, Preisser, in a 2004 interview, admits that he "understand[s] the feelings people have about the history of Welles's *Macbeth*: Welles came, did his thing, and he left. But we're not going anywhere. We're in this for the long run, and we're here to stay" (Rux 84). Much like Welles, however, at times Preisser seems to want to have it both ways, calling his production not "a deconstruction or a cultural transplant," yet admitting "Of course, when working uptown, another culture asserts itself. The dance and music score are Afro-Caribbean" (Copage).

The non-traditional *Macbeth*s that seem the least burdened by the icon of Welles are those that have studied it closely, and immersed themselves in it—and then have moved on to formulate their own production decisions. Meditating at length upon their distance from Welles's production permits them to draw upon its precedent without re-doing its downfalls. The extraordinarily ambitious undertaking that Lenwood Sloan details in the following essay, for all of its difficulties, aimed to embody some of the same communal artistic spirit that Welles managed to tap into, in large part by concentrating on the FTP production's process (of a multi-media, collaborative artistic creation), rather than exclusively its content (of an all-black *Macbeth*).

Notes

1. The founder of this company, which re-occupied the very Lafayette Theatre where the Harlem *Macbeth* premiered in 1936, was the influential black theatre impresario Robert Macbeth.
2. John Wilders's edition of the play for the *Shakespeare in Production* series mentions the 1948 Welles film (58) without any apparent knowledge of the FTP stage production (or any other North American production, for that matter).
3. As Russ McDonald notes, following dozens of critics before him:
   > The auditory pattern that overrides all others in *Macbeth* is unremitting repetition. This is a striking technique in such an exceedingly short play: echoing sounds register with unusual force because they reverberate in so short a space. Not only are words repeated...but consonants and vowels are doubled and trebled, rhythmic configurations repeated insistently, and phrases and images reiterated, not just immediately but memorably, across several scenes. (47)
4. Welles went on to describe how he did not direct the pronunciation of the 1936 production:
   > The blacks invented the whole diction of Shakespeare.... I didn't suggest anything about Shakespearean tradition or the white way of reciting Shakespeare. But their sense of rhythm and music is so great and their diction so good that they found their own way of reciting Shakespeare. It was astonishing! (Marienstras 154)

   For an account of the "the assumption that vocal timbre is an unmediated reality of the body" and "how this has played out historically in the cognitive reception of the African-American voice in the United States," see Eidsheim (1). A fuller examination of the racial legacy of the 1936 production in Welles's later career would need to explore further Welles's relationships with actors from the Negro Theatre Unit whom he recast in later productions, in particular Jack Carter (Macbeth) as Mephistopheles to his *Dr. Faustus* (1937), and Canada Lee (Banquo) as Bigger Thomas in Welles's adaptation of *Native Son* (1944).

11

The Vo-Du Macbeth!: Travels and Travails of a Choreo-Drama Inspired by the FTP Production

Lenwood Sloan

*T*he *Vo-Du Macbeth* (2001–2005) was a massively collaborative undertaking, intended to revive the original 1936 FTP production. Yet even more ambitiously, we sought to involve a wide range of artistic communities across the nation in this revival. Throughout the development of *The Vo-Du Macbeth*, over 500 performers participated in residencies and staged presentations across the country; more than thirty scholars conducted post-performance dialogues, workshops, and symposia; collaborators led more than 100 community workshops, master classes, and classroom presentations; and thousands of listeners experienced presentations via public radio. We achieved our goal of replicating the WPA's impact by reaching more than two dozen American communities—even though we never achieved the fruition of our vision on the main stage. In this essay, I relate my role as coordinator of this complex and fraught process, and detail the project's historical and cultural groundings.

The Call to Action

Between 1985 and 2000, the National Endowment for the Arts fostered an impressively interdisciplinary collaboration between visual and performing artists across the country. The result was an innovative body of new works for the American stage. Dubbed "Inter-Arts," the movement encouraged true co-creation through shared vision. Institutions, local communities, and artists' colonies became the incubators for such collaborations. Organizations, colleges, and universities revealed both the process *and* the product of making art to their audiences.

In 1999, I was invited by the Contemporary Arts Center of New Orleans to mount an historically-informed work inspired by the WPA period. While doing research in the Schomburg Library (New York), I found a photo of Katherine Dunham holding Orson Welles's palm as if she were giving him a reading; looking on were Alicia Alonso and Maurice Chevalier. The caption read: "Katherine Dunham and friends discuss the Voo-doo Macbeth." I soon learned that Dunham, choreographer of the Negro Dance Units, profoundly affected the young artists of the Federal Theatre Project. Her 1935 fieldwork in Haiti, which eventuated in her 1947 study *Dances of Haiti*, helped contribute to an American fascination with Haitian culture (see Aschenbrenner).

That was it. I was hooked. I had found my project: a "revival" of this famous all-African-American *Macbeth* production.

THE ROOTS

In 2000, New Orleans artists and community activists from Act One Theatre joined me in the project. I nicknamed them the "Emory Eight," after host Emory White, in whose living room we met each Wednesday evening. They included: White, Barbara Trevigne, Harold Evans, Kaia Livers, Ausettua Amenkum, Adela Gautier, Luther Gray, and Carol Sutton. We began by inviting a number of guests to our salons, including Joe Logsdon and Jerah Johnson of the University of New Orleans, Father Ledeau of St. Augustine's Church in Tremé, Laura Rouzan of Dillard University, and patron and mentor Gary Esolen, a renaissance man. These humanists introduced us to important historical and cultural relationships between *Macbeth* and events in nineteenth-century New Orleans.

Soon Carol Bebelle and Douglas Redd of the Ashe Cultural Center joined the circle and invited us to move the Emory Eight to their center to continue the dialogue. While meeting there, we learned that most of Orson Welles's cast had day jobs as domestic workers, elevator operators, taxi drivers, and factory workers. Because many worked two jobs, the crew could only rehearse late at night, so Welles held midnight suppers and provided early breakfast for all who showed up for rehearsals. The Ashe Cultural Center likewise created a regular Wednesday night buffet, salon, and open house in the style of Welles, blending the cultural community with the working-class neighborhood. After receiving a Louisiana Playwright's Fellowship to develop the script and libretto, and the Louisiana Bicentennial Commission Award to build a team to reconstruct the score and choreography, I was invited to serve as Ashe's Artist in Residence. Welles had found his patron in the Federal Theatre Project and his incubator in the Lafayette Theatre Group; I found mine in Ashe and the Act One Theatre.

My first step was to turn to composer and sound engineer Bill Turley, with whom I had previously worked. So far as we knew no original score was still extant. In order to begin our task, we decided to explore the relationship between the musical ensemble and the voice choir that we had read so much about. The Emory Eight invited composer Hannibal Lokumbe to join our Wednesday salon; he helped moderate dialogues on forms of new opera theatre. Later composer and conductor David Amrahm was invited to work with Bill Turley on an opera treatment for soloists, chamber ensemble, and chorus.

A second trip to the Schomburg uncovered Abe Feder's lighting plot with hanging and rigging notes, gel colors, gobo templates, and the layout for every lighting instrument. This was a true source document, and we decided to hang Feder's plot for a staged reading at the New Orleans Center for Creative Arts. Additional research at the Smithsonian revealed photographs of the Lafayette Theatre production, including important images of the costumes and set pieces. After studying all that we had discovered about the FTP production, we discussed candidly the clash of aesthetics of Welles's original team. While Welles and Houseman staged a classic proscenium production, diverse artistic perspectives contributed to the Voodoo aspects of the production (see the Rippy essay). The juxtaposition must have been as unnerving as it was overwhelming.

First Reading

For our first major reading we assembled a cast of forty readers, vocal soloists, chamber ensemble, and chorus. It was merely one-third of Welles's cast of 125, but still large for today's standards. A reader's theatre format featured voice and showcased a shortened script, aiming to remain true to the language, characterizations, and structure of Shakespeare's text. A rear projection screen was hung overhead. Throughout the reading, images transitioned from landscapes of Scotland, to images of Haiti, to settings of New Orleans and its bayous. Performers sat in an orchestral pattern, facing the audience before a platform with two conductors. Amrahm conducted vocalists and instrumentalists through his treatment, and I directed the readers. Bill called lighting, sound, and projection cues from the percussion section, and Luther Gray provided outstanding African drumming.

A spirited post-performance dialogue dealt with the critical elements of sense of place, viewer experience, and personal narrative. The audience asked questions like:

> Where are you supposed to be? It's confusing. We don't get the Scotland stuff! Why are these Scots people in Creole and Cajun settings anyway? If this is Louisiana, why's his name Macbeth, and his Banquo? Aren't Creole names exotic enough?

Lolis Ellie encouraged us to "cut bait with Scotland" and find our way home to New Orleans by asking:

> What was it like for freemen in 1863 in a secession state? How did the Creoles deal with mandatory freedman's papers? How did they hide their anger and frustration about their condition? How did they feel about the collision of French democratic ideals and American democracy? How did they feel about the infusion of other people of color into their system of race, class, and status?

But perhaps the question that stumped us most was from a woman in her mid-twenties holding a two-year-old child. "What about the baby?" she asked with a voice of alarm. "What baby?" we responded politely. She was persistent. "That woman said that she'd take her baby who felt so good on her nipple and bash its head against the wall. What kind of woman is she anyway? How could she even think of doing that?" These discussions helped us reposition ourselves in relation to *Macbeth*.

STARTING OVER

A few days later performance artist John O'Neill encouraged me to take the process outside of New Orleans to see if a larger dialogue could lead to a more compelling adaptation. Ashe secured funds from the Louisiana Endowment for Humanities to test a traveling production entitled "Meet the Macbeths." The Emory Eight decided to focus on the marriage bed and bond (and find out about that baby!). We selected a FTP treatment entitled "The 45 Minute Macbeth," and a collaborative team convened at Howard University. Choreographer Chuck Davis joined the team and encouraged us to return to the source word "*vo-du*," which meant "spirit." Thus, the National Spirit Project (NSP) was born. Our goal was to mirror the Negro Theatre Unit of the FTP by establishing an alliance of multicultural community arts organizations across the country.

NSP host communities would cast local reading groups and chamber ensembles to spend a week with the creative team, exploring the adaptation. Each host community would also recruit a Shakespearean scholar to conduct a series of panels on the project. At the end of the week we would stage a reading followed by a dialogue with the local community, utilizing the "animating democracies" process of investigation facilitated by psychotherapist Karla Jackson-Brewer.

HISTORICAL AND CULTURAL GROUNDING

Scholar and theatrical producer Vernell A. Lillie assembled a cast of readers and humanists at the University of Pittsburgh. They advised us to give up on the opera's rigid form in order to showcase the Afro-Franco culture that we embraced (see Adjaye). At San Antonio's Trinity College, members of the English Department went through the script scene by scene, freeing the pentameter. We explored cascading rhymes, Ebonics, colloquialisms, and open verse form. Then we evaluated each scene against its historical and cultural grounding, language, sound, rhythm, and nuance to reconstruct the drama and translate it from Scotland to Louisiana.

Americans confuse the distinctions between race and culture, generally dividing the issues into black and white, north and south, male and female. Doing so fails to capture the complexity of American life that we ought to understand in order to frame the dialogue that defines American culture. According to the 2003 New Orleans "Undoing Racism" workshop, "Racism is discrimination by [one racial] group against another for the purpose of subjugation or maintaining power. Colorism is the term for describing the prejudice that African Americans hold about each other and seemingly use against each other to advantage themselves based upon complexion" (People's Institute). Emphasis on light skin, straight hair, and sharp features allowed some individuals to attempt to distance themselves from the typical image of blacks. Nowhere is this more evident than in New Orleans.

Following Haiti's revolution, the United States experienced the largest migration of highly skilled freemen, freedmen, and enslaved individuals in its young history; New Orleans was the first and closest port of entry for these immigrants (G. Hall 81–85). During the Battle of Orleans, freemen of color endeared themselves to Louisiana by lending their famed Native Guard to Andrew Jackson, who would credit them with the victory there (Lachance 114–15). The Creoles of color were bound to the system of *plaçage* which placed a woman of color under the sponsorship of a white gentleman (Mills 72–81). He was responsible for her safety, her household, and the education of their offspring. This created generations of privileged mixed-blood children often schooled in Paris and more rooted in French culture than American. These children of *plaçage* had lineage as well as inheritance, and considered themselves far above freemen and freedmen alike. They measured class and status by the *Code Noir*—a seventeenth-century French legal decree that continued to define degrees of blackness. Thus, they enjoyed the status of a virtual third race.

By the onset of the Civil War, New Orleans's *gens de couleur libres* owned $15,000,000 worth of Louisiana's real estate. When General Butler's union troops came on shore, Louisiana divided. Shreveport became the capitol of the Confederate stronghold, and New Orleans served as the Union seat. Following the invasion of New Orleans, Union soldiers were left to their own devices for food and sustenance, becoming scavengers as well as conquerors (Gill 15–30). Most Scotch-Irish foot soldiers found the Creoles of color far too arrogant for losers. They felt it was their duty to challenge the status that Creoles claimed for themselves. Meanwhile, returning Confederate soldiers marched solemnly through the streets, wounded, broken, and in despair. Confederate sympathizers vowed never to forget the sacrifice these men gave to preserve their way of life, and imagined that the Creoles were beneficiaries of their military defeat. In short, the Civil War did not end in 1865 in New Orleans. Even though the troops eventually went away, and the men took off their uniforms, the conflicts of race, class, and status between the Napoleonic Creoles and the Northerners continued.

The Concept

The NSP transported Shakespeare's *Macbeth*—a parable about power and the manipulation of power—to New Orleans in 1863. It is the Feast of the Winter Solstice, yet not a time for celebration. The great houses of the Free Gentlemen of Color are Balkanized. Some choose to "cross over," and "pass" for white; some pledge allegiance to the Union cause; many take exile in Cuba. Maspero (Macbeth) has joined forces with the Confederacy to fight for his land and preserve the old way of life (see Bergeron and Rollins). At the crossroads in Congo Square, Maspero's power is revealed to him by the Mambos (Witches). Blinded by desire, he conspires to perform crimes against nature by murdering his kin—first Marigny (Duncan) after the Christmas Eve reveillon, then Oskar (Banquo) before an elaborate New Year's Eve masquerade ball (Saxon, Dreyer, and Tallant 164–65).

During the Civil War, old family societies did much of their subversive business through the membership of their secret "krewes" (hereditary clubs). Guarded by codes and oaths of allegiance, the "cover of mask" became a new political construct (Gill 45–55). Each krewe maintained its own benevolent and loan society, real estate power brokers, mercenaries, and militia. The masquerade ball, parade, and revel were provided new purposes, and became excellent façades for covert activity. Thus, the banquet scene is transformed into an elaborate and colorful tableau ball based upon the Minotaur legend, with Maspero dressed as Minotaur, Antoine as Icarus, and the ghost of Oskar entering as Daedelus. The Minotaur—part man, part bull—becomes a metaphor for miscegenation, and thereby provides a parable for the masquerade.

On New Year's Day, Maspero spills the blood of Charlotte (Lady Macduff) and her son. This offends the Orishas and the Ancestors, who turn their backs on him (Teish 190–91). Later, as he is drawn closer to his fate, the guns of the Twelfth Night revels are heard in the distance, and the songs to Hecate begin down by the river. Maspero fulfills the prophecy about the end of his kind and the beginning of a new people.

Constructing the Script and Libretto

Our two major tasks were to translate the geographic references of the original play to comparable locations in New Orleans. The *Open Place* became Jackson Square, a *Camp* became a barge on the Mississippi River, a *Heath* became the center altar in Congo Square, the *Forrese Palace* became the council chambers, *Inverness* became Tremé.

Alvin Baptiste and Bill Turley conducted an intensive residency at Southern University for soloists, chorus, and chamber ensemble. Key monologues, dialogue, and speeches were identified to construct two styles of song cycles. A French cycle of nine songs, representing the intellectual sophistication of Tremé, framed solo and melodic works for principle characters. The cycle included:

> Alas poor Crescent ("Alas poor country") (Ross, 4.3.165)
> Awake, Awake ("O horror, horror, horror!") (Macduff, 2.3.59)
> "Who could refrain / That had a heart to love" (Macbeth, 2.3.113–14)
> Walking Shadows ("She should have died hereafter") (Macbeth) (5.5.16)
> "To bed, to bed" (Lady Macbeth, 5.1.56)

"Double, double" and "Come, you spirits" (Witches, 4.1.10; Lady Macbeth, 1.5.38)
"But who did bid thee join with us?" (Murderers, 3.3.1)
"Whither shall I fly?" (Lady Macduff, 4.2.73)
"How does my wife?" (Macduff, reprised by Macbeth, 4.3.177)

A German cycle of five songs, representing the passion of the people of Congo Square, was interwoven into the French cycle. These works for "voicestra" (interactive group singing) were designed to support character dances that promoted aspects of nationalism, tradition, and Creole life. They included:

Hecate's descent—an opening blessing (ring shout) (3.5.1–35)
Macbeth and Banquo with the oracle of the Witches (labyrinth dance) (1.3)
Duncan's arrival to Macbeth's castle (junkanoo) (1.6)
The banquet scene (masquerade) (3.4)
The apparitions (ballet macabre) (4.1)

THE CHARACTERS

HECATE (MOTHER OF THE NIGHT): Literary scholars surmise that Hecate's speeches were added to Shakespeare's manuscript (see the Daileader essay). Because of this, many productions remove, or greatly alter, Hecate's role: Welles, for example, changed the gender of Hecate to a male character. Freed by this knowledge, we changed Hecate into the story's narrator. According to the customs of the African Diaspora, the Mother of the Waters is honored between December 21 and January 6, in a ceremony that celebrates the end of the year. The old fetishes are stored away, and the new ones are introduced (Teish 133–35). The synergy between Catholicism and the *Vodun* is exemplified in New Orleans by the relationship between the Blessed Mary and Cuba's Yemayá.[1] Both are embodied in Hecate, who collects tithes at the crossroads of the old year and the new. In Haiti, she is called Agwe Tawoyo, the Ocean deity, whose roots are in the ceremonies of the Water Spirits, Olokun; in Toga and Ghana, the Mother of the Waters is Mami Wata (Teish 39–42). All these cultures make up the people who gathered in Congo Square in nineteenth-century New Orleans (Jerah Johnson 11–25). The Mother of the Waters is envisioned as a mermaid whose skin is covered with elaborate chalk white drawings. The white chalk does not depict European skin but rather the pale enrapture of her divine spirit.

THE MAMBOS: Shakespeare's witches are transformed into three Mambos—daughters of Hecate. One is called to the service of the Mother of the Waters through dreams. Hecate appears in a dream with her snakes (represented by her locks of hair) and calls her daughters to initiation. Female worshippers called Mamisis tend to find prestige through devotion to Mami Wata (Teish 40–42). Their voices represent the winds that carry Hecate through time and space, weaving together the dramatic scenes that support the play's wordplay. They keep the secrets of the Spirit of the Waters and the

Mother of Night, but test Hecate with their pranks and abuses. YASHA-BAD (FIRST MAMBO): A Senegambian soothsayer, keeper of the fire; MARIE (SECOND MAMBO): A midwife and healer from Haiti, keeper of the charm pot; TITUBA (THIRD MAMBO): A Maroon, root sorceress from the Louisiana swamps, keeper of the crossroads (William Katz 137–42).

MASPERO (MACBETH): Maspero, an octoroon son of a *plaçage* household, is patterned after the nineteenth-century pirate of the same name (Saxon, Dreyer, and Tallant 275–76). Although he is highly favored, nature played a trick on him and he was born dark. A Captain in the Native Guard and usurper of power, Maspero must pass through a maze and solve the riddles in order to achieve his destiny and keep all that he has gained (Logsdon and Bell 217–20).

MADAME MASPERO (MADAME MACBETH): Madame is inspired by the infamous and legendary Madame LaLaurie of New Orleans (Saxon, Dreyer, and Tallant 236–38). An octoroon, she avoided the condition of subordination required in *plaçage* by marrying a darker man. She abuses the ancestors and forfeits her soul as she dances across the tight rope of the *Code Noir*.

MARIGNY (DUNCAN): Marigny's character is based upon New Orleans' Bernard Marigny, who understood the threat of the Creole culture to the Union and feared that Union victory would deprive Creoles of their identity (Saxon, Dreyer, and Tallant 284–88). The choreo-drama's Marigny is a "*passé blanc*," the son of a "*demi-meamelouc*" (1/32 black) woman and white gentleman. Living between the races, he is faced with the possibility of being exposed racially.

OSKAR (BANQUO): In 1872, the African American Lt. Governor Oscar Dunn of Louisiana received the popular vote for Governor (Logsdon and Bell 248–50). He celebrated too soon and died of a mysterious death on the eve of his inauguration. His history becomes the pattern for our Oskar, a quadroon freeman (Mills xiv). His death is meant to stop the inevitable events of his time, but it becomes a force that sets change in motion.

ANTOINE (MACDUFF): Patterned after the third of the black Lt. Governors of nineteenth-century Louisiana, C.C. Antoine, our Antoine is a Protestant mulatto landowner from Shreveport who attempts to enter New Orleans Creole society through his marriage to Charlotte (Desdunes 140–41). Freed, he represents righteous anger and retribution (Tregle 168–72).

MADAME CHARLOTTE (LADY MACDUFF): Madame's sister, Charlotte, a member of New Orleans Creole families, has been raised for the *plaçage* but marries outside her culture and her faith. Thus, Creole society turns its back on her.

YOUNG ANDRE (THE SON): Charlotte's son has been born out of *plaçage*. His mother has been shamed and shunned by society for breaking her contract and marrying his stepfather, Antoine.

DR. DE VOIR (THE DOCTOR): A northern-born and Victorian-bred Negro freedman who is not at all impressed with the Creole gentry or their customs and traditions.

LABAT/OLD UNCLE (THE GENTLEMAN'S SERVANT): Labat is a composite of several characters including messengers, the porter, and aspects of both Ross and Macduff. A Creole freed by emancipation, he is nonetheless bound to Maspero's estate. Labat is a wise old uncle whose skills of survival and philosophy of life are well worth following.

CELESTINE (THE HANDMAIDEN): A mulatto freed by manumission but bound in the service of Madame until her twenty-seventh birthday (Talty 20–22). She is a young novitiate of Hecate's sect.

GWINE AND NOTHIN': Two slaves from the Georgia cane fields are called into service as Maspero's hounds (Logsdon and Bell 201–04). They serve as murderers, spies, and thieves. Neither has any expectation of freedom or long life; they are in it for the profit and opportunity.

PRODUCTION AND DESIGN

The score is constructed for principle characters, a quartet of singers, and a chamber group of six, including keyboard, bass, cello, violin, drums, and horns. African drummers use a variety of traditional instruments and found objects to provide the drama its driving rhythm. Drummers also play prepared piano as a percussion instrument representing Maspero's inner voice. The sounds of nature (birds, crickets, frogs, wolves, owls, thunder, lighting, and rain) blend with sometimes haunting sound effects. Recorded voices wailing, shrieking, and intoning create a soundscape for the performance.

Character dances include bamboula, calinda, junkanoo, juba, ring shout dances, funeral dirges, and saint walks. They blend with quadrilles, pavanes, jigs, and waltzes to complete the nineteenth-century dances of New Orleans ballrooms, marketplaces, and Congo Square (Jerah Johnson 11–25). The choreographic plot also includes a "ballet macabre" (in apparition scenes) and totentanz at the time of Madame's suicide (Cohen 2–7).

The work is presented in arena style with entrances and exits through the audience as well as through the wings. A tri-leveled scaffold with visible stairs to the second tier and escape stairs to the third level allows performers to appear on all platforms. Set elements are made of cloth and serve as banners, flags, drapes, canopies, and rugs. They are rigged to fly, drop, and vanish. A soft sculpture drop hangs as the full moon. It is pulled down to become Hecate's grand cape and turned over to become the maze that the Mambos use in their rituals at the crossroads. Scrims assist the shadow play along with projections to create supernatural and allegorical time. Multi-image projection adds to the visual design, defining a sense of place.

THE CURSE OF *MACBETH*?

Early on we invited Karla Jackson-Brewer and Barbara Trevigne to assist us in coming to grips with the *Vodun* and the force that we were calling up. They would assist us in avoiding the curse of *Macbeth* many times. Yet events seemed set against us. While loading into Pittsburgh, the World Trade

Center Towers tumbled in flames, and the economy came tumbling after; September 11, 2001 was the beginning of the end for the NEA Inter-Arts Program. Like Maspero, we did not understand that the world we knew was drawing to a close. During the project, I was hospitalized three times; both Thelma Horton's and Gloria Powers's houses flooded, ruining manuscripts and records of the project. Along the way we mourned the passage of John Scott, the brilliant facilitator of our design team.

Bill Turley and I attempted one last strategy to continue our production. The Pennsylvania Council on the Arts co-commissioned a radio drama version for Philadelphia's New Freedom Theatre, broadcast before a live audience by Pennsylvania Public Radio. Bill performed on a multiple-sound-effects table in the style of a 1936 radio drama. The Southern Arts Federation planned to broadcast a four-part version in all thirteen southern states. On the eve of launching the radio show, we were nominated as one of the ten best plays of the New South for 2004, a positive omen.

As 2005 approached, an ambitious plan to mount the full production was developed during a design team intensive led by team member Philip Mallory Jones at the North Carolina School of the Arts (NCSA). NCSA established residencies to build the sets and costumes and to develop the lighting plot, production schedule, touring schedule, and shipping schedule during the summer of 2004. Upon building the sets and costumes for the full production at NCSA, a professional ensemble would tour a series of HBCUs across the country. The proposed costs were staggering, but it seemed like an achievable plan.

Then Hurricane Katrina came along, blowing the entire project to the four corners of the universe: collaborators, sponsors, Dillard, audiences, and all!

Note

1. *Vodun* is one of the many derivatives of the word for the spirit religion of Africa; others include *Vaudoux* in the Caribbean, and *Vou-dou* in Cuba and Haiti. In New Orleans and the Gulf south, *Vodun* was most commonly known as *Vo-du*, *Voo-doo*, or *Hoo-doo*.

4

Further Stages

Figure S.4 Weird Sisters from Teatro LA TEA's *Macbeth 2029*. Credit: Antoni Ruiz: aruizphotography.com

12

A BLACK ACTOR'S GUIDE TO THE SCOTTISH PLAY, OR, WHY *MACBETH* MATTERS

Harry J. Lennix

In 2007 I starred in an all-black production of *Macbeth* at the Lillian Theatre in Los Angeles. As an actor/producer on a production that very deliberately set out to utilize Shakespeare's play to target a potentially uninterested audience of urban blacks, I realized that our company had to find the means to make our production relevant to our time and audience. Although actors in classical productions always struggle to balance the universal with the particular, we struggled to make manifest a reason to perform this play in the present moment *and* to an experience specifically based in the African Diaspora. The act of producing an all-black *Macbeth* would put us squarely in the center of a current dilemma: When is it appropriate *and* important to have black companies mount productions of so-called "Classical Theatre"?

Once the theatrical color line was substantially broken here in the United States by pioneering commercial producers like Joseph Papp, most especially with productions of Shakespeare's plays, there was an almost wholesale indulgence in a kind of thoughtless, orgiastic zeal to ignore the endemic and problematical realities of race in the American theatre. Close on the heels of its commercial viability, "colorblind" productions of *Julius Caesar, Taming of the Shrew*, and so forth, were put up in theatres across the nation. So frenzied were the impresarios to throw racial considerations to the wind that there was even a colorblind production of *The Wiz* at a Chicago area theatre in which Dorothy was white (Lacher). The reasoning (if it can be called that) behind the larger movement is that the theatre represents a kind of transcendent reality in which one should be cast according to one's deeper humanity, irrespective of the realities of race. As a result, black actors implicitly accepted that there was a higher common denominator of humanity that was separate and apart from their own race and ethnicity. One can see this taken to an even greater extreme now in England, where the Royal Shakespeare Company has any number of black Brits who populate productions of Shakespeare's history plays. Of course, such choices are often accompanied by critical outrage both here and abroad.

As such, it seemed to us that the avoidance of the realities of race in favor of an idea about who we are as a people was demonstrably false and the very definition of irresponsibility. We took it as self-evident that a known correlation to our own collective experience must exist in order to justify the appropriation of any "classic" production worthy of dramatic rendering. The play must perforce have a recognizably syncretic dramatic line between two otherwise disparate realities. This could be either literal or poetical.

Armed with this position, it became a mandate that our production find a means to set *Macbeth* from the point of view of the African Diaspora. Early in the rehearsal process the subject of how we could approach the play in a way that is ethnically authentic, and remains true to Shakespeare, became integral to the process of production. Many of the actors in the play were passionate acolytes of August Wilson's insistence on equal consideration for a black aesthetic. As stated in *The Ground on Which I Stand*, Wilson makes clear that, while the Western Classics are incredibly great works of literature, African-American literature is not afforded the same consideration (Wilson 1996). Plays by serious black dramatists are neither funded with equal zeal (by blacks or whites) nor granted the same latitude in academic and critical circles.

Why then do we, who have been trained and educated from a decidedly traditional worldview, exhaust our resources on putting on a play by Shakespeare? What has he to do with the development of a black theatre?

Why *Macbeth*?

If I had a dollar for every occasion that question was asked, it might be possible to bail out Wall Street all by my lonesome. There is not a question in my mind that Orson Welles dealt with the same query when he first came up with the idea of appropriating the tragedy for the black experience seven decades ago.[1]

What lies behind the question is not nearly so much curiosity as revelation. It is safe to assume that nowhere near the number of people who ask this about *Macbeth* would ask "Why *Othello*?" In our case it was primarily black folks who were asking the question. These people were educated and generally familiar with the state of black theatre. Still, there is an assumption that black people have less agency when dealing with classical (read: universal) texts than white people: unless our blackness is being put into a context that whites deem worthy of observation (as in the case of Welles). This holds true even when black directors and producers undertake work they hope will be considered a "crosses over" success.

So why *Macbeth*? And equally, do we as black actors have any business remounting a production of a play dealing with Scottish history, and if so, what should the conditions of such a production be?

The answers are complicated. First, it may be the case that Africans and their descendents are uniquely empowered to bring a deeper resonance to this particular tragedy because of its interest in the metaphysical world. I do not know a black person who does not, on some level, believe in ghosts. The very earliest personal stories my family related involved the spirit world interacting with mortals. When reading *Asimov's Guide to Shakespeare*, it struck me as curious that Shakespeare's primary source, Raphael Holinshed, submitted *as fact* the existence and intervention of the "weird [weyward] sisters" (Asimov). While much of their participation in the received play is surely not directly from Shakespeare, we can say with certainty that their contributions to the events that Shakespeare relates are not completely apocryphal. They are there in Holinshed's *Chronicles* as factual entities. Similarly, we find in the works of twentieth-century African-American writers the introduction of a seamless relationship between the pedestrian and the ethereal. Examples abound from August Wilson's *Piano Lesson* to Toni Morrison's *Paradise*, *Beloved*, and *Love*. It is part of the black idiom that the spiritual realm informs and determines much of everyday life.

Second, *Macbeth* is a great work that does not carry the onerous burden of race. No course of action in its direct dramatic line depends upon race. It is mercifully free from any delineating considerations that lie outside the most basic human impulses. While plays like *Othello*, *Titus Andronicus*, and *The Merchant of Venice* deal directly with issues of cultural and racial difference, the characterizations of the "Other" are often burdened with racial, cultural, and religious stereotypes. Because there is an assumption that Shakespeare's philosophy, ethos, and human portraiture are infallible, black actors, for example, are often put in the position of fighting with Othello: trying to dignify the character, or denying the existence of a problem with what the play is saying about black men. These problems are entirely absent from *Macbeth*. The only wrinkle comes with the fact that when a black company puts on *Macbeth* (or anything else), it automatically becomes, at least partly, a race play.

Yet all appropriations of plays should be treated with equal circumspection. While I agree with the critics who bemoan the *laissez-faire* approach of modern theatre to non-traditional casting, I do not give a *prima facie* pass to theatre companies that produce the classics without serious consideration to the times in which we live, including contemporary issues of race. Given the hyper-real examination of morality in *Macbeth*, we must necessarily struggle to find explanations for the events that make sense to our contemporaries without the safety net of viewing it as a distant history. While Shakespeare may be many things, he was neither an historian nor a documentarian.

Third, in a time when politicians speak of the importance of an "Ownership Society," we must constantly make the philological bedrock of our way of thinking contemporary. We must own the language that so ably describes us. Of all of Shakespeare's plays, it seemed to the company that *Macbeth* had the deepest parallels with our everyday realities, both the tangible and emotional.

The World of the Play

When our company reached a few conclusions regarding the reasons for doing *Macbeth*, then it was time to attack *how* to do it.

1. We had to find a way to make Shakespeare's racially-transcendent truths dramatically specific to our African-American experiences.
2. We needed to find an historical template on which to lay the physical framework of the events of *Macbeth*.
3. We needed to define a means to illustrate Shakespeare's significance to our modern collective black experience, and reciprocally to correlate the experiences of the African Diaspora to Shakespeare's understanding of the political and sociological events in *Macbeth*.
4. We wanted our audience to understand that *Macbeth* is a play with contemporary importance by demonstrating the dangers of unchecked ambition.

We then realized that it was imperative to look at our plan to engage *Macbeth* through the broadest of telescopes. Our company did not believe that the play teaches us anything of importance as a function of race consciousness. Thus, we wondered when in the history of the Americas the descendents of Africa have experienced the larger human truths represented in the play.

It took a few weeks to grasp that a literal parallel was not to be found. To the extent that blacks have controlled an autonomous land of their own in the Americas, only certain Caribbean islands and, to a much lesser extent, the Gullah Islands on the eastern coast of the United States have experienced black political autonomy. While many of the actors in our production were in direct contact with their African and Caribbean ancestry, we also had those who had only known the experience of the contiguous 48 States.

This is not to say that there was no correlative with the besieged land of the Celtic Scots. Since emancipation there has been a divide in and between black people. It can be reduced to a single, unresolved question: are we black first or American? The practicalities of repatriating millions of dispossessed people proved too troublesome to accomplish in spite of the efforts of those like Marcus Garvey (who exhorted, "Up! Up! Ye mighty race! You can accomplish what you will") (quoted in Van Deburg 10–11). Nor was the dream of a black nation within the United States ever realized in terms of real estate, for which the Honorable Elijah Muhammad advocated. Our nation's inability to erect, or conceive of, a seamless integration resulted in a culturally, economically, and spiritually partitioned America. There have already been, and continue to be, virtual autonomous states—racially and culturally—within the United States of America. Gunnar Myrdal detailed the schizophrenic quality of American society over sixty years ago in *An American Dilemma: The Negro Problem and Modern Democracy* (1944).

Ours, then, is a profoundly analogous situation to the sociopolitical conditions set forth in *Macbeth*. Though we possess no land, our resources have been our labor, our arts, our physical being, and our "soul." Therefore, we set up a construct of a parallel reality where blacks were in control of a spiritual state. For our purposes, we referred to it as Dred Scotland after Dred Scott, the slave who unsuccessfully sued for his freedom in the Supreme Court case of 1857 (see Fehrenbacher). More than anything else, Dred Scotland was a state of mind. How we attempted to set the location of our vision for *Macbeth* relied greatly upon the performances of the company members.

For reasons of economy (our meager budget had ballooned to a whopping $32,000!), and to avoid any semblance of direct analogies, the set and costumes were not localized. In our reality, Dred Scotland was situated somewhere in the Americas. It, like historical Scotland, was an island. Our island, like the people of the Diaspora, floated in and out of various regions. Nonetheless, it retained its hard earned sovereignty. Those from the outside who sought to conquer the island did so for the spoils represented by the island's inhabitants. The idea was that if Dred Scotland were conquered, they would be returned to chattel slavery. Taking it a step further, we reasoned that the people of our land learned much of their English from the paucity of any literature outside of the King James Version of the Bible. The biblical element added a somewhat poetic twist given the fact that King James was a constant between the society we were erecting and the language that fleshed out the intellectual infrastructure of that world.

The characters in our production were regionalized by the use of accents—linguistically and aesthetically. As a company, we employed African, Caribbean, South American, and southern American rhythms and cadences. Of course, this is also true of most black, urban neighborhoods in America. If you were to drop by any block in Harlem, you would not immediately be able to distinguish it from Dred Scotland aurally. The actors' facility with dialects also helped to differentiate the multiple characters most of us played. With the exception of me (Macbeth) and Patrice Quinn (Lady Macbeth), all the other actors in the company played many different characters. There were ten of us playing over forty parts.

Playing the Part

For me, it all started with finding an approach to the character of Macbeth itself, and discovering what traits I could activate in myself to bring out his true character. As it turns out the appearance of the weird sisters in *Macbeth* was the first aspect of approaching the part that I had to get my head around. They are cumbersome figures to deal with. Could it be that they are figments of his imagination (though Banquo sees them, too)? This inkling was reinforced by our director. Adhering to Harley Granville-Barker's reading of *Macbeth*, he suggested the opening scene with the witches was not written by Shakespeare (Granville-Barker).

Macbeth and Banquo are shocked by the appearance of the creatures, the ambiguity of their gender, and the purpose of their message. It seemed terribly unfair that Macbeth could be doomed by the predetermination made by the weird sisters in their revelation of his imminent rise to the throne. Granville-Barker's deeply held view of the apocryphal nature of the sister's first appearance helped us in our attempt to establish the fact that it is Macbeth himself who has been thinking of a means to rise to power all along (see also the Daileader essay).

When the rehearsals began, I had to struggle with the following questions: When does Macbeth decide to kill Duncan to assume the title of king for himself? What/who decides it for him: the weird sisters, his wife, or his own errant nature? The answers form the bedrock of the emotional and intellectual approach to the character.

Thanks to the excellent contribution of Horace Howard Furness, we are all beneficiaries of a treasure chest of thinkers who have tackled just such questions. Included in Furness's *A New Variorum Edition of Shakespeare* is a section on performance history with excerpts from Edmund Kean, Edwin Booth, and a host of other players who arrive at widely divergent conclusions about crucial dramatic moments. Here is William Charles Macready on *Macbeth*:

> Acted Macbeth in my very best manner, positively improving several passages, but sustaining the character in a most satisfactory manner.... I have improved Macbeth. The general tone of the character was lofty, manly, or, indeed, as it should be, heroic, that of one living to command. The whole view of the character was constantly in sight: the grief, the care, the doubt was not that of a weak person, but of a strong mind and of a strong man. (quoted in Furness 496)

In some ways reading another actor's interpretation of a part is a bit like spying. In modern times actors are careful to be as different as possible from a predecessor in the hopes of distinguishing themselves and "discovering" some heretofore unearthed gem that will definitively illuminate the character for an audience of their contemporaries. Yet I found great value in being able to examine what elements of an actor's rendering surprised and illuminated the viewers of an earlier age. Here, Macready clearly read Macbeth as a manly character (a "strong man") whose ambition motivated him.

To extend this motivation a bit further, it became clear to me through the *Variorum* that Shakespeareans of antiquity considered the possibility that Lady Macbeth was as much an aspect of Macbeth as an entity unto herself. This was, of course, not to be taken literally as there is no way to dramatize such a concept without inviting scorn or radically altering the play. It is at times such as these that it helps to have a friend. Happily for me Sarah Siddons came to the rescue.

> It was my custom to study my characters at night, when all the domestic cares and business of the day were over. On the night preceding that in which I was to appear in this part for the first time, I shut myself up, as usual, when all the family were retired, and commenced my study of Lady Macbeth. As the character is very short, I thought I should soon accomplish it.... I went on with tolerable composure, in the silence of the night (a night I can never forget), till

I came to the assassination scene, when the horrors of the scene rose to a degree that made it impossible for me to get farther. I snatched up my candle and hurried out of the room in a paroxysm of terror. (quoted in Furness 476)

Reading Sarah Siddons's approach to Lady Macbeth, I found a way in to my own approach to Macbeth. Her fear, her "paroxysm of terror," helped me to understand the sheer horror of the scene. It became clear that Macbeth is as frightened and incredulous as any modern man would be when he sees the ghost of Banquo in the banquet scene. In addition to eliciting some of the most horrifyingly beautiful poetry of the play, this singularly shocking paranormal event also serves to illustrate a point of no return in his relationship with his wife. It signals a transfer of power, if you will.

My experiences with Shakespeare have taught me to try to find anchoring points. To be clearer, I am inclined to approach Shakespeare in a manner similar to a musician's approach to a piece of music. Whether the genre is classical or jazz, one looks for an entry point. Somewhere in the piece is a point of relative stability. There exists in any piece a small section that is to be the one position from whence the performer can venture forth into the extremities of the rest of the work. A tactic used by jazz pianists is a "pedal point," and it is a base tone—and usually in the bass—from which position other tones or chords can modulate. The underlying mood of a kind of predetermined doom referenced above seemed the most practical one to me. Macbeth's mood (tone) is largely self-invented, and it is constantly reinforced by his wife and the weird sisters. After announcing and fulfilling his intention to visit the weird sisters again, however, Macbeth's character changes and becomes more isolated. Although the Macbeth that is described by the Captain in 1.2 is not substantially different from the man we see at the end of the play, his motives become more self-directed. What *is* different is the absence of fear that seems to have fallen away after his last encounter with the sisters.

Thus, although I was invested in anchoring this production of *Macbeth* in a distinctly black aesthetic, contextually, culturally, and racially, I found my own personal inspiration coming from nineteenth-century white, British actors. Again, I think this speaks to the odd balance all actors, but especially actors of color, must face when approaching classical texts. In order to create the specific cultural context I wanted to portray, I needed to explore the "universal" aspects of the character that previous actors had developed.

MACBETH'S DIRECTION

Under the direction of my college friend and Shakespearean colleague, Steve Marvel, we were able to reduce and combine various characters to keep down expense and maximize space in the 99-seat, Equity-waiver theatre. Many of his cuts proved to be judicious and helped keep the line of action of clean. Yet all of the considerations that resided outside of the immediate action of the play (its cultural, social, and racial components) were debated only by the black company and our predominantly black audience. Steve never really engaged in the debate as to whether or not our production needed any consideration of identifiable setting, even though it is typical for directors to do so.

Steve was, though, keenly interested in the schematics of the set; the physical dimensions of the dramatic structures which allowed the action to proceed with fluidity. After all, he was directing *Macbeth*, and not moderating a social debate that used a tragedy as a platform. It was his considered opinion that the play speaks for itself, and the very act of mounting it endows the audience with the power to create its own superstructure and concept. Racial dynamics seemed a tertiary issue for him. I shall never be certain how this can be possible for white artists. I appreciate (to the extent that I can) Steve's delicate situation in directing a company that was entirely black, replete with highly trained and educated actors with strong opinions and clear aims. Though he never betrayed any sense of discomfort, there had to be some sense of avoiding potential landmines given the troublesome history we were exploring through *Macbeth*. Did he ever find himself reluctant to say something that he may have said to a white actor? Did he refuse some insight that may have opened up another channel of thought or emotion? Could it be the reverse? Did he uncover some truth about a line or moment that was given poignancy as a function of our insistence on relating *Macbeth* to our unique American experience? We will never know because he did not engage in these types of racial and cultural questions.

Most actors, irrespective of race, simply want to perform in good plays. For me this is never enough. It will always be my first item of business to find a way to present great works of literature in ways that are consistent with the realities of the contemporary world. In order to do this, we actors, directors, and producers will be at the mercy of those who are saying what we believe to be true about the people we know. Again and again, Shakespeare gets quite a great deal of this right (because he was writing about humanity), even though he was not necessarily writing *for* black people.

For this specific case, though, I will always believe that it worked to our advantage to have a director who could concentrate simply on the play. It is a luxury that I envy, and will ever aspire to. I shall also be plagued by never being able to know with any certainty if my eternal awareness of race lessens or increases my own experience of a play such as *Macbeth*. If scientists are right about the existence of parallel universes, perhaps it can do both.

Note

1. See the essays in the "Federal Theatre Project(s)" section of this collection for further details on Welles's 1936 production.

13

ASIAN-AMERICAN THEATRE REIMAGINED: *SHOGUN MACBETH* IN NEW YORK

Alexander C. Y. Huang

Macbeth has a long and varied history of Asian-style enactments in the United States. Among the best known performances are Akira Kurosawa's film *Throne of Blood* (1957), Yukio Ninagawa's *kabuki*-style *Macbeth* (1985), Wu Hsing-kuo's Peking opera *The Kingdom of Desire* (1987), Tadashi Suzuki's all-male *The Chronicle of Macbeth* (1988; English and Japanese versions), Shozo Sato's *Kabuki Macbeth* (1997), Sato and Karen Sunde's *Kabuki Lady Macbeth* (2005), and Charles Fee's *kabuki*-inflected *Macbeth* (2008) (these last three in English), all screened or staged multiple times in North America. These works either map the English imaginary of Scottish incivility onto what is perceived to be equivalent Asian contexts (as Kurosawa's film does), or create a new performance idiom from amalgamated elements from various traditional Asian theatre styles (as Sato's production does). It is not uncommon for artists to combine unfamiliar styles of presentation with English and even Shakespeare's lines, of which the American audience tend to assume ownership.

John R. Briggs[1] combines both approaches when he brings the Scottish play, Kurosawa, and Asian America together in his *Shogun Macbeth* (1985), a play in English (interspersed with a great number of Shakespearean lines) set on the island of Honshu in Kamakura Japan (1192–1333). Japanese titles or forms of address replace Shakespeare's originals: the Maruyama [Birnam] Wood moves to Higashiyama [Dunsinane]; Thane of Cawdor becomes Ryoshu of Akita; bottom-lit *bunraku* puppets as ghosts are summoned by the witches; "the best of the *ninja*" are the Murderers; and the Porter scene morphs into a *kyôgen* comic interlude performed by a pair of drunken gatekeepers. The emergence of a work such as *Shogun Macbeth* coincided with Japan's rising economic influence in the United States in the 1980s and American theatre's continued interest in select Japanese cultural tokens in the new millennium. Briggs notes with enthusiasm the "world-wide rebirth" of Japanese culture that fed into American fascination with "all things Japanese, especially things samurai" (J. R. Briggs 2009). As a "samurai" film, *Throne of Blood* has been so successful that it has been cited as inspiration for new works beyond Asia, including Briggs's play, Alwin Bully's Jamaican adaptation (1998), Arne Zaslove's stage production (1990), and Aleta Chappelle's proposed Caribbean film *Macbett* (2010) (see Appendix). In contrast to *The Throne of Blood*, which uses *Macbeth* as a launching pad for cinematic experiments, *Shogun Macbeth*, as an American play, deploys fragmented Japanese performance culture to rescue Shakespeare from what Peter Brook has called "cold, correct, literary, untheatrical" interpretations that make "no emotional impact on the average spectator" in a time when "far too large a proportion of intelligent playgoers know their Shakespeare too well [to be willing to] suspend disbelief which any naïve spectator can bring" (quoted in J. R. Briggs 1988, 8–9).

Regarded as a "Kurosawa-lite adaptation," in both the positive and negative senses of the phrase (Gussow), *Shogun Macbeth* was first staged at the Shakespeare Festival of Dallas in 1985 and then by the Pan Asian Repertory Theatre in New York in 1986 (also directed by J. R. Briggs). The play has also been performed by non-Asian groups, including a San Francisco State University production directed by Yukihiro Goto in 1992 and a Woodlands High School production directed by Carlen Gilseth at the Edinburgh Fringe Festival in 2003 (see Muto and Mire). It was revived in New York in November 2008, again by the Pan Asian Repertory (directed by Ernest Abuba, who had played the title role in 1986) with a cast of white and Asian-American performers.

The curtain opened to reveal a minimalist set cast in ominous green light. All of the action took place in front of an eight-foot statue of Buddha behind a Torii gate (a traditional gate commonly found at the entry to a Shinto shrine), a statue that seemed to look down at the dramatic events with a sense of aloofness, a transcendental indifference; this Buddhist reference informed the framework of narrative. Biwa Hoshi (Tom Matsusaka), an itinerant blind priest and narrator, opened the production with a powerful delivery of lines from the *Sutra* that echoed the ideas of fourteenth-century *noh* playwright Zeami (*Atsumori*) and anticipated some of Macbeth's later lines:

> BIWA: Life is a lying dream, he only wakes who casts the world aside. The bell of the Gion Temple tolls into every man's heart to warn him that all is vanity and evanescence. (J. R. Briggs 1988, 11)
>
> MACBETH: Life's but a walking shadow, a poor player
> That struts and frets his hour upon the stage
> And then is heard no more.
> It is a tale told by an idiot; full of sound and fury, signifying nothing.
> (J. R. Briggs 1988, 64; compare *Macbeth* 5.5.23–27)

Fujin Macbeth (Lady Macbeth; Rosanne Ma) walked the stage in a *kimono*, while most characters carry a *katana*. The three white-faced *yojos* (weird sisters) in Day-Glo wigs and *kabuki* makeup played a major role throughout the performance, as they manipulated and channeled events and news. Briggs intended "all horrible events or negativity" to spring from the *yojos* (J. R. Briggs 2009). They manipulated the *shoto* (dagger) to which Macbeth reacted ("Is this a *shoto* I see before me, the handle toward my hand?" [see *Macbeth* 2.1.33–34]). Their appearance marked as androgynously "Japanese" through *kabuki* makeup, movements, and chants, the three *yojos* were played by both male and female performers (Claro Austria, Shigeko Suga, and Emi F. Jones, who doubled as *isha*, the doctor). The *yojos*' cultivation and channeling of various characters' desires and behaviors were enacted by the kinetic energy of their presence in many scenes—seen or unseen by the characters. They delivered armor and a helmet to an agitated Macbeth (Kaipo Schwab) in the final battle scene, and remained on stage as enchantresses and indifferent observers of the end of his story.

Unseen by Fujin Macbeth in the sleepwalking scene but exerting a felt presence, the *yojos* followed her every step of the way, creating the impression that these creatures were both the cause and result of her nightmarish imaginations. Fujin Macbeth's *suri-ashi* ("slide her feet and shuffle along") gait articulated well with the *yojos*' presence in this dreamscape. The doubling of the enchantress (witch) and the healer (doctor), facilitated by a *noh* mask, exemplifies Briggs's investment in the capacity of Japanese signs to generate new meanings from this famous scene (5.1):

> ISHA / YOJO ONE: When was it last she walked?
> *[The lights come up to reveal Yojo One wearing a* noh *mask.]*
> GENTLE WOMAN: Since Shogun [Macbeth] went into the field…
> ISHA / YOJO ONE: In this slumbery agitation, besides her walking and other actual performances, what, at any time, have you heard her say?
> GENTLE WOMAN: That sir, which I will not report after her.
> ISHA / YOJO ONE: You may to me, and 'tis most right you should.
> (J. R. Briggs 1988, 58; compare *Macbeth* 5.1.2–13)

The otherwise benign diagnostic conversation between the doctor and the gentle woman bears a more malignant undertone as the *yojo* (Shigeko Suga) speaks from behind her mask. The very reference to Hell by Fujin Macbeth brings forth the other two *yojos* as she washes her hands:

> FUJIN MACBETH: Yet here's a spot. Out damned spot! Out, I say! One; two; why, then 'tis time to do't. Hell is murky!
> YOJO TWO: Fujin Macbeth!
> YOJO THREE: Fujin Macbeth!
> FUJIN MACBETH: [She does not see or hear the *yojos*] What, will these hands ne'er be clean?!
> YOJO THREE: Fujin Macbeth!
> FUJIN MACBETH: No more o' that, my lord, no more o' that!

YOJO TWO: Fujin Macbeth!
FUJIN MACBETH: You mar all this with starting.
ISHA / YOJO ONE: You have known what you should not.
GENTLE WOMAN: She has spoke what she should not, I am sure of that.
...
YOJO TWO AND THREE: Oh! Oh! [echo]
ISHA / YOJO ONE: What a sigh is there!
(J. R. Briggs 1988, 59–60; compare *Macbeth* 5.1.26–44)

Even though Fujin Macbeth did not interact directly with the *yojos*, the way her lines coincide with those of the *yojo*s creates a suggestive layer of intertextuality, as she unconsciously danced to the rhythmic hissing and growling of the *yojo*s toward the end of this scene. Briggs envisioned an aesthetic structure and "solipsistic" philosophical framework that allowed violence to "scream its horrors beneath the fragrant cherry blossoms" (J. R. Briggs 1988, 8). As both observers and instigators, the witches are given substantially more agency than they have in Shakespeare's play.

The artistic and critical focus of *Shogun Macbeth* has thus far rested upon the production's capacity to test Shakespeare's universality and liberate *Macbeth* from variously defined traditionalist interpretations; as one critic wrote in 1986, "Though language and character are altered, the soliloquies remain and...at its bloody heart, the play is still *Macbeth*, albeit an exotic one, with a universality transcending time, place and performance style" (Gussow). And yet as a unique English-language adaptation exploiting Japanese sensibilities, *Shogun Macbeth*, as director Abuba points out, also provides an opportunity for exploring what it means to be Asian American: "One of the major intents of re-visioning *Shogun Macbeth* is to demonstrate the exceptional talent of the new generation of Asian American actors" (Abuba). This vision is in line with the Pan Asian Repertory Theatre's stated mission to "bring Asian American Theatre to the general theatre-going public and deepen their appreciation and understanding of the Asian American cultural heritage" (Pan Asian).

Yet is performing in the style of a culture (*kabuki*, for example) actually embodying the culture itself—"not just visiting or importing [it] but actually doing [it]" (Schechner 4)? Despite its popularity, *Macbeth* is not typically associated with racial questions for Asian-American theatre, a racially-defined theatre that was established in 1965 with the founding of the East West Players. *Shogun Macbeth* negotiates challenging cultural terrains as it deploys various elements of Japanese culture to interpret *Macbeth* and expand the Pan Asian Repertory's repertory of otherwise Asian or Asian-American plays. The founder of the group, Tisa Chang, has been criticized for commercializing "Asianness" as foreign and exotic. Plays such as *Shogun Macbeth* address the younger generation of performers' resistance to her request to "keep focusing on their Asian identity," which, they believe, limit the creative possibilities even as it promotes Asian American solidarity (E. Lee 91).

Chang herself seems to be resisting the same concept: "I was so tired of Westerners using Asian-ness as an exotic characteristic" (quoted in Griboff). The identity of *Shogun Macbeth* remains unclear, as the blending of different cultures does not necessarily lead to a hybrid one, though the identity of the lead actor, Kaipo Schwab (Macbeth) embodies this ideal: born in Honolulu, he is of Hawai'ian-Chinese-German-Irish ancestry. Critics such as Leonard Pronko consider *Shogun Macbeth* a "non-Shakespearean play," albeit with "many of the famous speeches [from *Macbeth*]," but Briggs maintains that the play is still "a Shakespearean play, in the best traditions of what that means" (Pronko 29; J. R. Briggs 1988, 9). One may wonder whether *Shogun Macbeth*, despite its repackaged Asian cultures and Asian bodies, might not harbor an investment in the notion of "a self-consciously white expression of minority empowerment" (Hsu 52).

Briggs sought to produce a work that rediscovers Shakespeare's insights by "displacing the audience, forcing involvement in his language, creating an atmosphere that is new and different and capable of spontaneous surprise" (J. R. Briggs 1988, 8). Perhaps his displacement succeeded too well; I attended a Sunday matinee performance in the 2008 revival, where more than half of the seats were empty, and quite a few audience members did not sit through the entire performance. Despite their respectable effort, a number of performers appeared to be lost in both the Japanese *mise-en-scène* and the Shakespearean lines, making for an atmospheric but uneven performance. Perhaps the playwright's "white" identity and the racially-mixed cast distracted audience and critics; the reception history of Asian productions of Shakespeare in North America and Europe suggests that reviewers are often more tolerant of cultural differences and artistic innovations when these works are written and directed by artists from Asia (Huang). As much as the company, Abuba, and Briggs wanted to break out of the stereotypical association of Asian-American theatre with a necessarily Asian-American repertory which is defined by plays such as David Henry Hwang's *M Butterfly*, Shakespeare—however Asian—is always "white." However, *Shogun Macbeth* has successfully constructed a contact zone that remains open for future inscription.

NOTE

1. Note that playwright John R. Briggs is no relation to my fellow contributor John C. Briggs.

14

THETLINGIT PLAY: *MACBETH* AND NATIVE AMERICANISM

Anita Maynard-Losh

Following the opening battle scene, we see three creatures creep onstage through fog. We hear drumming, and these creatures come together in a frenzied dance. Replete, they hunker down together; we hear:

> Gwátgeen sá nás'gináx wooch kagéi haa kaguxdaháa?
> Xeitl tóox', séew kaa ch'u k'eeljáa gé?

On the screen above the stage, we read the words:

> When shall we three meet again?
> In thunder, lightning, or in rain?

(1.1.1–2)

Beginning in 2003, I directed the Perseverance Theatre (Juneau, AK) production of *Macbeth* that set Shakespeare's story within the cultural context of Southeast Alaska's indigenous people, the Tlingit (non-Tlingit speakers pronounce it as "KLINK-it"). Other than a couple of local productions that combined *The Tempest* with a traditional story about Naatsilanéi (a revenge plot that includes the origin of the killer whale), a play by Shakespeare had never been performed in a full-scale Tlingit version until then. The idea for such a production began to take shape during the eleven-year period that I lived in Hoonah, Alaska, a fishing village on Chichagof Island that had a population of less than 700 at the time, at least 90 percent of them Tlingit. The community had good reasons to be distrusting and unwelcoming of non-Natives—most of their experience with non-Natives historically had ended badly—and I was not accepted right away. (I am sure some community members never accepted me.) It seemed that the defining ethnic difference between people was whether one was Native or non-Native; for example, a white person and an African American would be seen as members of the same non-Native group, a marked difference from how people are categorized in the lower 48. The cultural differences between the dominant culture and the Tlingit culture span areas that many are not even aware of: tempo, space between questions and answers, relationship to animals, view of senior citizens, what can be owned, and so forth. It is very easy for the group that I frequently call "well-meaning white people" to make egregious mistakes without even knowing it.

I was not looking for an interpretation of *Macbeth*, so the idea floated to the surface over the years—idly at first as I began noticing things like the Tlingit clan systems that are still in place, when the clan context I was familiar with was from Scotland. As I learned more about the Tlingit culture that surrounded me, I saw further connections between this complex ancient culture and the world of *Macbeth*: both possess a strong connection with the supernatural; both were feared for their fierce warfare; and both value the needs of the group over the individual. In this interpretation, Macbeth's life begins to disintegrate when he breaks this cultural imperative and places his personal ambition over the good of the tribe.

Our use of the Tlingit culture was intended to be a celebration rather than exploitation. The actors, designers, and I wanted to make sure that we had the support of the elders of the community. Respected leaders such as Nora and Richard Dauenhauer and Paul Jackson came to my house and talked with us as a group, and eventually they gave us their blessing to proceed. Original music, drumming, and songs in Tlingit were written for the piece. There are cultural restrictions on traditional music, based on clan ownership of music and songs, a form of oral copyright—for instance, only members of a certain clan are permitted to tell certain stories. In addition, the existing Tlingit clans use particular animals as a kind of crest/symbol. Thus, the visual symbolism we used came from Tlingit traditions without employing restricted images which belong to individual clans. For example, designer Robert Davis Hoffman created totemic crab ("shaming") images that were used as Macbeth's crest; deer ("peace") became Duncan's crest—neither of which belong to actual Tlingit clans. Dances were choreographed, most often accompanied by drumming. The witches were interpreted as shape-shifting creatures traditionally known as the *kooshdaa kaa*—or land otter people; they became the three murderers (wearing totemic wolf masks, and dancing their killing of Banquo and the Macduff family) as well as other characters that hurry Macbeth towards his fate. Original masks were used extensively within these roles. All of the actors, as well as the composer, lyricist, costume designer, and set/prop designer were Alaskan Natives. Most were at least part Tlingit, but other Alaska Native tribes included Yup'ik (Lady Macbeth), Gwich'in (Duncan), and Inupiaq (Macduff).

There were eventually three incarnations of this *Macbeth* production, all marked by differences in the balance of English and Tlingit language. The original two incarnations of the production were all in the original English, except for occasional uses of Tlingit. The one part I desperately wanted to be in Tlingit was the "Double, double, toil and trouble" section, as that has become such a cliché in English. Before mounting a third version at the National Museum of the American Indian in Washington, DC, we started thinking of how we might use the Tlingit language to represent the culture.

In this production the conceit of Macbeth's alienation was highlighted by having the characters speak Tlingit when observing traditional group values, and English when following individual desires. The entire beginning of the play was in Tlingit until Macbeth meets the witches for the first time and begins to contemplate the murder of Duncan. The first English we hear is the beginning of his soliloquy:

> Two truths are told
> As happy prologues to the swelling act
> Of the imperial theme.
> (1.3.126–28)

After that moment the characters speak Tlingit when obeying cultural rules, English when not. Thus, most of the scenes between Macbeth and Lady Macbeth are in English, as well as the asides that they make in scenes with others. English supertitles were projected above the stage at the NMAI in March 2007.

Translating so much more of the play into Tlingit was a monumental task, which began with support from the then artistic director, PJ Paparelli, and the Sealaska Heritage Institute. Tlingit is an extremely complex language, with sounds that English speakers find difficult to hear, much less pronounce. When the U.S. government colonized Alaska and established compulsory education, Native children were punished for speaking Tlingit in school, a policy that was instrumental in threatening its survival. For years Tlingit was considered a dying language, with perhaps only sixty living fluent speakers. Today the language is experiencing a resurgence due to efforts of Native leaders, cultural organizations, Tlingit school programs, and dedicated individuals. The translator for our production was Tlingit elder Johnny Marks. Lance Twitchell, the young actor who played Ross, was studying Tlingit at the university and helped us understand the images used within the translation. It was deeply emotional and difficult for the actors to learn the text in Tlingit, as most of them spoke no more than a few words or phrases in what many of them believe should be their Native tongue. They struggled and eventually triumphed. It was fascinating to see what happened to the actors when they spoke in Tlingit—they inhabited their bodies in a much fuller, more grounded way. Performing in another language forced us to be very clear with what action was being played on individual lines or beats. The change in tactics became so clear in working with the actors in Tlingit that even the English scenes seemed much too general, forcing us to apply comparable rigor to them. There was a fierce joy present when the actors spoke this language in Washington—this language which they felt the American government had once tried to eradicate.

The production was collaborative in a much deeper way than a usual theatre piece, given the generous contributions of the actors and the rest of the creative team. There was no way that I, particularly as a non-Native, could have made this play by myself. The rehearsal process mirrored the tribal value of everyone working for the good of the whole. For example, Gene Tagaban, who played Banquo and created the drumming and dancing for the production, had the idea to stage the final duel between Macbeth and Macduff by using drums as shields, and striking with beaters instead of swords. This became one of the most exciting, physical moments of the play: the actors reached out and hit each other's drum/shields with all the fierceness of combat, accompanied by the percussive rhythm of the drummed hits themselves. George Holly, who wrote all the original Tlingit songs, was inspired to write a song for the "Double, double, toil and trouble" section of the piece. This was fairly late in the rehearsal period of the last incarnation of the play, and we had been using a Tlingit chant which was working well, so I must admit that my impulse was to decline when he brought up the idea. However, I had learned through this process that saying "yes" to the ideas that the cast generated was always right, even if it was not always easy. George created a powerful song that was eerily beautiful and haunting—as well as seeming so evil that the cast and crew held a cleansing ceremony in the theatre every night after the show!

I had previously heard tell of the "Voodoo" *Macbeth* when studying Shakespeare, and thought it sounded like an intriguing idea, but was not consciously responding to it. (I also was aware of Kurosawa's film, *Throne of Blood*, although I had never seen it until I was already working on our production.) Like Welles, I was a white director of a non-traditional production that deployed "native" elements in revising Shakespeare. However, I believe that Welles did not collaborate with the people of color that performed in his production to the extent that we did. If I learned anything in my time in Hoonah, it was how humble and respectful one must be in working with a culture that is not one's own. It was important to me that it be the actors' play more than mine. It might have been more "right" if there had been a Tlingit director, but I was fortunate to have had the training in Shakespeare, the relationship with Native actors and artists, the experience of living in Hoonah, an established relationship with Perseverance Theatre, and an artistic director (Peter DuBois, now at the Huntington) who made it possible for me to move forward with what I would most like to do. So I was a catalyst who could never have created the production without the generous contributions of the people working with me. I was always conscious that I did not own the culture; as I said to the actors in rehearsal, "Shakespeare belongs to everyone, all of us; but your culture does not belong to me, it belongs to you. You own both."

Shakespeare could be considered the ultimate "white" playwright, and thus a symbol of the dominant culture that stripped away so much from the Tlingit. We all felt the transformational power of art shifting our world view as the play was illuminated by setting it in this cultural context. It became more than a play to us, more than art—it became a mission, an invocation of ancestors, an event that bridged art, culture, history, and community. Native audience members spoke of how moving and healing it was to them to see that their culture "made Shakespeare better." Tlingit teenagers said that they now realized that Shakespeare "is about us," whereas previously they had felt disconnected. Early in rehearsal, Allan Hayton, who played Duncan, said that while he had been in several of Shakespeare's plays before, he was moved to tears by the power of seeing all the brown faces in the room. There was a point in rehearsal when we felt overwhelmed by the rigorous task we had undertaken. Jake Waid, the actor playing Macbeth, stirred everyone in the room to renew their commitment to the challenge when he declared, "We need to claim this play for Native American people."

15

A Post-Apocalyptic *Macbeth*: Teatro LA TEA's *Macbeth 2029*

José A. Esquea

Macbeth and multicultural repertory intersected in our recent adaptation, *Macbeth 2029* (2008). Drawing upon the rich multi-ethnic traditions of New York City, Teatro LA TEA seeks to produce Shakespearean productions that provide access to Latinos and others traditionally excluded from the enriching experience of the performing arts—artists and audiences alike. Producing Shakespeare presents us with the challenge of mastering the highest standard in theatre arts, what I consider a kind of performative "measure of achievement" for theatre professionals. Yet at the same time, we find ourselves producing plays first created in an Elizabethan world in which people of color rarely existed. Not surprisingly, Shakespearean productions have historically been closed to dramaturges, directors, production designers, and, most notably, actors of color. This was my immediate reality after my graduation from Skidmore College in 1993. Like most kids still green from undergraduate optimism, I was ready to change the world with my training as an actor. Yet I found myself frustrated, at times even angry, with an industry that had very little room for me in it.

After many casting disappointments, I turned to directing and producing. I believed that as the director/producer, I would be able to *create* the world in which the story took place, thereby freeing me and my fellow actors of color from the racist constraints of our industry. For years I paid my dues, working in just about every theatre in New York City. Finally my friend and mentor Veronica Caicedo helped me earn my shot as a director, first at the Songs from Coconut Hill Latino Playwrights Festival, and then later at Latin American Theatre Experiment & Associates (or LA TEA). Its founders were all members of the famed Repertorio Español: Nelson Tamayo (Ecuadorian), Mateo Gomez (Dominican), and Nelson Landrieu (Uruguayan). In 1982, urged by a desire to develop their own voices, they began Teatro LA TEA in the Lower East Side. The theatre has a philosophy of inclusion that allows the theatrical world of Shakespeare's plays to be as diverse and grand as New York. As theatre historian Joanne Pottlitzer notes, "Hispanic theatre in the

United States, which dates to the sixteenth century, is as diverse as the people it portrays"; we feel strongly that Shakespeare should be part of that diversity too (1).

My first Shakespearean play for LA TEA was *Hamlet* in 2006. With Jaime Luce as the lead and Gil Ron as Claudius, we set the story in a present day locale much like New York: Hamlet and his family were a very wealthy fair-skinned Latino family; Ophelia was black, though her father was white; the players were led by an Asian Shakespearean actor. In 2007 we put on a groundbreaking production of *Othello*, in which the lead role alternated between a male actor (David Beckles, Jr.) and a female actor (Cheryl D. Hescott). We were once more fusing experimental and classical theatre, and sought to explore racial prejudice as well as homophobia. Strikingly, audiences seemed more unsettled by acts of violence committed by the female Othello than by any of the more overtly "queer" aspects of the production. The play was set in contemporary Venice with a diverse, cosmopolitan cast: both Othellos were black; Desdemona (Mariko Barajas) was an actress of mixed Asian descent; and Cassio alternated between a Donimincan actor (Jesus Martinez) and one with Chinese heritage (Chester Poon). We won four HOLA (Hispanic Organization of Latin Actors) Awards in two categories: Ricardo Hinoa (Roderigo) and Will Sierra (Iago) for their performances; Jesus Martinez and Chester Poon for their fight choreography.

2008 brought us to *Macbeth*. We were certainly not the first Latino company to stage *Macbeth*—indeed, in 1966 it was produced in the New York Shakespeare Festival's Spanish Mobile Theatre, in a Spanish-language production directed by Osvaldo Riofrancos.[1] Raul Julia played Macduff in that production; he had earlier played the title role for Tapia Theatre (San Juan, 1963) and would again in 1989 (New York, Public/Anspacher Theatre). In 1985, David Richard Jones directed *Macbeth: A Modern Mestizo Story Set in Central America* for La Compañía de Teatro de Alburquerque. More recently, the British Kelvin Players adapted *Macbeth* to a 1970s Latin American military regime (Roche). We differentiated our production by casting leads who were not Latino, although there were Latino principals throughout the rest of the cast. Furthermore, rather than setting the play in some vaguely "Latin American" context, we envisioned it taking place after an apocalypse, in the year 2029, with the only semblance of order coming from observing ancient Scottish rites. This production was our most immense challenge to date, forcing us to make sure that all elements—costuming, setting, lighting, sound, weapons, choreography, etc.—simultaneously evoked a sense of medieval Scotland, the grim reality of the apocalypse, and passing remembrances of the twentieth century. Such a setting appealed to us in part because it made the multiracial cast more plausible for a Shakespearean play[2]: our Macbeth, Malcolm, and Duncan were white; Banquo was Asian; Macduff was black and Lady Macbeth was Latina; and the witches were white, Hispanic, and black (see Figure S.4).

The production of *Macbeth 2029* exemplified how the collaborative process of theatre is greatly enhanced when the participants have vastly diverse backgrounds, racially, technically, culturally, economically, etc. The execution of a pluralistic theatre must be carried out not only on our stage but also *behind* it. To that end, our production crew is as diverse as our actors. Assistant Director Dara Marsh, Lighting Designer Alex Moore, and Costume Designer Tracy McClaire are African American. Set designer Yanko Bacoulick is Chilean by way of Eastern Europe. Makeup artist Ingrid "Skull" Dubberke is Dominican and Italian, while our photographer Anthony Ruiz and our Videographer Ruben Dario Cruz are Puerto Rican. Our fight choreography team—noted above as the Cassios from our *Othello*—is a seemingly unlikely duo: Jesus Martinez is Dominican and Chester Poon Chinese. Our guest lecturers, Professors of Romance Languages Patricia Akhimie and Manu Chander, are themselves a diverse couple: she is biracial of South African and German decent, and he is Indian by way of Detroit.

This amazing diversity allows our team to address both production concerns *and* human concerns. We have dealt with every "ism" you can imagine. We have suspended our audience's preconceived ideas about our backgrounds and made the *story* their main concern. We strive to transcend the limitations of oppression through the theatrical experience *and* we strive to interpret Shakespeare's work successfully. They go hand in hand for LA TEA. We anticipate continuing with one Shakespearean production per season. Our 2009 *Romeo and Juliet* will be set on the Lower East Side, with a Chinese Romeo (Poon) and a Latino Juliet (Carrissa Jocett Toro); parts of the production will be in Cantonese and Spanish, accentuating language as well as ethnic difference.

Power is not transferred without revolution, which begins with a mastery of the oppressor's language. The next step in revolution is to proclaim that art belongs to everyone. We cannot wait for our oppressors to hand us the opportunity to perform in the classics. We have to create these opportunities for ourselves. The one thing artists in this collection have in common is our decision to produce any work that our interests, imaginations, and ambitions will allow. We have produced these works to the highest creative and technical standards. It is our right as artists, and as humans.

Notes

1. In 1979, Joseph Papp attempted to establish a permanent black-Hispanic acting theatre troupe in New York, beginning with *Julius Caesar* before proceeding to an all-black *Coriolanus* (Kalem).
2. The 1981 Los Angeles Actors' Theatre Production, in which Danny Glover starred, was also set in a post-apocalyptic world (Wilds). For another take on a *Macbeth* set in the future, see Robert Alexander's play, *Alien Motel 29: Secret Outtakes of the Ebony Lady Macbeth*, "an African-American fantasia in space" (2005).

16

Multicultural, Multilingual *Macbeth*

William C. Carroll

Two weeks to the day after Barack Obama was elected President, on a flight to Hawai'i to see the Kennedy Theatre production of *Macbeth* at the University of Hawai'i at Mānoa, I came across this sentence in David Remnick's *New Yorker* article, "The Joshua Generation." Remnick, quoting Randall Kennedy, a professor at Obama's alma mater, the Harvard Law School, writes, "'He's operating outside the precincts of black America,' Kennedy said. 'He is growing up in Hawai'i, for God's sake'" (72). The phrasing suggests the manifest absurdity of an African American from Hawai'i being informed in, or shaped by, the racial discourse of the "mainland," or, as it is now commonly referred to by residents of Hawai'i, the "continent." And it is certainly true that Obama's background—white mother from Kansas, black father from Kenya, a childhood lived in Indonesia and Hawai'i, a grandfather who was a medicine man—differs radically from the stereotype of the experience of those on the continent. What Kennedy's comment implies, however, is the irrelevance, more or less, of the racial politics of Hawai'i—but the *Macbeth* production reflected racial divisions and linkages that are not to be dismissed. What constitutes "race" looks rather different in Hawai'i than it may appear in the heart of the mainland.

According to the U.S. Census Bureau, in 2006 the racial composition of the inhabitants of Hawai'i presents a complex picture. Where 80.1 percent of the United States as a whole self-identifies as "White persons," only 28.6 percent of residents of Hawai'i do. On the other hand, 40.0 percent—a plurality—of residents self-identify as "Asian persons," as opposed to 4.4 percent for the United States as a whole. "Hawai'ian and Other Pacific Islanders" make up 9.1 percent of residents, to just 0.2 percent for the U.S. as a whole. And while 12.8 percent of the U.S. as a whole self-identifies as "Black persons," just 2.5 percent do in Hawai'i.[1] Perhaps even more relevant is the fact that Hawai'i has "the largest multiracial population [of any state in the United States] with 24.1 percent of its population identifying with two or more races. Alaska follows a distant second with 5.4 percent identifying as multiracial."[2] These statistics suggest, first, that there are fluid boundaries to several racial categories, and second, that Obama was even more of a minority in Hawaiʻi than he would have been growing up in, say, Virginia.

The Kennedy Theatre *Macbeth* both reflects and takes advantage of the islands' mixed racial heritage. Directed by Paul T. Mitri of the Department of Theatre, University of Hawaiʻi at Mānoa, the "student" production had real ambition. The set consisted of multilevel ramps, platforms, and sewer grates, from which the witches slithered in and out, and into which the play's many corpses conveniently fell (no troublesome body-removal staging required). Banquo's ghost rose and fell eerily without the distraction of awkward stage business. Fumes and harsh sounds belched forth occasionally. The above-ground view was post-apocalyptic: eleventh-century Scotland by way of Samuel Beckett and *Blade Runner*, with some of the costumes as apparent castoffs from *Mad Max*. The costumes indicated that the inhabitants of this strange world were Scots, samurai, and Russians, among others. Eerie sound effects added to the audience's disorientation. The casting included students of Anglo-Saxon, Japanese, and Chinese descent, among others, and a single African American.[3]

Perhaps the most distinctive thing about Mitri's production was that it was not only multiethnic, but also elaborately multilingual, as a way of further creating a multicultural world. The witches spoke several different languages; Duncan and his clan spoke a great deal of Japanese, and were of Japanese descent; Macduff and his family spoke Russian. There was even a smattering of Gaelic (the Gentlewoman in 5.1) and Arabic (the Doctor). The Macbeths spoke Spanish as well as English. The choices, Mitri says in his "Director's Notes," "create a cohesive world populated by remnants of our present world." Notice the fluidity of the language shifts for the entrance of the witches in 1.3 in Mitri's Working Script (these were parti-colored in the original):

> Enter the three Witches muttering *"fair is foul and foul is fair"* (1.1.10) (*Lo amable es siniestro y lo siniestro es amable, Kagayaku hikari wa fukai yamiyo, ukande iko, yogoreta kiri no naka wo, prekrasnye graznye y graznye prekrasnye*).

And for the entrance of Macbeth and Banquo:

> MACBETH: So foul and fair a day I have not seen.
> BANQUO: How far is't call'd to Forres? *Nanda are wa,*
> So wither'd and so wild in their attire,
> *Kono yo no monotomo omowarenga,*
> *Tashika ni soko ni oru?*—Live you or are you aught
> That man may question?
> MACBETH: *Habla*, if you can: *¿qué eres?*
> FIRST WITCH: *¡Gloria a Macbeth! gloria a ti, noble de Glamis!*
> ALL: Hail to thee, thane of Glamis!
> SECOND WITCH: *¡Gloria a Macbeth! ¡gloria a ti, noble de Cawdor!*

ALL: Hail to thee, thane of Cawdor!
THIRD WITCH: *Gloria a Macbeth, tu serás rey de aquí en delante.*
ALL: All hail, Macbeth, that shalt be king here after.

Banquo's response continues the mixture of alienation and familiarity:

> Good sir, why do you start; and seem to fear
> Things that do sound so fair? *Tashika ni kikugana,*
> *Omae tachi wa maboroshika,*
> *Soretomo me ni mieru tohri no mononanoka?* My noble partner,
> You greet with present grace
> That he seems rapt withal
> *Dono tane wa me o fuki dore wa minoranto, wakaru mono nara*
> Speak then to me...
>
> (Mitri, Working Script)

The results, for an English speaker familiar with the play, are fascinating: an alienation effect that somehow forces greater attention to the dynamics of power on display, and the otherness of the experience. (On the other hand, "Tomorrow, and tomorrow, *y mañana*" is an unfortunate choice, in my view.) For those not as familiar with the text, and for traditionalists, the linguistic experiment is less successful; one reviewer comments that "Serious Shakespeare fans will be disappointed that much of the play's poetry is literally lost in translation" (DeRego). Such comments of course reflect the ethnocentric nature of the speaker, since the audience for Mitri's production—no doubt, *any* theatre audience in Hawai'i—is multicultural and multilingual; for some in the audience, then, the English would have been the distraction. Even such critics as the one quoted above, however, admit that the power of the actors made the production comprehensible (in different ways).

Mitri, who self-identifies as "of Egyptian descent," has directed multicultural productions prior to this one (Mitri 2009). At the Minnesota State University campus in Akita, Japan, he took on *Macbeth* in 1997; the actors were Japanese students studying English and American students learning Japanese: the Japanese students "were given roles with predominantly Japanese lines with English lines that were deemed slightly above their language level," while for the American students, "most of their lines would be in the original English, but some would also be in Japanese." The result, he notes, was that audience members with either language capacity "would be able to follow the story clearly" (Mitri 2007). Mitri also directed a similar version of *A Midsummer Night's Dream* in Japanese and English, and somewhat later, a production of *The Comedy of Errors* at the American University in Cairo, Egypt, in both Arabic and English. In reflecting the multiple cultures of the Pacific, Mitri follows some other directors, such as Ken Campbell in his 1999 pidgin *Makbed*; that production's language was a pidgin Bislama, a pidgin English originating from Pacific Islander slaves of late-nineteenth-century Australian plantations.[4]

Mitri's purpose in staging his multilingual versions was threefold: (1) he strove "to make a play that could be understood by any audience member" (i.e. in a multicultural, multilingual audience) (Mitri 2007); (2) he wished to provide an "experiment for actors to test actor-training techniques in different forms/languages and discover how those techniques translate and carry over into other forms and languages" (Mitri Grant Proposal); and (3) he hoped that both students and audience would "use Shakespeare's works as a way of exploring their changing global commitments" (Mitri 2007). Or, put another way, "in an ideal world, part of what I was also hoping to attain is some questioning by the audience of what their own beliefs/views of theatre are" (Mitri 2009).

In this last aim, Mitri is close to his early modern source. For Shakespeare's London audiences in 1606, the Scots were, if not a separate race, certainly thought of as an inferior people: uncivilized, too often wild and savage (revealing their descent from the Picts, as opposed to the English descent, predominantly Anglo-Saxon-Norman).[5] A few years earlier, Shakespeare himself staged two history plays that were multilingual in much the same way that Mitri's production is. In *1 Henry IV*, Hotspur, Glendower, Hal, and Falstaff all seem to speak different languages, even though all are in English. But Lady Mortimer, Glendower's daughter, "can speak no English" (3.1.189), and in fact "*speaks in Welsh*" (3.1.196sd) and sings in Welsh—which most of Shakespeare's audience would not have understood. *Henry V* goes much further, with the multi-national officers of Henry's army—Captains Fluellen (Welsh), Gower (English), MacMorris (Irish), and Jamy (Scots)—all speaking in different stage dialects; even the common soldiers, such as Williams, sound different. The capstone, of course, is that other scenes include French, from partial lines and exchanges (4.2, 4.4, 5.2) to the English lesson conducted in French (3.4). In the Henry plays, especially, Shakespeare investigates the ways in which language divides us, but also how such divisions can, in some cases, be overcome: thus Mortimer loves his wife though he cannot understand her words, and Henry successfully woos Katharine of France across the language barrier.[6] So too Mitri's production of *Macbeth* erects language barriers as a way of revealing cultural and racial divides, yet is able, for most members of his audience, to bridge those barriers even while emphasizing them. The Scottish play has rarely if ever seemed so cosmopolitan, nor have *some* contemporary racial issues been so directly represented as in this production.

Notes

1. Statistics cited from http://quickfacts.census.gov/qfd/states/15000.html.
2. http://www.censusscope.org/us/s15/chart_multi.html.
3. My attempt to identify racial categories by external appearance and program-book names is, I realize, doomed to failure, particularly in Hawai'i.

4. "It's a huge improvement" over Shakespeare's language, Campbell claims, "there's none of that iambic pentameter rubbish which is Shakespeare's main drawback" (quoted in Smurthwaite).
5. For primary documents on the early modern English perceptions of the Scots, see Loomba and Burton.
6. For a slightly more dissonant reading of the multilingual dynamics of this wooing, see Blank (165–67).

5

Music

Figure S.5 Duke Ellington, sheet music for "Lady Mac," c. 1956. Courtesy of Duke Ellington Collection, Archives Center, National Museum of American History, Smithsonian Institution.

17

REFLECTIONS ON VERDI, *MACBETH*, AND NON-TRADITIONAL CASTING IN OPERA

Wallace McClain Cheatham

There are still many roles within the frequently produced lyric-theatre repertoire that have not been undertaken by African-American singers. Each operatic culture is historically equipped with its own unique prejudicial mindset about casting and programming. There are often audiences who find noticing the race of singers in classical opera a jarring experience. One viewer of the 2008 Opera Memphis production of Verdi's *Macbeth*, for example, complained in a letter to the local newspaper that it required "too much suspension of disbelief" to view "vocally qualified black singers playing white characters such as Macbeth and Lady Macbeth." He then suggested that "black performers [should] avail themselves of the expertise of makeup artists and conform to the requirements of character, time and place" (Feazell).

Yet many African-American singers have been cast with seemingly little or no controversy in productions of Verdi's *Macbeth*. Opera historian Eric Ledell Smith identifies Simon Estes and Mark Rucker as having performed as Macbeth; Grace Bumbry, Gail Gilmore, Margaret Tynes, and Shirley Verrett as Lady Macbeth; Brian Gibson and Kevin Short as Banco; and George Shirley as Macduff, who also performed as Malcolm, as I have previously documented (Cheatham 7). Smith further notes that Curtis Rayam, Jr. has done various roles in Verdi's opera. Rayam has confirmed to me that those roles include both Malcolm and Macduff (Rayam). Smith's volume singles out Willie Anthony Waters as the only black conductor of Verdi's *Macbeth*; Waters has also conducted full-scale productions of Verdi's *Otello* and *Falstaff*.

To Smith's catalog, I add Donnie Ray Albert (Oper der Stadt Köln, 1998; Opera Columbus, 1999), Gordon Hawkins (Seattle Opera, 2006), and Lester Lynch (Dayton Opera, 2008) as Macbeth, and Russell Thomas as Malcolm in the Metropolitan Opera's 2008 production. Martina Arroyo performed Lady Macbeth at the Met in 1973 (Abdul 21; Schonberg 15). I myself attended the January 2008 Opera Memphis production (discussed above), which had Gregg Baker in the title role and Marquita Lister as Lady Macbeth. This production marks a watershed moment in non-traditional operatic casting because I believe it was the opera's first staging with a black couple in the leading roles. Likewise, the production did not shy away from another controversial casting practice: Jason McKinney, a light-skinned African American, performed as Banco, and his son was performed by a white child. The chorus in this production was also quite racially diverse.

Many performance theorists have noted that erotic intimacy, more than anything else, complicates the casting of African-American performers in classical productions. As Shirley Verrett observed in an interview that gained her a reputation for outspokenness: "When we can see an African-American man standing next to a white woman, singing a love duet, without cringing, then I think we will have made it into the human race" (Gruen 11). Strikingly, Verdi's *Macbeth* gives casting directors little dilemma with regards to intimacy because it is not a love story. In some ways this appears to make it more amenable to casting black singers. Neither the hero nor the heroine is a sympathetic character, and there is no suggestion of physical attraction or emotional affection within the libretto.[1] Although some, like the former opera conductor Burton D. Fisher, suggest that committing homicide was for the Macbeths a sublimated mode of erotic expression (13), the sublimation of the actual expression of erotic desire seems to have opened the doors to the inclusion of African-American performers in the opera. In other words, Verdi's *Macbeth* is often interpreted as offering timeless psychological issues that need not be isolated to any particular population, nationality, or race.

Although most discussions of non-traditional casting focus on the issues of visual believability, aural quality is an equally important issue when casting opera. So one must ask whether the roles Verdi created in *Macbeth* are particularly suited to a "black" or even "Negroid" sound. Some might object to the notion that there are specific, racial aural qualities. As an African-American composer and musician, I have no problem with the notion that there is a "black" or "Negroid" sound because I hear and define that sound as a vocal instrument of great richness, deep resonance, and abundant color. When someone says, "That's a black sound," or "That's a Negroid sound," I do not think it has to be attributed to a stereotypical putdown. In fact, I am convinced that this sound is not a natural birthright of race. Not all black singers have this sound; and some white singers do. Rosa Ponselle, soprano, the first American-born singer to become a member of the Metropolitan Opera in 1918, stands as one striking example. Likewise, Marilyn Horne is another white singer who had the rich tones often associated with African-American voice. In fact, she dubbed the singing voice for the African-American actress Dorothy Dandrige in the 1954 film *Carmen Jones*.

While Verdi could have heard a figuratively "black" or "Negroid" sound in his native Italy from an Italian singer, it is certain that Verdi never heard this from African-American singers.[2] One can nevertheless speculate that Verdi heard in Lady Macbeth's music something that approximated what is in today's American culture called a "black" or "Negroid" sound. The "ideal sound for this role," Verdi said, "is a harsh, stifled, dark sound" (cited in C. Osborne 8–9). In his identification of Lady Macbeth's "ideal sound," Verdi was not addressing vocal technique, or aesthetic beauty. He was alluding to an identifiable vocal *personality* that should be present whenever Lady Macbeth's music is brought to life. Although it is clear that Verdi intended something far more comprehensive than what one might call a "black" or "Negroid" sound, his notion of this "ideal," "dark" sound has created a space for the inclusion of black Lady Macbeths.

Perhaps more so than any other contemporary opera singer, Shirley Verrett has been closely associated with her interpretation of Lady Macbeth. As Richard Osborne lauded her 1976 recording, Verrett was "the finest Lady Macbeth of the age, the most formidable and absorbing performer of the role on stage since [Maria] Callas and [Leonie] Rysanek" (9). *I Never Walked Alone*, Verrett's autobiography, documents the long and fruitful relationship that she had with "Lady," beginning at La Scala in 1975. Verrett initially felt "it was too dark a drama for me. I was afraid of Lady Macbeth. She was power hungry" (Verrett and Brooks 193). Nevertheless, Verrett reprised the role in the first La Scala Opera performance in the United States during the 1976 Bicentennial at the Kennedy Center. *People* magazine featured her that year, inquiring why an Italian opera company had used an American in its production, to which Verrett replied, "if they liked you and appreciated your talent, as far as they were concerned, you could be from Mars" (Moore 68). She later performed the role in Boston, San Francisco, a disastrous Rome production (where, unable to sing, she "decided to *speak* the role" [Verrett and Brooks 263]), a European film version, and finally a Japanese production in 1988. Arias from the role were favorites for her recitals, including Marian Anderson's eightieth birthday celebration in 1982. Verrett reflects that, contrary to the theatrical superstitions associated with the play, "it has been a good luck role for me"; her biographer calls it one of her "crowning roles" (Verrett and Brooks 269, 309).

The reasons why the Metropolitan Opera did not follow through with a production of *Macbeth* by showcasing Verrett as "Lady" in 1976 remain open for speculation; she confirmed, through my correspondence with her biographer Christopher Brooks, that she knew that "negotiations had begun, but the production never materialized." My take is the production did not materialize because there remains some truth in the long-held belief within the black classical music community: the "powers that be" will make room at the top for only one or two black stars. Verrett's autobiography confesses how this belief brought personal anxiety about how far she might be permitted to go in her career (87). At the Metropolitan that "one" was Leontyne Price—and to this day remains Price, even in her retirement. I posit that as long as Price is alive there will be no other comparably promoted "black star" at the Met; indeed, it is quite possible that Price will remain thus even after her death. My speculations are not intended to take anything away from the artistry or musicianship of either Price or Verrett. If Verrett had actually performed her "Lady" at the Met to the acclaim she had received at La Scala,

I do not believe that the "Price star" would have been at all tarnished. But I think this serves as yet another example of racism's most debilitating legacy: the tremendous waste of resources that it causes, perpetuates, and sustains. Price and Verrett were both excellent Verdian singers, but unfortunately the need for a singular black singer at the Met likely impeded Verrett's acceptance there.[3]

How then did race play into the *Macbeth* paradigm with Shirley Verrett? Much of *I Never Walked Alone* meditates upon the racial dynamics of her career—from her being unable to perform in Houston early in her career, to controversial responses to her hairstyle for *Aida*. Yet except for a brief comment about stage lighting in the original La Scala production—"The one complaint I had with the production was the lighting. I didn't think there was enough of it for my skin color" (197)—Verrett does not seem to consider her work with this role as being particularly groundbreaking for an African-American female. Her work in another Shakespearean role, as Desdemona in Verdi's *Otello*, somewhat complicates this picture. For the 1981 Boston Opera production, she wore makeup to "lighten my complexion slightly" (240)—just as James McCracken wore dark makeup for Otello. One critic noted favorably: "makeup was not the key to her portrayal; singing was, abetted by tremendous dramatic integrity" (Tommasini). However, a double standard regarding African-American performance emerged during an interview: "the question of a black woman playing a white character in an opera in which race is a central issue came up. I answered the question with a question, 'Did you pose that question to James McCracken?'" (Verrett and Brooks 239). Verrett saw this as being no different than white sopranos darkening themselves to play Aida.

Verrett even believes that the existence of typecasting is often overplayed. *The Autobiography of an American Singer*, the subtitle of Verrett's account of her career, is a forceful statement about how she sees herself as a person, and how she saw herself as a performing artist. Discussing an offhand comment she once made to a reporter about the visual unsuitability of Marilyn Horne for a role, she notes that "In retrospect, I also see that it was wrong for me to even offer an opinion on who should be staged in what operas, because similar arguments have been used to justify not staging African Americans in certain operatic roles" (Verrett and Brooks 210). While Verrett moves toward a greater sense of the fluency of casting, one still must register the unusual history of Verdi's *Macbeth* and African-American performances. This opera more than others has enabled non-traditional casting, opened dialogues about casting practices on the operatic stage, and broadened the casting options for African-American singers.

Notes

1. Opera Memphis stage director Larry Marshall's insistence on intimacy between Macbeth and Lady Macbeth did not work for me; it contradicts my reading of the libretto. It felt incongruous to have Macbeth and Lady Macbeth locked in a

passionate embrace after they plot to commit multiple murders. Nevertheless, the director's decision to have Lady Macbeth place the king's crown on Macbeth at his coronation ceremony was a touch of creative genius.
2. This lack of personal familiarity with African Americans sadly did not mean Verdi was immune to racist slurs; see his correspondence with Léon Escudier when discussing working on *Macbeth*: "I have been back from Turin some days, and now I am at close grips with *Macbeth*. You think I work only at the last minute? No, I'm working like a nigger even now—I won't say I do much, but I work, work, work" (Dec. 2, 1864, quoted in Prodhomme). Opera scholar Naomi André confirms that "Unfortunately, this is not a one-time occurrence and similar references are used in letters related to the genesis of his opera *Aida* and *Otello*" (André).
3. While Price recorded the Lady's "Sleepwalking Scene," she never did the complete role, and, so far as I know, appears not to have even contemplated undertaking it.

18

Ellington's Dark Lady

Douglas Lanier

S*uch Sweet Thunder*, Duke Ellington's 1957 suite of twelve musical vignettes based on Shakespearean themes, is a landmark of adaptation of Shakespeare to contemporary popular styles. Written after Ellington played at the Stratford Shakespeare Festival in Canada, hard on the heels of his triumphant comeback at the Newport Jazz Festival in 1956, *Such Sweet Thunder* seeks to consolidate Ellington's status as a "popular classic" artist by suggesting affinities between his own art and Shakespeare's, as I have previously argued (see Lanier). It self-consciously pushes against the way Shakespeare had been used in nineteenth-century minstrel shows to ridicule the cultural aspirations of black performers and, by extension, black art. What is more, *Such Sweet Thunder* inaugurated a dominant concern in Ellington's late compositional career: showing points of contact between jazz and many other cultural idioms and subjects. Ellington sought to demonstrate the potential for jazz, a quintessentially African-American idiom, to be a cosmopolitan *lingua franca* that extends well beyond its humble and once-denigrated roots. This view of *Such Sweet Thunder* is decidedly not universally shared by Ellington scholars and fans. One widely-held perspective on Ellington's career remains that the high-water mark of his work was the early 1940s, when his band was filled with his most famous players, and Billy Strayhorn, Ellington's longtime collaborator, contributed some of his most memorable tunes and arrangements. Judged against this period in Ellington's output, many argue, most of his later suites seem collections of lesser quality music, long-winded, pretentious, unmemorable, or recycled exercises in Ellington's signature style. Indeed, some critics complain that the themes that unified the later suites sometimes ran no deeper than Ellington's witty song titles. Ellington's suites, then, raise two related issues that bear upon the cultural ambitions of his later career: how genuinely unified are the suites? And how profound is Ellington's engagement with the cultures his extended works purport to take up?

These questions hang over the history and critical reception of *Such Sweet Thunder*. From the start many have been suspicious of how perceptively Ellington addressed his Shakespearean sources. Though Don George claims that Ellington owned a copy of Shakespeare in which he had "underlined parts that appealed to him, not only to be set to music but to be performed by him" (136), and Ellington asserts that he and Strayhorn read through the plays in preparation for composing, a survey of Ellington's effects at the Smithsonian found no copy of Shakespeare (Hudson 21). Several critics have expressed skepticism about the nature of Ellington's musical portrayal of Shakespearean characters, finding them "a collection of self-indulgent fragments that are tied to Shakespeare by great leaps of logic and that show very little understanding of what the plays are actually about" (Collier 285). Related to this issue are Ellington's principles of selection from the panoply of Shakespeare characters he might portray musically. The prominence of Othello and Cleopatra makes sense, given their associations with African origins and cultural sophistication. So do selections featuring Hamlet, Henry V, Julius Caesar, Puck, and Romeo and Juliet, given the pride of place these plays had at the time in the Shakespearean canon. But other choices seem peripheral to the suite's predominantly affirmational tone: why Kate from *The Taming of the Shrew*? Why pair Iago with the witches from *Macbeth* in a single piece? Why Lady Macbeth? The last of these selections, titled "Lady Mac," seems especially odd given Ellington's curiously cheerful treatment of her character and her popular status as a Shakespearean *femme fatale*. Despite its anomalous place within the suite, this musical portrait is congruent with and, perhaps surprisingly, central to Ellington's concerns in *Such Sweet Thunder*; analyzing "Lady Mac" helps reveal the conceptual unity of the suite as a whole as well as the sophisticated, actively revisionary nature of Ellington's engagement with Shakespeare.

"Lady Mac" is distinctive within *Such Sweet Thunder* in several respects. It is the only jazz waltz in the suite, though an interlude is played in 4/4 time before the solo melody line by Clark Terry on trumpet is played over a 3/4 rhythm; in fact, the tension between the two rhythmic structures (3/4 and 4/4) makes the piece slightly off-kilter (see Figure S.5). It opens with a piano solo by Ellington (one of only two in the suite), in effect stating its main musical theme, though in somewhat tentative, modified form, identifying the roots of the melody in the blues. Ellington dubbed most of the pieces portraying single characters "sonnets," and as Bill Dobbins has noted, they are constructed in terms of fourteen phrases of ten notes each, emulating the fourteen iambic pentameter lines of Shakespeare's sonnets (cited in Crouch 441). "Lady Mac," by contrast, is one of two pieces portraying single characters, which does not adopt this form, though the main musical theme is divided into two phrases of five notes each, a structure that is somewhat reminiscent of the ten-note lines of Ellington's musical "sonnets." Formally, then, "Lady Mac" is both integrated with and distinguished from other sections in *Such Sweet Thunder*. The placement of "Lady Mac" as fourth of twelve suggests that it is the structural counterpart of the lament "Sonnet to Sister Kate," eighth in the sequence. As the only two isolated women in the suite (Cleopatra and Juliet are referenced in conjunction with their lovers), this pairing is noteworthy, for both are preoccupied with mastery of their respective male partners. Perhaps most curious about "Lady Mac" is its upbeat tone, quite surprising given the ultimate tragic fall of Lady Macbeth and the predominantly negative portrayal of her onstage. Set in a major key, "Lady Mac" is sandwiched between a parodic portrayal of Henry V and a portrait of the dejected Othello, so that its placement tends to magnify the seemingly incongruously jaunty tone.

In the liner notes to the original LP release of *Such Sweet Thunder*, Ellington observed of "Lady Mac": "Though she was of noble birth, we suspect there was a little ragtime in her soul." This rather cryptic comment—for the music in *Lady Mac* does not evoke ragtime—has only confirmed some commentators' suspicions that Ellington's knowledge of Shakespeare was slight. Collier singles out "Lady Mac" as a prime example of Ellington's failure to engage Shakespeare's work in any depth, for "Lady Macbeth...does not have a little ragtime in her soul and is not reflected in the lighthearted piece" (285). Other critics have addressed the jauntiness of "Lady Mac" by noting that Ellington, in a press release for the Stratford festival, described his technique as "parallel[ing] the vignettes of some of the Shakespearean characters in miniature...*sometimes to the point of caricature*" (emphasis added; quoted in Edwards 9). Thus the lightheartedness of "Lady Mac" might be understood as an irreverent portrait of an aristocratic wannabe, something also hinted at in the wry informal nickname he gives to Lady Macbeth (the same irreverence toward aristocrats surfaces in the title "Hank Cinq").

However, "Lady Mac" may not be a general evocation of Lady Macbeth's personality, as has been almost universally assumed by commentators.[1] Rather, I propose that the song refers to a specific scene in *Macbeth*, 1.7, in which Lady Macbeth bolsters Macbeth's flagging resolve to kill Duncan. This is arguably her most memorably forceful moment in the play, where her sensual seductiveness and undaunted mettle are most on display, qualities that may be reflected in the feminine-masculine name "Lady Mac." Musically, the song resembles a dialogue far more than a monologue. After Ellington's bluesy solo opener and a statement of the main musical theme by a saxophone ensemble, the orchestra breaks into different sections for the second statement of the theme, each playing against the other in syncopated counterpoint. It is as if we are hearing the centrifugal fragmentation of a once-unified musical idea, the aural parallel to Macbeth's opening soliloquy in which he moves from fixed purpose to self-division (1.7.1–28). This second statement of the first theme ends inconclusively, without the expected cadence. Instead, we hear the first statement of a second theme, a sequence of ascending motive figures first taken by the saxophones and then by the brass, growing in volume and intensity as the melody moves higher. This musical sequence—what we might call the "aspirational" theme—identifies the subject matter—ambition—over which the two solo sections of the piece have a dialogue.

The first solo, stated by Russell Procope's alto sax, begins a somewhat tentative (if lyrical) melody line in the subdominant key, not the tonic, backed up by a low-key saxophone ensemble, as if to musically portray Macbeth's irresolution (elsewhere in *Such Sweet Thunder* Ellington assigns specific instruments to certain characters). When the melody returns to the tonic, Clark Terry on trumpet, the voice of Lady Macbeth, begins his/her solo, the longest of the piece. At first mirroring the tentativeness of Procope's melody line as if in answer to it, Terry's trumpet solo, constructed on the song's opening theme, eventually becomes more virtuosic, fluid and forceful, in effect out-talking Procope's Macbeth and returning the piece to the melody with which it started. Significantly, Ellington's piano is newly assertive in this section as he emphasizes the waltz rhythm with syncopated comping during Terry's solo. This sequence ends with an emphatic restatement of the "aspirational" theme, as if after the conversation between soloing voices this theme can be reasserted with renewed vigor. The piece ends with a forceful final statement of the initial musical theme, now with brass and saxophones (and piano) playing together in well-harmonized counterpoint, a musical parallel to Macbeth and Lady Macbeth's coming to a shared resolve at the end of the scene. The only indication of tragedy to come is a suddenly dissonant cadence at song's end.

In its positive, exultant tone, Ellington's musical portrait of Lady Mac seems neither parodic nor uninformed. Instead, "Lady Mac" offers a fascinatingly revisionary reading of *Macbeth*, addressed to a specific emblematic moment in Shakespeare's text. That moment engages with a central concern of Ellington's late suites, African-American cultural aspirations, particularly as Ellington pursued those aspirations through his tireless promotion of jazz as an African-American idiom with potentially international reach. It is in *Macbeth* 1.7 where the ambitions of the Macbeths are most threatened by Macbeth's failure of nerve and moral qualms, and though Lady Macbeth has long been regarded as inappropriately self-assertive, heartless, gender-crossing, and ethically monstrous, it is her single-minded devotion to attaining greatness that triumphs in this scene. Given Ellington's commitment to full artistic enfranchisement for African Americans still living in the shadow of Jim Crow and the minstrel show, Lady Macbeth's discussion of aspiration and courage might have held special resonance (1.7.39–44, 49–51). In "Lady Mac," I am arguing, Ellington reads this key scene in *Macbeth* through the prism of post-war black history in which the cultural ambitions of African Americans, once systematically ridiculed, were being awakened. Lady Macbeth, a Shakespearean voice of aspiration traditionally evaluated as a figure of moral darkness, is transposed by Ellington into an affirmational key, making her into an icon for the strivings of those who are racially dark.

Lady Macbeth thereby becomes the "dark lady" of Ellington's musical sonnets, read against the grain of the white mainstream. In "Sonnet to Sister Kate," Katherine Minola becomes Lady Mac's tragic counterpart, a figure whose desire for self-assertion is crushed. Rather than seeing Kate's final speech and fate as a victory for marital accord, Ellington portrays her as a pitifully sad woman, defeated by the man and singing the blues. Significantly, he calls her "sister," identifying her with the African-American community, and he uses the name Petruccio prefers for her, "Kate." In an irony unnoticed by commentators, Ellington reverses the generic logic of *Macbeth* and *The*

Taming of the Shrew, recasting the former in terms of comic triumph and the latter in terms of tragic pathos. "The Telecasters," the other piece in *Such Sweet Thunder* that references *Macbeth*, brings together the three witches (a trio of trombones) with Iago (Harry Carrey on baritone sax), the only movement which combines characters from different plays. In this halting ballad, Iago first states the theme with accompaniment from the witches, then the witches take the theme with Iago in accompanying descant. This evocatively brings together seductive, verbally tricky figures who bring low two Shakespearean heroes linked by their aspirational natures, Macbeth and Othello; Ellington revealingly places "The Telecasters" immediately after the vignettes devoted to *Macbeth* ("Lady Mac") and *Othello* ("Sonnet in Search of A Moor"), as if to clarify who eventually destroys these male protagonists. Considered in the context of "The Telecasters," then, the portrayal of Lady Macbeth in "Lady Mac" is all the more triumphal, for unlike conventional readings, which stress her monstrous affinity with the witches, Ellington seems to stress her difference from them.

There is another dimension to Ellington's revisionary strategy in "Lady Mac," one that comes into view if we consider the relationship of *Such Sweet Thunder* to *A Drum is a Woman*, the other major work that Ellington and Strayhorn were composing as they were putting together the Shakespearean suite in 1956. *A Drum is a Woman* was an ambitious, now largely forgotten multimedia piece, involving music, original lyrics, narration, dance, and theatre all by Ellington and Strayhorn; originally it was destined for broadcast on network television with a largely black cast, a milestone in itself. It is the most fanciful of several large-scale pieces Ellington constructed throughout his career which seek to portray the history of African-American culture and music. It is a grown-up fairytale, which sketches out a loosely-structured history of jazz from its roots in African rhythm through its transplantation to the Caribbean and New Orleans, its ripening in New York urban culture, space-age modernity, bebop, and its return to African rhythm. At the heart of the fairytale is the story of the tense relationship between Carribee Joe, a primitive *naif* who is most at home in the jungle, and Madam Zajj ("jazz" spelled backwards), the sensual spirit of jazz rhythm who reappears throughout various times and locales. When Carribee Joe discovers an elaborate drum in the jungle and reaches out to touch it, the drum speaks to him, revealing that it is in fact Madam Zajj, an "African enchantress" who tries to coax Joe to travel with her and make beautiful music together.

Rather surprisingly, Joe resists her blandishments, preferring to remain "in love with the jungle, the virgin jungle God made untouched"; his refusal sends Madam Zajj on a quest through time to find another musician, another "Joe," though she soon confesses that she is "beat and blue for Joe," a longing she repeatedly expresses by singing the descant "Carribee Joe" throughout the suite. Eventually Joe joins her in contemporary New York, and briefly samples the city's bebop scene; but he leaves soon thereafter, and upon his return to the jungle he tells his "jungle clique" about the modern music he has witnessed. At the end of the suite, Joe asks "his favorite new shiny drum" to tell him a story, and in circular fashion the tale begins again: "Once there was a boy named Carribee Joe, and one day he found an elaborately fabricated drum, and when he touched it, it actually spoke to him, saying I am not a drum, I am a woman." This ending does much to encourage us to identify Ellington at some level with Carribee Joe, since it is Ellington who begins the narrative of *A Drum is a Woman* early in the suite by uttering these very words.

Ellington's characterization of Madam Zajj throughout *A Drum is a Woman* is strikingly complex and relevant to his reconceptualization of *Macbeth* in "Lady Mac." Several commentators have noted that she has many qualities of a *femme fatale* and bitch-goddess, the stereotypes in which Lady Macbeth has traditionally been read and, arguably, to which she gave definitive articulation in popular culture. In her opening song, "A Drum is a Woman," Madam Zajj establishes herself as an African siren whose seductiveness and sensual exoticism is entrancing but potentially engulfing to men. On the other hand, Madam Zajj symbolizes great power and worldly success. In her earliest encounter with Carribee Joe, she offers to make him "rich and famous," and in several vignettes she dresses in flashy jewelry and fine clothing and is surrounded by adoring multitudes. She exemplifies threatening enchantment and incredible power throughout the suite. What is more, Zajj is a figure of sheer speed and restless innovation, moving effortlessly from locale to locale and era to era, associated especially with the kind of urban cosmopolitanism that Carribee Joe, devoted to his primitive jungle, resists. Despite her association with elemental, even primal rhythm, Zajj becomes the voice of mid-century city life and modernity—notably in "Rhumbop" she is a "tropical, nuclear vision," a telling conjunction of adjectives.

Given these qualities, Carribee Joe's attitude toward Madam Zajj is rather surprising. One might expect Carribee Joe, Ellington's alter-ego in the piece, to enthusiastically embrace this universal, entrancing, passionate spirit of rhythm (after all, Ellington titles his memoir *Music is My Mistress*). But Joe actively resists her blandishments at every turn, unwilling to surrender his connection with the "the virgin jungle God made untouched" and its sounds and animal language. As Zajj notes, in reality there are many Joes—Joe is clearly a figure for the jazz musician throughout history—but noteworthy are the ways in which many of them express resistance and antagonism toward Madam Zajj. One of the most notorious stanzas in *A Drum is a Woman*, a much-quoted passage, which has in large measure shaped the frosty critical reception given the suite, offers this troublingly misogynistic lyric which pushes the governing metaphor too far:

> It isn't civilized to beat women
> No matter what they do or they say
> But can someone tell me
> What else you can do with a drum?

Rather than reject *Drum* in its entirety for this one regrettable passage (as nearly all contemporary commentators do—e.g. Hadju [159] and Bradbury [89]), it is important to situate the lyric within the suite's overarching thematic concerns and to recall that these words are uttered by a character, "a man who lived in Barbados," not necessarily by Ellington *in propria persona*. Even so, the attitude of this speaker, though extreme, is consistent with moments of hostility toward and fear of Zajj throughout the suite. "You Better Know It," a male crooner's reply to Zajj's reprise of "A Drum is a Woman," opens with the couplet, "Zajj, darling, we're in love and I surely want to thank you, / But if you get ideas, I'll surely have to spank you," and indeed the song's title line, "you better know it," hovers ambiguously between emphatic devotion, fear of abandonment, and a vague threat, an odd tone for an otherwise routine romantic ballad. In the end, Madam Zajj, despite her fame, wealth, and beauty, is left without Joe, and Joe, despite his modest success "giving drum lessons," remains safely in the jungle surrounded by his drum collection, until the cycle of temptation and resistance begins again when his newest drum begins to retell the story.

It may be tempting to dismiss this as garden-variety misogyny, and certainly the complexities of Ellington's romantic life might provide ample fodder for such a reading. But *A Drum is a Woman* is far more than a window into Ellington's erotic psychology, for its divided attitude toward Madam Zajj conveys a certain ambivalence toward changes in the jazz idiom in the 1950s, when jazz moved decisively away from Ellington's big band sound to emphasize solo improvisation in small ensembles and a more cliquish, overtly political, and less broadly "popular" style. *Drum* acknowledges the currency of Latin rhythms and bebop while still standing at some distance from them. Indeed, Carribee Joe's uneasy relationship with Madam Zajj portrays not only a pocket history of jazz, but also a central tension within Ellington's own compositional ambitions: the urge to pursue wide popular appeal versus the obligation to remain true to the "pure," "primitive" African rhythms at the heart of jazz and with it to the distinctly African-American nature of the style, the tension between modern assimilative innovation, which offers status and international recognition, and fidelity to one's own cultural traditions, the "jungle music" with which Ellington began his career. As Ellington presents it, the history of jazz is the history of tension between Zajj's restlessly modern, potentially ecstatic nature, and Carribee Joe's faithfulness to an elemental, African tradition, a tension which *Drum* in the end does not resolve so much as explore and exploit. This tension runs throughout Ellington's late extended works, which struggle to find ways to hold faith with Ellington's distinctively African-American musical idiom while extending it to all manner of subjects and styles.

This tension likewise animates *Such Sweet Thunder*, as the title of the piece, taken from Hippolyta's comments about hunting hounds in *A Midsummer Night's Dream*, makes clear: "I never heard / So musical a discord, such sweet thunder" (4.1.114–15). The oxymoronic combination of "sweetness," an essential quality of Ellington's elegant late style, and "thunder," an elemental quality of rhythm from nature, captures the ideal balance between Ellington's "cool" and "hot" styles in his own suites as well as the meeting of "sweet" "universal" Shakespeare and "primitive" jazz Ellington as a "musical discord" in which both elements are preserved without compromising the other. According to Ellington, Shakespeare and jazz have deep affinities because they are at their heart popular artforms which involve a "combination of team spirit and informality, of academic knowledge and humor" (193). But both are threatened, Ellington continues, by their becoming the property of "experts," those too bound to tradition, academicism, or hidebound genre boundaries to appreciate their immediate, visceral appeal.

This tension too informs Ellington's unconventional portrayal of Lady Macbeth in "Lady Mac." As should be apparent, the traditional image of Lady Macbeth has many affinities to Ellington's Madam Zajj: both are seductive, ambitious, potentially castrating, simultaneously enticing and threatening to men. Ellington's Lady Mac is yet another avatar of Madam Zajj, a figure who embodies Ellington's ideals and anxieties about his aspirations as a jazz composer and African-American artist. Though many critics have focused on Ellington's comment that Lady Macbeth has "a little ragtime in her soul," they have missed the fundamental contrast in his formulation: "*Though she was a lady of noble birth*...we suspect there was a little ragtime in her soul" (emphasis added). Ellington insists upon the operative *tension* between Lady Macbeth's aristocratic status and her jazzy, syncopated restlessness, her refusal to remain within established patterns of rhythm and social and cultural hierarchy. Ellington intensifies this tension with his nickname for her, "Lady Mac," a combination of formality and informality, fidelity and infidelity. In other words, his Lady Mac embodies the tension that runs throughout the entirety of Ellington's Shakespearean suite, a tension between Ellington's own cultural aspirations embodied in the very act of a black man adapting Shakespeare, and his desire to remain true to the African-American roots of his music.

It is for this reason, I believe, that Ellington includes a solo for himself in the two songs devoted to aspirational characters with which he associates his own plight, Macbeth and Othello. As previously noted, it is Ellington who first states the theme of "Lady Mac" on the piano, an indication of his close identification with the aesthetic and ideological issues raised by the scene from *Macbeth* he is portraying. His solo opens with an elegant statement of the theme ornamented by flourishes, but ends with a repeated minor-key blues motif, as if we are witnessing Ellington's own musical struggle with the friction between "sweetness" and "thunder." When Ellington's piano next becomes prominent in the mix, it is to accompany Terry's portrait of Lady Macbeth, so that as his trumpet solo begins, it becomes at first essentially a duet between Ellington and the voice of Lady Mac. And it is Ellington's piano that introduces the recognition of tragedy to come in the piece's dissonant cadence. What I am arguing, in short, is that the dialogue of blues and aspiration that emerges in "Lady Mac" springs not only from Ellington's reading of *Macbeth* through the history of African-Americans' denigrated aspirations and cautious hope, but also from Ellington's own personal aesthetic struggle to craft an artistic style at once widely popular and culturally specific, capable of embracing all topics and cultures but remaining constant to its African-American ancestry. And that aesthetic struggle is at the heart of Ellington's decision to adapt Shakespeare to jazz form.

Despite the fact that several commentators have recently sought to rehabilitate the critical status of Ellington's extended suites within his canon, by and large those efforts have taken the form of broad assertion of their value. Yet close analysis of their form and immediate context, such as that undertaken here, can better tease out their ideological purposes and aesthetic nuances. Ellington may have composed too quickly sometimes and in his later years he did a fair amount of recycling of earlier material, but *Such Sweet Thunder* and its counterpart *A Drum is a Woman* reveal careful, deliberate, imaginative, ambitious design. Though at first glance it may not seem congruent with Ellington's apparent plan for *Such Sweet Thunder*, "Lady Mac" is indeed crucial to the project, and reveals much of the suite's extraordinary artistry and intellectual energy. Far from being a casually chosen component or, worse, a ham-handed misunderstanding of Shakespeare, "Lady Mac" in fact offers a masterfully revisionary perspective on a key scene and character from *Macbeth*. Ellington recasts the play so that it speaks more directly to the African-American experience and to his own aesthetic endeavor to extend the reach of jazz without compromising its essential "ragtime soul." His approach to *Macbeth* is not dilettantish or idiosyncratic, but rather profoundly considered; it suffuses not only the overall conceptualization of "Lady Mac" but its many musical details. Critical discussion that treats the piece as evidence of *Such Sweet Thunder*'s status as bad Shakespeare spectacularly misses Ellington's point. Irving Townsend, Ellington's producer at Columbia Records, surely has it right when he observes that

> to appreciate the [*Such Sweet Thunder*] Suite in its final form, one must listen to the music as a statement of what Duke got out of Shakespeare, a uniquely personal, deeply perceptive encounter. Duke likes Lady Macbeth, whether you're supposed to like her or not, and he treats her right. (321)

Note

1. Critics of all stripes have been hostile to the idea that Ellington's music has the referential dimensions of programme music, likely because programme music occupies a lower status in compositional circles, since its referentiality compromises the "pure" musical values of the writing. Yet Mark Tucker notes that in a list of his classical music favorites, Ellington showed "a distinct taste for early twentieth-century programmatic works" (268).

19

Hip-Hop *Macbeth*s, "Digitized Blackness," and the Millennial Minstrel: Illegal Culture Sharing in the Virtual Classroom

Todd Landon Barnes

A surprising number of hip-hop *Macbeth*s currently circulate in our globally networked, multimedia environment, and this constellation of performances constantly grows and changes. This coincidence of *Macbeth* and hip-hop culture has been structured by motion pictures like Gerald Barclay's *Bloody Streetz* (2003) and Greg Salman's *Mad Dawg* (2004), as well as stage performances such as Ayodele Nzinga's *Mac, A Gangsta's Tale* (2006), or Victoria Evans Erville's NEA-sponsored *MacB: The MacBeth Project* (2002, 2008). Even "mainstream" theatrical performances have begun incorporating hip-hop aesthetics: for example, Rupert Goold's recent *Macbeth* (2008), staring Patrick Stewart and Kate Fleetwood, featured rapping witches (Brantley 2008). Perhaps the most productive site for the intersection of hip hop and *Macbeth*, though, has been within pedagogical programs like Flocabulary's Stephen Greenblatt-endorsed CD *Shakespeare is Hip Hop* (2007), Tonia Lee's *Macbeth in Urban Slang* (2008), or Aaron Jafferis and Gihieh Lee's commissioned book/rap/play *Shakespeare: The Remix* (2004). In this essay, I examine the strange effects of local, culturally specific pedagogical practices fusing Shakespeare and hip hop which—like the music itself—have been cut, copied, pasted, and practiced outside of what was once their "proper" domain.

Confusion Now Hath Made His Masterpiece (2.3.62)

Like thousands of other YouTube members, high-schooler "MadskillzMan" (Matt Dacek) uploaded his senior English project, *Macbeth Act V: Revenge of the Ghetto*, to share it with the global community. Of the hundreds of student performance projects I have seen, which rework Shakespeare through hip hop, *Revenge of the Ghetto* serves as a representative example, not an exception. The 23-minute video remixes and re-contextualizes the final act of *Macbeth*, setting it in the present-day "streets of Cleveland." The soundtrack moves between the hip-hop beats of Chamillionaire, the jazz guitar riffs of Marcus Miller, and the nasal lyricism of Cypress Hill's B-Real. Elizabethan verse is transformed into hip-hop vernacular. Martin Luther King Jr. is evoked alongside other African-American cultural icons. The political drama of Macbeth's Scotland is uprooted and transplanted into the inner-city economies of drugs and violence. In these respects, MadskillzMan's video is not unlike other productions aiming to trouble the line between high and low culture, or between an isolated, European early modernity and Black Atlantic postmodernity.

But the video should give us pause, as *Revenge of the Ghetto* quickly reveals itself to be a minstrel show. The all-white cast awkwardly cites hip-hop vernacular in ways definitive of performative infelicity: affective gestures seem contrived, awkwardly timed, or "unnatural," illustrating how "the performance of blackness backfires when it finds itself in unwitting or unaware hands" (Godfrey 19). Perhaps most striking is how *Revenge of the Ghetto*'s performers enact their cultural remix against the virtual backdrop of a green-screen digital milieu. This space-collapsing technology allows them to walk alongside the black pedestrians of an anonymous inner city. At one point in the video, Macduff engages in antiphonic dialogue with a digital crowd of hip-hop superstars, which he casts as his soldiers; the crowd is a collage that pulls together images of Ja Rule, DMX, Method Man, Nelly, and Eminem. Their mouths are animated, and they signal their approval: "We'll do it! Yeah!" Unlike the rest of the video's dialogue, the approval ventriloquized through these constrained and authenticating "black" bodies is not spoken in what Mark Anthony Neal terms "sonic blackface" (quoted in Ogbar 31).

In *Revenge of the Ghetto* signifiers of African-American cultural heritage are sampled, casually tossed around, and recontextualized, making the struggles of the Civil Rights movement seem as fungible as the digital media through which they are now represented. King's "I Have a Dream" speech is evoked alongside an image of Gary Coleman (TV's Arnold from *Different Strokes*) with a caption reading, "Keep yo pimp hand strong" (a citation of 2007's *Date Movie*, itself a parodic pastiche of pop culture). With no lived memory of the struggles of the previous generations, the millennial generation parodically pillages the icons of the past, creating strange mashups of the present. Matt Dacek is not the only artist desecrating the culturally sacred, stealing lives, and creating masterpieces of confusion with these "mad skillz." YouTube abounds with hundreds of hip-hop Shakespeares, many of which resonate to great effect, but most of which, to invoke Eric Lott, seem products more of theft than love. In many ways, *Revenge of the Ghetto* is illustrative of a strange sea change in the politics of racialized representation in the age of digital reproduction.

When the Battle's Lost and Won (1.1.4)

The increasing ubiquity of the phrase "Hip-hop Shakespeare" might sound like a triumph for critical pedagogy, a rich spoil paraded home from the culture wars. I know that in the years I spent learning and teaching hip hop and Shakespeare with low-income, "at-risk" youth, forging connections between these disparate cultural performance practices was a key critical pedagogical practice, which allowed my students to connect language to life. Many of the aforementioned hip-hop *Macbeth*s succeed in accomplishing important cultural work, and this critique is in no way meant to undermine, *in toto*, the labor of valiant frontline cultural warriors such as Nzinga, Jafferis, and Flocabulary. Yet the slippery fact remains that the culture war is often simultaneously a battle both lost and won. As *Revenge of the Ghetto* illustrates, aligning Shakespeare and hip hop does not guarantee the *ways* in which this alignment will be received or reproduced. As Lila Abu-Lughod reminds us: "If the systems of power are multiple, then resisting at one level may catch people up at other levels" (53). I explore other, potentially more productive modes of understanding encounters between *Macbeth* and hip hop, ways that might keep us from jumping Jim Crow while playing the Upstart Crow.

When faced with the apparently novel connection between *Macbeth* and hip hop, one is tempted to ask, "Why Shakespeare and hip hop?" The first order of business we must attend to when looking at hip-hop *Macbeth*s would be to dismiss any astonishment we may feel at the novelty of this encounter. Both hip hop and Shakespeare occupy massive territories within the increasingly refined contours of our highly networked cultural imagination. We should express neither dismay nor giddy delight at their coincidence within a techno-symbolic landscape wherein conjunction has become the norm, disjunction the exception. Such reactions, apart from being naïve, only work to hide the numerous techno-structural and semantic forces at work in their increasing interaction. Audiences are often polarized in response to these cultural intersections, responding with either wide-eyed astonishment or dismissive disdain; however, both reactions arise out of a desire for *historical* and *cultural analogy*. The analogy either works or it fails. When it does "work," the dangerously reductive premise and conclusion remain intact: Shakespeare and hip hop are the *same*, hip hop is the Shakespeare of today, or, to cite the title of Flocabulary's program and school tour, "Shakespeare is hip hop." Nearly all the hip-hop *Macbeth*s I have seen on YouTube have been structured by this overzealous analogy. These analogical performances produce visions of a static history in which only the *mise en scène* changes: Kings become kingpins, horses become Hummers, and knives become nines. Unfortunately, most YouTube performances (and their corresponding pedagogical sourcebooks) mobilize the worst of urban clichés in order to signify their recontextualization.

In *Revenge of the Ghetto*, Macbeth asks Seyton for his "bling" (jewelry) instead of his "armor." Compared to more typical analogies (with their desire for an impossibly transparent equality), this analogy's exceptional clumsiness contains a productive disequilibrium. The bling/armor analogy illustrates how these performances might potentially activate a similitude that preserves and even highlights—rather than erases—difference. This preserved difference might function as a way of inviting classroom discussions around historical change and cultural singularity. How, for instance, is the "bling" logic of late-capitalism operating differently than the heraldry of feudal armor? More often than not, however, hip-hop Shakespeares operate according to an analogic of the Same that irons out seams of difference between two discrete historical identities in order to make old robes as habitable as new ones; however, this desire ultimately only further buttresses arguments for Shakespeare's universal or transhistorical relevance.

These transhistorical equations collapse the thick duration of history, reducing it to nothing more than a continuous and progressive repetition of the Same. These discourses seem to be telling our students: "Yo groundlings! Shakespeare has already discovered *moral stories* which, if *translated properly* today, might empower or teach you about the postmodern complexity of your world!" This breed of analogical fallacy pervades many a "liberatory," "progressive" pedagogy. Many well-intentioned teachers utilize this transhistorical, analogical moral valuation in the service of a multicultural canon-reformation project that seeks to erase the line between high and low culture. This project, more often than not, seeks to make strange bedfellows of the postmodern and the early modern—but it is always a procrustean bed. And sometimes students respond with *Revenge of the Ghetto*. Hip-hop Shakespeares, however, might radically change shape if educators instead focus on performance's ability to register and rehearse historical change and cultural difference. Hip-hop Shakespeare pedagogies might instead seek to explore with students how the historical analogies enacted by these performances falsely connect struggles of the past continuously with those of the present. Instead of teaching students that all things are equal (adding an historical blindness to rival attempts at race blindness), educators might help students explore the changing *difference* between the early modern and the postmodern, between elite and popular culture.

This focus on hip-hop-as-youth culture, and its relation to culturally relevant pedagogies, has its own institutional history. Paulo Freire, critical pedagogy's seminal philosopher, is often narrowly remembered as an advocate for pedagogical frameworks "constituted and organized by the students' view of the world, where their own generative themes are found" (90). Hip-hop culture, as a result, has been naively touted by many critical educators as the authentic "ur-text of cultural resistance" within which one finds the authentic "thematic universe" of marginalized youth (McLaren 169; Freire 90). However, Freire's avocation plays out of tune unless one also considers that the role of the critical educator is to "'re-present' that universe to the *people from whom she or he first received it*—and 'represent' it not as a lecture, but as a *problem*" (emphasis added; 90). When hip hop enters the Shakespeare curriculum today, however, it does so from globally mediated mass-cultural sources, not from experience with students or local communities; furthermore, hip hop enters as the ostensible *solution* to a canonical problem, not as what it is: a cultural formation equally fraught by its constitution within problematic historical, sociopolitical forces. Freire's progressive humanism, expressed through the broad strokes of his Hegelian dialectic, is too often understood as an injunction to reverse the cultural direction of what he described as "cultural invasion":

> [T]he invaders are the authors of, and actors in, the process; those they invade are objects. The invaders mold; those they invade are molded. The invaders choose; those they invade follow that choice—or are expected to follow it. The invaders act; those they invade have only the illusion of acting, through the action of the invaders. (133)

Although Freire subtly undercuts the absoluteness of this formulation throughout his work, this notion of invader's active authority and the oppressed's passive, illusory existence nevertheless continues to persist in pedagogical practices that imagine Shakespeare's equality with hip hop as the *telos* of a progressive narrative steadily approaching wider racial and economic equality. But if we no longer take Shakespeare to represent transcendent, absolute authority and power, and if we no longer take hip hop to represent pure, local authenticity untouched by power, how else might we understand the balance of their interaction? For this reason, instead of asking "Why *Macbeth* and hip hop?" we must start asking ourselves and our students "*How Macbeth* and hip hop?" What is the nature of this mix?

Bloody Instructions... Return to Plague th'Inventor (1.7.9–10)

In the 1980s and 1990s, while Shakespeare studies was interrogating its origins and the corpus' constitution within the matrix of authorship and authenticity, hip-hop studies was struggling to legitimize itself as a field. In the wake of the author's death and Foucauldian projects stressing the subject's contingent formation, Shakespearean cultural studies worked to de-authorize, de-valorize, and de-mystify (even de-humanize) the idea of a metaphysical, transhistorical Shakespeare. Meanwhile, hip-hop studies preoccupied itself with humanizing a history of dehumanized representations, recuperating authentic folk cultural forms ("keeping it real"), tracing, valorizing or debating origins across time and space, and legitimizing and lending authority to the aesthetic products of marginalized communities. The two projects did not seem to be in tension, but instead collaborated along the unified revolving axis of cultural revolution. Now this revolution is coming to an end, but its end has left cultural studies looking for something "real" or "vital" after the death of theory, while hip-hop studies—in the wake of hip hop's oft-proclaimed death at the hands of mass culture—critically examines its own complicity with the auratic cult of authenticity it now strives to disrupt. In the age of *Alternative Shakespeares* and the global mainstream commodification of hip-hop culture, we need to adjust the terms of the debate in order to understand *how* hip hop meets *Macbeth*.

We need to find ways to critically destabilize authority and authenticity in both high *and* low culture instead of just, hurly-burly, replacing one authoritative commodity with another while remaining within the same economy of authenticity. But the discourses of the culture war packaged these cultural artifacts as commodities to be traded or protected, appropriated or re-appropriated within a unified and coherent economy of value. Shakespeare and hip hop were understood as cultural properties within a restrictive market of exchange. They could be stolen, *or* they could be preserved. They could be public, *or* they could be private. Shakespearean cultural studies attempted to *open* and *share* high cultural treasures with the masses while hip-hop culture attempted to *close off* and *protect* hip hop from the mainstream. The title of the Folger Library's *Shakespeare Set Free* series testifies to the desire for open access to Shakespeare. Halifu Osumare, in her fascinating article on hip-hop dance appropriation in Hawai'i, illustrates hip-hop protectionism's complicity with the property logic it seeks to oppose: "underground hip-hop positions itself in *proprietary* opposition to the commercialization of rap music and hip-hop dance...which seeks to *protect* itself from the all-encompassing field of late capitalism in the postmodern era" (emphasis added; 41). But the law of property is never undone by an oppositional ownership.

Instead of framing transactions between Shakespeare and hip hop within a "free market" of fungible commodities, which move unchanged across differences of class, race, and time, we might ask our students to interrogate the changing logic of an appropriation game that attempts to determine what types of culture are deemed "proper," who gets "props," and who gets to own or reproduce whose cultural "property." Contained etymologically within the idea of *appropriation* is the structuring principle of property ownership. Indeed, we might ask our students and ourselves how the very term "appropriation" might be leading us astray. As technological transformations redefine how cultural property is created and exchanged, so too do laws restricting the ownership and reproduction of cultural property. As the "copyright wars" feed off of meats baked at the funeral of the culture wars, Stanford legal scholar and spokesman for copyleft culture Lawrence Lessig highlights the tenacity of the either/or logic subtending both wars: "The 'copyright wars' have [led] many to believe that the choice we face is all or nothing.... Either we're about to lose something important that we've been, or we're going to kill something valuable that we could be. Whoever wins, the other must lose. This simple framing creates a profound confusion" (34). How might Shakespeare studies and hip-hop studies escape this confusion?

Mingle Mingle, Mingle, You That Mingle May (4.1.45)[1]

Revenge of the Ghetto's "sonic blackface," without a doubt, participates in the American minstrel tradition. Undoubtedly, our gorges rise at the video's horrid racism; we are unsettled as we watch the video strengthen and continue a tradition of violent, racialized misrepresentation. However, we can learn from this video if we refuse to dismiss its performance as a "mere racist appropriation." Esther Godfrey points out how "[t]o a new millennium American society increasingly ambivalent about the existence of racial categories, the theatricality of minstrelsy and other metaphorical blackface performances serves dual purposes—dismantling stereotypical notions of racial identity while recreating and affirming them in the process" (3). Not even African-American hip-hop artists escape playing into this duality. Many hip-hop scholars, and a whole host of hip-hop *artists*, take it as a *given* that many African-American hip-hop stars continue in the tradition of the minstrel (Ogbar 12). One could argue that the "ambivalence" Godfrey points to is nothing new and has always been the case with minstrelsy. In fact, Eric Lott bemoans the way studies of blackface have oscillated between mutually exclusive notions of "wholly authentic or wholly hegemonic," and his study goes a long way towards escaping this reductive dualism.

Similarly, W. T. Lhamon stresses minstrelsy's ability to occupy both sides of the dualism by noting that while blackface performance "enacted an identification of whites with blacks, it also encouraged racialist disidentification. While both could go on simultaneously, they might also go on separately" (1998, 139). The mutual exclusivity of these oppositions rarely holds when blackface is embodied and put into practice. Lhamon's genealogy of blackface performance highlights Catherine Market at the turn of the eighteenth century in Manhattan as a site wherein black and white bodies were highly mixed and "there was an eagerness to combine, share, join, draw from opposites, play on opposition" (3). Lhamon points out how the "relative integration of these streets was not usual," and how it produced the "mingled behavior," which marked the market as a site where difference could be displayed and transgressed (1998, 3; 19). Likewise, Lott stresses how "frontier towns" were "not coincidentally the most important centers of blackface innovation" (47). Where are today's "frontier towns"? Where might we find today's Catherine Market?

What Seem'd Corporal Melted as Breath into Wind (1.3.79–80)

The virtual frontiers of cyberspace are producing, like frontier towns before them, a virtual mingling of behaviors, a new way of viewing, performing, and modifying the habits of the body in relation to race. Acknowledging that millennials' *habitus* is "profoundly impacted by the virtual space of the Internet," Osumare notes that "the synthesis of globally proliferating popular culture body styles with local movement predilections" forms what she calls the "Intercultural Body" (38–39). This fusion of the global and the local enabled by the virtual spaces of the Internet produces intercultural, "glocal" bodies that take on an increasing ability to incorporate gestures that are "not indigenous, but assumed, yet not contrived" (38). When Macbeth gets news of his promotion, Banquo comments that "New honours come upon him, / Like our strange [i.e., foreign] garments, cleave not to their mould / But with the aid of use" (1.3.143–45). Likewise, these intercultural gestures, as "strange garments" or "borrowed robes" escape their local "mould" and are globally naturalized through the "aid of use" (1.3.107). Thus, the natural mother is bypassed, and when a local space can easily "unfix [its] earth-bound root," we see the birth of new, timely bodies illegally "ripped" from digital wombs (4.1.112; 5.10.16).

These virtual frontiers, like Catherine Market, blur distinctions between private and public cultural property, but they also disrupt the distinction between private and public space. The private spaces and properties of the Internet have become radically charged with public potential. As a result, the priority given to the "local" and the regional in hip-hop and Shakespearean cultural studies needs to adjust. We can no longer fetishize "local" cultural tradition while innocently ignoring the implication of this valuation within the broader logic of private property. The private is no longer clearly and always distinct from the public; it always maintains the potential to "go public"; this publicized privacy, in turn, can just as easily return to a new local privacy. A student's English project finds its way into an edited collection of essays as easily as anonymous inner-city pedestrians find their way into a high-schooler's English project.

The Internet repeats Catherine Market's liminality, but it does so with a difference. This difference comes as a result of the shift from analog appropriation to digital sampling and the kinds of copying and sharing each makes possible. In describing the "copyright wars," Lessig's pertinent analysis articulates a conflict between what he calls RO ("read only") culture and RW ("Read/Write") culture—terms taken from "permissions" attached to computer files (28). A phonograph's grooves analogically hold the trace of sound waves once written upon it, and when played, the needle is moved along the inverse trace of these vibrations. This RO technology, because it is analog, is difficult and expensive to rewrite. Nevertheless, hip hop arguably begins its life when Grandmaster Flash, through technical know-how, invents the cross fader and places his hands on a pair of records. Fusing tactile dexterity and electronic expertise, Flash moves his hands back and forth on ostensibly RO records. This analog rewriting, however, still remains connected to the labor of Flash's *hands*. With the shift to digital technologies, the body's labor is reduced to a minimum, and its immediate contact with the medium with which it mingles may not exist at all.

Lessig points out how "the 'natural' constraints of the analog world were abolished by the birth of digital technology. What before was both impossible and illegal is now just illegal" (38). In the same way, what geography, segregation, and fears of miscegenation once made impossible, the birth of glocal, digital performance now makes possible. Digital characters in *Revenge of the Ghetto* illustrate this dissociation from the analog world as they walk through digital alleyways inscribed with graffiti pieces that still retain the anonymous traces of the movement of the writer's hand, which carefully rewrote his or her environment. Madskillzman "throws up" his own pieces on digital walls: they read "crack house" and "left coast." But this rewriting is different. The writer is less exposed to the history of blood and bodies that once labored to write.

An important distinction between the effects of RO and RW culture is best explained by Lessig: "One emphasizes learning. The other emphasizes learning by speaking. One preserves its integrity. The other teaches integrity. One emphasizes hierarchy. The other hides the hierarchy" (87–88). Lessig's analysis stresses both the freedoms and the dangers presented by digital RW culture. *Revenge of the Ghetto* pretends to an integrity found by transcending hierarchical difference, but the difference, paved over through analogical fallacy, persists in hidden form. Digital integrity will only be learned when this difference can be revealed, preserved, and highlighted. In the strangest moment in *Revenge of the Ghetto*, this difference peeks through with all the power of the repressed in its unruly return. In a flash, Madskillzman seems to acknowledge the analog conditions that allow for his digital tale. His Macbeth cries, "Not even all the perfumes made by all the sweatshops in India could wash away that smell off my hands." The video then cuts to an image of sweatshop workers with a caption that reads "This film was brought to you by sweatshops!!" Real bodies still exist, and they are filled with blood. Blood flows in the history of an analog labor that produces the freedom of the digital. Unlike Macbeth, Madskillzman never drives his knife into the body of those occupying the position he usurps. Instead, Madskillzman's virtual *Macbeth* exists in a space wherein he can assume a "digitized blackness" as he wipes the filthy whit[e]ness from his hands and continues surfing on the incarnadine virtual sea (Ogbar 32).

Revenge of the Ghetto begins by introducing its characters to the beat of Chamillionaire's "Ridin'," a song about racial profiling. The refrain repeats, "Try to catch me ridin' dirty." But this is not Chamillionaire. It is the famed parodist "Weird Al" Yankovic singing "White and Nerdy." The lines go, "I wanna roll with the gangsters, but so far they all just think I'm white and nerdy." Yankovic goes on to list every stereotypical white and nerdy activity, from being computer savvy to spending every weekend at the Renaissance Faire. According to these criteria, *Revenge of the Ghetto*'s digital Shakespeareans certainly qualify as "white and nerdy." *Revenge of the Ghetto* claims to be "A Production by the Whitest Kids in Your Class," a citation of the popular HBO race parody *The Whitest Kids U'Know*. Does Madskillzman imagine his project as a parody? What he does not understand is what Judith Butler explains: "Parody by itself is not subversive, and there must be a way to understand what makes certain kinds of parodic repetitions effectively disruptive, truly troubling, and which repetitions become domesticated as instruments of cultural hegemony" (quoted in Godfrey 18).

As we continue renegotiating the relationship between RO and RO/RW culture, between copyright and copyleft, between the *de jure* restrictions and *de facto* freedoms to copy, we must remain focused on *how* Shakespeare and hip hop intersect. Now, more than ever, we need to focus our attention on the *ethics* of copying the anonymous, constitutive forces of history without enslaving these cultural-historical forces within the logic of private property, authenticity, or the authority of origins. We will have to experiment with mixing culture without yoking historical and cultural difference to the homogenous temporality of a "progressive" Sameness. Only then will we, along with our students, learn to mashup culture without shedding blood. Only then can we begin to teach and understand how the proprieties of intercultural performance pedagogies stake out what constitutes "fair use" in and between the digital and the analog world. Our students will continue to copy and share regardless of how they understand this practice. This is what terrifies the Recording Industry Association of America and motivates its war on copyright infringement. Let it not terrify cultural studies.

Note

1. As these lines are an interpolation from Middleton's *The Witch* (5.2), they thus enact the intertextual "mingle"ing of which they speak (see the Daileader essay). One way to destabilize the origins of *Macbeth* and hip hop would be to point out *Macbeth*'s collaborative identity and textual instability (all those interpolations!) alongside the idea of the "studio gangster" in hip hop. We could talk about how both studio gangsters and Shakespeare remix history in the service of authority. As Dr. Dre invents a criminal origin in order to construct street authority, Shakespeare remixes Scottish history in order to construct the origins of King James's authority.

6

SCREEN

Figure S.6 Harold Perrineau, *Macbeth in Manhattan*, 1999.

20

RIDDLING WHITENESS, RIDDLING CERTAINTY: ROMAN POLANSKI'S *MACBETH*

Francesca Royster

AGAINST READING AS WHITE IN SHAKESPEARE STUDIES

In her poem "Passing" from her book *The Land of Look Behind*, Michelle Cliff writes:

> Isolate yourself. If they find out about you it's all over. Forget about your great-grandfather with the darkest skin—until you're back "home" where they joke about how he climbed a coconut tree when he was eighty. Go to college. Go to England to study. Learn about the Italian Renaissance and forget that they kept slaves. Ignore the tears of the Indians. Black Americans don't understand us either. We are—after all—British. If anyone asks you, talk about sugar plantations and the Maroons—not the landscape of downtown Kingston or the children at the roadside. Be selective. Cultivate normalcy. Stress sameness. Blend in. For God's sake don't pile difference upon difference. It's not safe. (Cliff 23)

Though I am not a black Jamaican, Cliff's is the position that I know intimately, the "Black-Eyed Squint," to borrow Ghanaian writer Ama Ata Aidoo's subtitle from her book, *Our Sister KillJoy*. Perhaps this is why I have been particularly interested in images of whiteness gone wrong, not-quite whiteness, moments where we see the gaps in a unified sense of white identity, from *Titus Andronicus* to *Antony and Cleopatra*. My view of Shakespearean studies has always been from the outside of whiteness and Britishness, through a gaze that has become (for the necessity of my health) oppositional. I may have been told by my professors, by editors, and anonymous readers to produce a view that is "historicized" and objective, but I have always understood that directive deep down as meaning that I should bring some forms of knowledge to the table while silencing others. I learned this lesson in my first year of graduate school, steered away from choosing Shakespeare's *Titus Andronicus* as my focus text in our bibliographic methods class, lest my "feelings" on my own blackness corrupt my understanding of the archive. The lesson resurfaced at my first conference paper, when I was told by audience members that I was too angry about Laurence Olivier's use of blackface in his National Theatre production of *Othello*. After the Q and A, an exasperated audience member came up to me, wanting to know "What white people should do—just not play Othello at all?" These lessons were not subtle; I did not have to scratch far beneath the surface to find them. That the "ideal" reading position of Shakespeare's text is gendered as male and raced as white, or should be in the service of a race-neutral (read "white") point of view has inspired two generations of field-changing scholarship in early modern feminist, anti-racist, and postcolonial studies. Coming of age as a Shakespeare scholar in the early 1990s, works like Ania Loomba's *Gender, Race, Renaissance Drama* (1989), Margo Hendricks and Patricia Parker's edited collection, *Women, "Race" and Writing in the Early Modern Period* (1994), and Kim F. Hall's *Things of Darkness: Economies of Race and Gender* (1995) were formative to my thinking. We now have a considerable body of texts that document the reading position of the "other" or "altern" in Shakespeare studies, as well as a growing body of work on critical whiteness studies in the field.[1] This work is still under siege by scholars who assail it as too "presentist" or as not sufficiently occupied with "aesthetics" or beauty. And because of this assault, it is important to keep producing work that reads against assumed whiteness in ways that expand its archive of the past while also linking with new fields of knowledge, including interdisciplinary and transhistorical links.

In this essay, I bring an oppositional reading of whiteness to Roman Polanski's 1971 film, *Macbeth*, a film that makes whiteness not only visible, but also downright strange. I am reading against the grain of many critical approaches of the film to consider Polanski's treatment of embodiment, idealized beauty, civilization, and power in terms that highlight whiteness as their focus.

Reviving a critical discussion of Polanski's *Macbeth* in terms of its treatment of white power, civilization, and bodies is complicated by the controversies around Polanski's life, particularly around abuses of power. Polanski's life, as a survivor of the Nazi regime, may well inform the film's complicated constructions of nationalism and racial purity. Surely it informs his interests in the ugliness of abused power.[2] At the same time, Polanski himself has been a figure of controversy around issues of power and its abuse—most notoriously, his charge for "unlawful sexual intercourse" with a thirteen-year old girl, and his flee to France in 1978. Certainly sexuality and power are germane to Polanski's treatment of whiteness, embodiment, violence, and vulnerability in *Macbeth*, and we cannot fully sever the artist from his art or times. The point here is less to recover Polanski's intentions than to interrogate his filmmaking in the context of the changing status of whiteness in the 1971 cultural moment.

Whether intentionally or not, Polanski's *Macbeth* represents whiteness in a way that lends itself to an "oppositional gaze," to borrow bell hooks's term.[3] Here, I owe a lot to Bryan Reynolds's trenchant reading of the film in his essay, "Untimely Ripped: Mediating Witchcraft in Polanski and Shakespeare," in which he brings together Foucault's notion of Power/ Knowledge, the Deleuzian crystalline image, and the Theatre of Cruelty to think about the ways the film riddles certainty, including certainty about the presence of witchcraft and the certainty of Shakespeare as an image of high culture. Reynolds draws on Polanski's 1971 interview, in which he says, "I show people something so obviously impossible as witchcraft, and I say to them, 'Are you certain it is not true?' Too many people accept things for certainty.... I want people to question certainties" (quoted in Reynolds 146). I would like to argue that one of the results of this interrogation of certainty is to disrupt the white embodiment and white culture as a neutral category. Reynolds suggests that Polanski's film constitutes "a terrorist intervention into a discursive and cultural and ideological struggle," that brings together "the philosophies of the peace-loving-revolution hippies of the 1960's, Vietnam War protesters and civil rights activists, the reified mainstream American populace, and the ruling conservatives, as well as society's preoccupation with aestheticization" (144). Reynolds argues that the film does so via its treatment of violence and power. However, while Reynolds argues that Polanski's film arrives at this critical juncture in terms of social certainty and change, he does not address specifically the ways the film addresses or fails to address outmoded forms of national or racial allegiance. Instead, Reynolds looks more loosely at the film as it situates itself in terms of hippie subculture, via the witches and witchcraft practice. Restricting himself to the film's treatment of witchcraft and the supernatural, and the violence that malefacurum might produce, Reynolds argues that the film may be read ultimately as ambiguous in its oppostionality. He suggests that even as the film seeks to resensitize its viewers to a violent world that has been commodified and fetishized, Polanski's film might invite viewers to question the dangers of the unconventionality of marginalized populations (147). Reynolds's analysis rests on the feeling of extreme terror and disease that permeates the film, that notion that "fair is foul and foul is fair" (1.1.10) as it takes particular graphic and bloody form. But I would like to draw our attention to additional and subtle countercultural moves that the film makes by undermining the notion of a "stable" or idealized fairness or whiteness. Whiteness is "indexable" in Polanski's Macbeth through its casting, lighting, costuming, and framing of white bodies, and through its use of the material culture and practices of white civility and society.

WHITENESS AND CIVILITY IN POLANSKI'S *MACBETH*

Though several critical studies of Polanski's films explore the resonance of Jewish identity and especially the terror of regimes of ethnic cleansing in *The Pianist*, I have found only one study interested in the racial politics of his other films, and none that discuss the issue of race in his *Macbeth*.[4] Ewa Mazierska's book-length study, *Roman Polanski: The Cinema of a Cultural Traveller* is one of the few that explores the racial ideologies of Polanski's work in some detail, though Mazierska does not discuss his treatment of the category of whiteness in *Macbeth* or in any of his films.[5]

Perhaps one might be tempted to suggest that there have not been critical discussions of Polanski's treatment of race in *Macbeth* because there is no representation of racial difference in the film. But this line of argument assumes that whiteness is only visible when it is seen in relation to other groups, and by extension, that whiteness is, in and of itself, a neutral category.

Historically, whiteness has, of course, traveled as unmarked; yet, it carries with it a system of values. Whiteness's functionality as supreme and privileged, in fact, depends upon its position as the unmarked center, as both neutral and transparent—and therefore unremarkable.[6] Yet, as Daniel Bernardi suggests, to not interrogate whiteness's claims to invisibility or inchoateness, or even to replicate it by labeling all discourses of authority and power, such as individuality, liberty, civility, or patriarchy as "white" without looking in detail at its forms, shifts, and contradictions within it, is both to collapse its specificity and to lose sight of how whiteness perpetuates itself as a racial formation (xx). One way that whiteness operates, particularly in academia, is through the myth of disinterested objectivity. In his book *Buying Whiteness*, Gary Taylor identifies what he calls "the myth of white disinterestedness," through which "whites have been systematically taught to believe in their own (unique) objectivity around racial and other matters, as beginning in the 1660's, with the scientific revolution" (Taylor 2005, 14). By extension, the neutralization of whiteness values and privileges certain kinds of information over others as "historical," as well as objective, and therefore, worthy of note. To participate in the privileged space of whiteness is to have one's tastes, desires, and modes of identification reinforced by these terms of authenticity, objectivity, and historicity.

As "the Scottish play," *Macbeth* is always already marked as a national and potentially a racialized project. First performed by The King's Men, *Macbeth* both reflects and negotiates the ascendancy of King James of Scotland to the English throne, and the need for the unification of these very different entities.[7] Coming at a new chapter of Britain's founding myth, the play reflects the potential volatility of a unified British (Scottish and English—and burgeoning white) identity—a discourse that becomes increasingly racialized in the backdrop of a barbaric and often disunified English past of invasions and cultural mixture.[8]

Polanski's film takes advantage of a contemporary nostalgia for ancient and medieval British "folk" culture that might be thought of as an extension of this early modern project; yet at the same time, it presents a version of this culture as significantly conflicted and troubling. Filmed in rural Wales, the film's location, costume, and *mis en scene* reflect the larger aesthetic influence of medieval English, Scottish, and Celtic folk traditions on late 1960s and early 1970s music, fashion, and literature. We might examine, for example, the obsession with medieval symbol, theme, lyrical style, and instrumentation in the early work of Led Zeppelin and Black Sabbath, the rise to cult status of Tolkein's medievalist *The Hobbit* and *The Lord of the Rings* (following multiple reissues of the texts in the United States throughout the 1960s), the echoes of the medieval hair and fashion in the late psychedelic robes, heavy rings, and medallions, and even the androgynously long hair of the period's countercultures. (Am I the only one to notice an echo between The Beatles' haircut and the traditional monk's hairstyle, minus the tonsure?) Perhaps this is why Polanski's *Macbeth* captures the uncanny feeling of being both contemporary and authentically medieval. In fact, Nicholas Selby's Duncan, caught sleeping naked under the covers by Macbeth, resembles the hirsute Sean Connery of James Bond fame—arguably one of the most famous Scots in the 1971 cultural moment. In addition to its use of medieval and vaguely early 1970's robes, tapestries, and fabrics, heavy beards, and long, disheveled hair, the film includes medieval dancing, feasting, and celebration, including a simulation of the Morris Dance with swords. Yet "tradition" is not represented as neutral or value free. The music by Third Ear Band is both ancient and modern in its use of cacophony with the chaotic sound of bagpipes and strings adding to the terror of the scene of Duncan's stabbing, the eerie meeting of the witches in 4.1, and the epilogue, in which Donalbain encounters the witches once more. The film also telegraphs its historical authenticity in its use of historically accurate bludgeoning instruments and other tools of war, bear baiting as entertainment, and methods of execution. The communal meeting halls, sleeping, and other living arrangements in the interior of the Macbeth castle that structure social life in the film do not provide tranquility or hominess, but rather are claustrophobic, dark, and become the sites of invasion and brutality. If we expect a revival of medieval sound and style to present an idealistic view of a white cultural past, these scenes reveal deep corruption and violence, and ultimate strains in the social structures of honor, hospitality, leadership, and marriage.

By providing this authentic and also troublingly visceral depiction of the Scottish medieval past, Polanski engages a vital change in the filming of Shakespeare plays. As with Franco Zeffirelli's 1968 *Romeo and Juliet*, Polanski widens the world of the stage by filming on location, including local ritual and music, attention to local architecture, as well as attention to the nude body. Yet I would argue that Polanski's approach to authentic European spaces and places is significantly less idealized and celebratory than Zeffirelli's. Indeed, Polanski's depiction of a historically white culture here might be contrasted with the approach of American blaxploitation filmmakers from the same time period. It is important to note the arrival of blaxploitation's high moment in the same year of Polanski's *Macbeth*, with the release of Melvin Van Peebles's *Sweet Sweetback's Baadasssss Song*. If one of the goals of blaxploitation films was to foster a stronger sense of black identity in part through the celebratory embrace of black style, music, and the often still segregated spaces of black urban ghettos, Polanski too gives us a film that depicts in tangible ways white spaces, white styles, and white cultural practices (see Guerrero). But Polanski gives us a much more troubling portrait of white culture that is nevertheless also indexable in its clothing, practices, and homes. Certainly we might see both Polanski's *Macbeth* and the flowering of the blaxploitation film as coming out of a larger crisis in white, mainstream U.S. culture, including the seemingly never-ending Vietnam War, the assassinations of leaders like King and Kennedy, and, as Reynolds details, the rise of multiple countercultures.

WHITE WITCHES AND THE WHITE DEVIL: MORALITY AND POLANSKI'S *MACBETH*

Another way that Polanski makes the functioning of whiteness and white supremacy visible is by calling into question the centering of whiteness as the space of measure for spiritual, moral, and civic health. Often this centering is conveyed by the white body itself, which becomes the site of innocence and idealized beauty, its beauty and strength the sign of an internal (spiritual, moral, and civic) stability.[9]

In Shakespeare's *Macbeth*, the witch's equivocating refrain, "Fair is foul and foul is fair" (1.1.10), would seem to put into question the reliability of fairness, or whiteness, as a marker of goodness in the play, even as it frequently uses the language of blackness to convey evil and the desire to create more evil. The linking between blackness and evil and whiteness/fairness with good is dramatically conveyed in Macbeth's asides: "Stars, hide your fires, / Let not light see my black and deep desires" (1.4.50–51); "Good things of day begin to droop and drowse, / Whiles night's black agents to their preys do rouse" (3.2.53–54); and "How now, you secret, black, and midnight hags, / What is't you do" (4.1.64–65).[10] But it is also clear that the inversion of fairness and foulness in the person of the fair-skinned (and often quakingly pale) Macbeth and Lady Macbeth is powerful enough to set nature off its course, loosing a whole cycle of unnatural acts (including, we learn from Ross and the Old Man's speech in 2.4, an eclipse of the sun, predators being eaten by their prey, and horses attacking their owners and eating one another). Ultimately, as the world of the play struggles to set itself on course, this language of hidden blackness becomes exposed for its hypocrisy. And we, the audience, already know, through our privileged position as listeners to Lady Macbeth and Macbeth's private dialogues and soliloquies, that the "innocent flower" hides the serpent. We might note, then, the ways that *Macbeth* reinforces the same dynamic of the early modern commonplace of the "white devil," put into motion in John Webster's 1612 play of the same name. In Webster's play, as well as in *Macbeth*, whiteness is a veil for the hypocrisy of politically and religiously powerful men, as well as the duplicity of women, particularly the painted hypocrisy of sexually powerful women. The white devil image speaks to the power of evil to take the disguise of innocence; yet it ultimately replicates the binary opposition of black versus white.

At first glance, Polanski's film seems to reproduce the ways that Shakespeare's play only temporarily suspends the binary of white over black.

Lady Macbeth (Francesca Annis) and Macbeth (Jon Finch) are cast as traditionally white beauties: both fair skinned, both with long hair and delicate, even "feminine" features. This beauty becomes the sign of their success in court and their charisma on screen. Douglas Brode suggests that Finch and Annis "resembled Zeffirelli's Romeo and Juliet, but were a little older and transformed from peace-and-love Yippies to selfish, ambitious Yuppies" (187). Annis's Lady Macbeth combines rhetorical with sexual persuasion in 1.7, raising Macbeth to greatness in both the metaphorical and sexual sense. Through means of voiceovers and close-ups in parts of 1.4, 1.5, 1.6, Polanski gives us insight into Lady Macbeth's and Macbeth's fair foulness in an even more intimate way, recreating us as an audience of conspirators. Yet by showing their psychic breakdown as evidenced in the changes of their bodies, we see the pair's emotional isolation and dissolution in visceral ways. Both are, by film's end, haggard, with widened, grey-socketed eyes. Macbeth has grown a full beard. Lady Macbeth's sleep walking scene, here, performed naked in front of the Doctor and waiting woman—a decision, as many have noted, that was the condition of Hugh Hefner's bankrolling of the film for its $2,400,000—makes her even more clearly cut off from her former self, breaching the rules of decorum and class. Finally, our last look at Lady Macbeth is as a corpse, dusty and left on the side of the road, legs splayed into a haphazard swastika, head hastily covered with a piece of burlap by a passerby. She could be anybody. To further our sense of the white body as vulnerable, we get a few moments of Macbeth's point of view after he has been beheaded, the world turned upside down and chaotic. The result is a view of the white tragedy as neither heroic nor truly condemning.

To further the slippery depiction of white morality, Polanski and his co-writer Kenneth Tynan enhance the figure of Ross, also played by blond-haired, blued-eyed John Stride. Brode sees this version of Ross as a Machiavellian figure of political slipperiness, one culled from Shakespeare's own culture's concerns with power as well as those of Polanski's, as a kind of a "pre-Watergate" figure (Brode 191). Ross slips easily from one leader to another, never fully committed to one moral position, but instead watches all: the celebration of Duncan in the first act, the crowning of Macbeth, rapes, murders, and then the return of Malcolm to power, all with a vague smile on his face. I would argue that Polanski uses his handsomeness to both mask his immorality and to help him slip into the background.

To question or even make explicit the idealization of whiteness is to risk one's continued belonging in white civilization. According to Thandeka, to participate in white culture, and ultimately, to receive white privilege, one must demonstrate one's belonging by following the codes of whiteness—not speaking about or naming whiteness being one of them (unless it is in reference to others): "We must remember the actual emotional content of the term white—specifically, the feeling of being at risk within one's own community because one has committed (or might commit) a communally proscribed act. Such an act threatens emotional perdition: the loss of the affection of one's caretakers and/or community of peers" (Thandeka 16). Whiteness,

Thandeka argues, is legislated by shame, guilt, and exile. Such can be the stuff of horror.

As a film that is one-part horror film, one-part Shakespearean tragedy, Polanski's *Macbeth* is haunted and fascinated by the liminal, the ways that categories are complicated by particular kinds of bodies, and potentially by all bodies. The opening scene, for example, gives us all of the bodily fluids that threaten to go AWOL: we watch the three witches on the Scottish shore, burying a hand, a dagger, and a noose. They chant their incantation, "Fair is foul and foul is fair" (1.1.10), then seal their curse with a vial of blood and a group spit. But these body parts will not stay buried. Indeed, nothing stays buried in this play. Nothing rests in its safe compartment, but will be severed, animated, or gouged out. This scene is the visual or fleshly equivalent to the riddled language of equivocation. If blood and hands are the signs of heroism, by the end of the play, this sense of a heroic standard is lost. This loss affects the ways we read the white body in the film, civilized and uncivilized alike.

Polanski makes us repeatedly aware of white bodies as they are fleshy and vulnerable—their skin the important sign in which we see gained and lost symbolic power, the witches, one hovering over another, scratching their backs; the horror movie soldiers, dripping blood from battle, exhaustedly reporting the battle he's just witnessed until released by the king; Duncan, stabbed repeatedly by Macbeth while sleeping naked in his bed (we see their eyes meet, see their moment of recognition); Donalbain, limpingly following his brother, ultimately disenfranchised; Macduff's child, murdered naked in his bath; and the servant, dress raised and thighs exposed unceremoniously by her rapists and then Lady Macbeth and Macbeth themselves.

Generally, the film shows us multiple white and imperfect bodies. Despite the production credit by Hugh Hefner, this is not a play that glosses over the bodily effects of age, war, and violence. The most obvious example of this is the witches, who are very human in their imperfections: boils and pockmarks, eyes that apparently have been gouged from their sockets by some past violent act, hair and skin that reveal both the trials of exposure to the elements and malnutrition, missing teeth, gnarled hands, and arms, breasts, and stomachs that sag with middle age. In their humanity, they are also indexably white: we note markers like skin and hair—they are culled from the real world of bodies around us. This is very different from, for example, Kurosawa's witches in *Throne of Blood*. While they are ghostly white, Kurosawa's witches are otherworldly, informed by Noh performance's aesthetics of artifice. All of these bodily flaws are within the realm of human experience and possibility for a population that is living hard and on the verge of war. Donalbain, son of royalty, is not an example of the body beautiful either. His limp could have come from birth or from the ravages of war. While the conventionally handsome Malcolm speaks the lines of apparent resolution that close Shakespeare's play, it is Donalbain, uncrowned, hobbling through the rocky terrain of the Scottish countryside to the theme song of the witches, who closes the play, and he becomes the sign of the unclear fate of things to come.

NOTES

1. For recent examples of early modern whiteness studies, see Erickson 2000, G. Taylor 2005, Callaghan, Iyengar 2005, Floyd-Wilson, and Royster.
2. Rothwell points out that the orchestration of the slaughter of Lady Macduff, "which begins with Ian Hogg contemptuously sweeping trinkets off the fireplace mantle in a terrified Lady Macduff's private quarters" was an echo of a childhood memory where "A Nazi bully apparently once subjected Polanski to a similar treatment when he was a child" (160).
3. hooks defines an oppositional gaze as one of critical spectatorship, pleasure, and refusal, in which "[b]lack female spectators actively cho[o]se not to identity with the film's imaginary subject because such identification [is] disenabling" (122).
4. For an analysis of *The Pianist*, see Stevenson.
5. Mazierska suggests that Polanski's films may give glimpsing looks at stereotype and dehumanization (in *Chinatown*, for example), though he also sometimes makes uncritical use of racial stereotypes, such as the image of black men as "brainless virility" in his film *Bitter Moon* (136).
6. As Dyer suggests, "Whites are everywhere in representation. Yet precisely because of this and their placing as norm they seem not to be represented to themselves as whites but as people who are variously gendered, classed, sexualized and abled. At the level of racial representation, in other words, whites are not of a certain race, they are just the human race" (Dyer 2005, 11).
7. For critical debates on *Macbeth*'s position on Anglo-Scottish unification, see Alker and Nelson, and Kinney (1991).
8. On whiteness and early modern English founding myths, see Erickson 2000 and Floyd-Wilson.
9. Floyd-Wilson has uncovered fascinating discourses by early modern Scottish writers, like Hector Boece, who, in his *Scotorum Historia* (1526), defends Scottish racial purity as it is demonstrated in bodily practices like breast feeding one's own children (to inhibit racial mixing through the wet nurse's milk), and the ability to survive in bold and harsh weather (56–57). This text situates Scotland's strength and racial purity against that of both the English and the Irish, who are said to intermingle too much with outside influence.
10. For an early discussion of Shakespeare's paradoxical use of whiteness, see Babcock.

21

Semper *Die:* Marines Incarnadine in Nina Menkes's *The Bloody Child:* An Interior of Violence

Courtney Lehmann

> The *Macbeth* idea was related to the women in *The Bloody Child.* One, played by Tinka [Menkes], is the alienated Marine captain in charge of the arrest. The other female character is the murdered woman. She is very young, childish, and super "feminine." Her solution to the problem of being a woman is to be cutesy, and small. And she ends up dead. So when her spirit is hovering around, all of her pent up creative energy comes out in a malicious form, like the witches in *Macbeth.*
>
> —Nina Menkes (Privett)

It's the desert at sunrise, and the only narrative is the howling wind. White cars come and go; white soldiers and white MPs dot the landscape of Twentynine Palms Military Base. The whiteness of this Mojave Desert outpost and its pale, tattoo-dappled occupants is ruptured only by shots of a bloody female body stuffed into the back of a car. These shots are, in turn, interspersed with glimpses of another woman; naked, her white skin is made all the whiter by an unidentifiable powdery substance that covers her body as she sits in the middle of a thick jungle, writing on her arm with a stick. The film is Nina Menkes's *The Bloody Child: An Interior of Violence,* the title of which obliquely invokes *Macbeth,* a play that is referenced directly in the film's intermittent repetitions of the witches' lines, which are intoned from an offscreen space in a distinctly non-narrative fashion. But for all its stunning, stark white imagery and its masculinist military setting, Menkes's film—like *Macbeth*'s own fearful summons of the dark, desperately disavowed "woman-within" every man of woman born—is very much about the intersection of blackness and femininity. Rather than externalizing this conjunction of blackness and femininity into the "weird" or "weyward" sisters (aptly described as Shakespeare's "*black* and midnight *hags*"), Menkes's spin-off pushes the viewer toward this connection through a series of mind-bending montage effects that force the spectator into an encounter with the "*interior* of violence," a phrase that, in addition to providing Menkes with her subtitle, also refers to the post-traumatic stress that is acted out by the Gulf War veterans featured in the film. In so doing, *The Bloody Child* performs a psychological autopsy of war, while exposing the "black and deep desires" (1.4.51) that maintain the fragile morale of a few good men, lost in the paradox of what it means to be *homo sacer*.

According to Giorgio Agamben's *Homo Sacer: Sovereign Power and Bare Life*, it is precisely the dual inflection of *homo sacer* as "sacred" and "damned" that accounts for its disturbing, nonsensical status as a life that can be killed but is unworthy of sacrifice. Long removed from its redemptive role in ancient socio-religious rituals such as scapegoating, *homo sacer* has evolved strictly into a juridico-political entity, functioning as the "bare life" that is "the referent of the sovereign decision," or the means by which sovereign power reinforces its own exceptionalism vis-à-vis the law (Agamben 85). The conversion of this once sacrosanct entity to a purely secular identity is, according to Agamben, the founding gesture of modern political power. Producing a subject whose murder can neither be punished nor celebrated, *homo sacer* thus refers to a category of being that is purely performative, invoking a species of life that has been engineered, paradoxically, for death.

Who embodies the role of bare life—of *homo sacer*—in *The Bloody Child*? At first glimpse, the film unequivocally implies that it is the pointlessly slaughtered woman who, perhaps inspired by Macbeth's treatment of the traitor Macdonald, has been "unseamed...from the nave to th' chops" by her husband (1.2.22). The total absence of meaning—the lack of ritual value or social significance (as in a revenge killing)—surrounding this act of violence is reinforced by the sergeant, who continually shoves the suspect's head into his wife's mutilated groin and screams: "You fuckin' like that?! Huh? Do—you—fucking—like—that?!...*Start fucking talking*!" Menkes repeatedly juxtaposes this repulsive image with a shot of the ashen woman in the jungle, who is composing illegible symbols on her arm. Intentionally ambiguous, Menkes adopts an Eisensteinian approach to editing or, more specifically, montage, employing obvious cuts to facilitate collisions of individual frames, rather than, as in classic narrative film, using montage "invisibly"— that is, to create continuity between frames. For Eisenstein, the collision of two radically different images produces a third, and not necessarily dialectical but almost always subversive, "dynamic concept" (45).

Although Jean Petrolle argues that *The Bloody Child* "invites allegorical reading" (94), I argue that Menkes's work—like Macbeth's witches—is as *anticipatory* as it is allegorical, for the film is imbued with an uncanny predictive capacity, foreseeing the ongoing "sequels" in the Gulf known euphemistically as Operation Enduring Freedom and Operation Iraqi Freedom. In this context, I read *The Bloody Child* as a proleptic—albeit elliptical—political statement about "sovereign power and bare life" in the era of the Bush doctrine. Focusing on three discrete incarnations of *homo sacer* in the figures of (1) the women, (2) "Africa," and, finally, (3) the marines themselves, I argue that *The Bloody Child* ultimately seeks to connect the Mojave to the Sahara, "us" to "them," in ways that have radical implications for the future of life—from bare to unbearable.

Bare Naked Ladies

On perhaps what is its most fundamental level, *The Bloody Child* is about the status of the female body as *homo sacer*. As a film that traces, in reverse, the arrest of a marine who has viciously slaughtered his wife and is apprehended by a female captain along with a male sergeant, Menkes's adaptation of this real event is, more specifically, a radical statement about the misogyny that haunts the "new," sexually-integrated U.S. military of the 1990s. More than just a straightforward indictment, however, *The Bloody Child* also fictionalizes a story about a female captain's response to the entirety of the situation. Adding further significance to the captain's part, Menkes casts her sister Tinka in this central role; in fact, with the lone exception of the dead woman, Tinka Menkes plays all of the female characters. A provocative spin on the concept of the weird sisters featured in *Macbeth*, this casting decision creates a sisterhood both within and outside of the film, as Shakespeare's trio is revisited in the combination of the film's director, its principal actor, and its subject, namely, the female murder victim. These ties are reinforced in the film's first line, which draws from the opening of *Macbeth* and is uttered by the distinctly girlish voice of the actress who plays the dead woman: "When shall we three meet again?" Indeed, as Susan Linville observes, *The Bloody Child* clearly establishes the murder victim and Shakespeare's witches—from *Macbeth* to *The Tempest*'s Sycorax—as "kindred spirits," and, in so doing, "resuscitates its slain woman-witch not to give us her story or to trigger our tears," but, paradoxically, "to remind us about her historyless status" (120n45). Linville adds that whereas typical Hollywood films employ the concept of "women and children" as "the shortest route to easy tears," "*The Bloody Child* makes 'women and children' strange in order to redeem them both from melodramatic manipulation and from historical oblivion" (120n45).

Yet estrangement is also the shortest route *through* the film's otherwise unbearable violence. For how are we to handle the many scenes in which the sergeant metaphorically rapes the murderer by repeatedly thrusting his head into the crotch of his bleeding wife as he yells: "Fucking look at that pussy!...Do you like that?! How's that fuckin' smell, huh buddy? Huh?" There are so many reprises of this scene that the audience cannot help but feel bludgeoned by the incessant "thump" of skull on flesh, as the sergeant uses the marine's head as a kind of sexual prosthesis, mimicking the rhythmic violence of hardcore porn films. However, Menkes denies both the on- and offscreen audiences any form of visual pleasure or narrative trade-off from these scenes. In the most literal, crudest sense of the phrase, this repeated encounter is a direct confrontation with "the *interior* of violence" that *The Bloody Child* seeks to depict.

But what happens when, in Linda Williams's well-worn phrase, "the woman looks," that is, when the film shifts the perspective of the "rape" scenes from no one's POV to that of the female Marine Corps captain, who watches her reflection as it watches the sergeant's abuse of both the dead woman and his fellow marine? The framing of these scenes later in the film is so careful and deliberate that we cannot help but wonder what she is thinking. Why does she permit the sergeant to interrogate the suspect in this way? Is she remembering sexual violence committed against her, before she became the Commanding Officer? It is not a stretch to answer this question in the affirmative, given the fact that the Nineties was a decade rife with the subject of violence against women in or by the military—so much so that Hollywood turned a profit on the subject with films like *Courage Under Fire* (1996), about the heroism and murder of Gulf War Captain Karen Walden (played by Meg Ryan), and *G. I. Jane* (1997), a fictionalized account of a female Marine (Demi Moore) who eventually succeeds in becoming "one of the boys"—an accomplishment efficiently summarized in her comment to a male superior to "suck my dick." After all, the Gulf War of 1991 was the first war in which men and women were fully integrated for months on end, with women frequently offering cover and supplies for combat troops on the front lines. Moreover, as Lynda Boose explains, the instances of "reported rape and sexual assault of women recruits at U.S. military training installations escalated so dramatically in the months leading up to the war that both the Pentagon and the chairman of the Senate Armed Services Committee were finally embarrassed into ordering investigations" (592).

Before we can return to Menkes's revisionist response to Linda Williams's question about the female gaze, we need to consider the ways in which, throughout the first half of *The Bloody Child*, various incarnations of the dead wife punctuate the audience's encounters with the marines, in both on- and off-duty contexts. Following the murder victim's representation as the hyper-white woman of the jungle, who is, significantly, almost always approached from the classic horror film angle of being stalked from behind, the dead woman's spirit is resurrected in the form of another woman—one who appears to have the privilege of looking both at the murderer and out at us—a privilege that is never once afforded the woman in the jungle. Simulating the shot-reverse-shot camerawork typically used to relay dialogue and convey reactions between characters within the same frame, Menkes juxtaposes a shot of yet another ghostly white woman, who wears heavy, oppressive eye makeup as she looks out at some offscreen space, with an image of the murderer. If we interpret this juxtaposition in terms of cinematic grammar, then the woman appears to be looking—with an utterly incredulous expression—at the murderer himself, as if to say: "How could you do what you're about to do?"

Finally the woman looks at herself in an unseen mirror, indicated by the offscreen voice of the girl-witch, who now taunts, "Mirror, mirror, on the wall, who's the fairest one of all?" That this woman (or incarnation of the murder victim), too, is rendered *homo sacer*, or a person who can—and will—be killed without purpose is suggested by the sudden discursive shift from the witches' lines in *Macbeth* to the words spoken by the witch in *Sleeping Beauty*.

A classic "fairytale," *Sleeping Beauty* stands in for all those countless variations on the theme through which little girls are indoctrinated into believing that passivity (by "sleeping," or some other form of waiting for one's "prince" to come) is the best way to reach self-actualization.

Adding one last element to this behavioral prescription, Menkes demonstrates that women must also—above all—be "in *distress*," as the sickly, swollen-eyed, sleepless look of the woman who peers into the mirror implies. Henceforth, we come to understand that answering the mirror cannot, in all fairness, confirm that this woman—rendered ugly by some unspeakable, "interior wound" (to use Menkes's phrasing)—is the fairest woman in all the land *except in terms of her skin color*, for like the "jungle woman" before her, she is also corpse-white. Indeed, the notion that "fair" in this case connotes a very specific, racialized conception of beauty-as-whiteness is made apparent by the subsequent juxtaposition of the woman aboard a bus in Africa. The racial inflection is apparent not only from the location but also from the strange woman's clothing: the harlequin, a stock figure typically outfitted in "motley," or the joyful, multicolored hues of the court comedian and *Commedia del' Arte*, is here represented in a stark, black-and-white diamond-patterned garment. This totally unexpected shift of *mise-en-scene* subtly establishes a critical, thematic connection between the pursuit of "fair" women and "foul" land—that is, the exploitation and conquest of the "Third World," which becomes central to the film.

BLACK AND DEEP DESIRES (1.4.51)

In a setting where "foul is fair and fair is foul" (1.1.10) ("foul," in the case of Shakespeare's language, also signifies "black" in Renaissance slang), we might expect racial relationships to be as confusing as—not to mention deeply intertwined with—the gender dynamics of *The Bloody Child*. The connection between the dead body of the murder victim, the invisible, "interior wound" of the Harlequin, and visions of a spiritually and materially impoverished "Africa" begin to emerge in the second half of *The Bloody Child*, wherein we see an entire continent represented as *homo sacer* in order to, as one marine puts it in the film, "protect[t] our [U.S.] interests." Through dramatic montage effects, the second half of the film introduces the violent contrast between the lives of U.S. citizens and the "bare life" that characterizes the inhabitants of the countries bordering the Red Sea (specifically, on the western shore, Egypt, Sudan, Eritrea, and Djibouti, and, on the eastern shore, Saudi Arabia and Yemen).

The first Eisensteinian montage is generated when a male marine is shown "sleeping in," lazily snuggled under the bedcovers in a faux ornate hotel room. When the female captain emerges from the bathroom in full uniform ready to report for duty, the marine asks her if he can stay in the room until he has to go to work, because "it's pretty noisy at the barracks." Having the option of calling more than one place "home" is typically a sign of entitlement. Although at first it is unclear why the captain spends the night at the motel, the answer, as we learn in the ensuing juxtaposed image, is *because she can*. Indeed Menkes makes this indulgent rationale blatant by contrasting the lavishly decorated hotel room with a cargo ship filled from stem to stern with black people in traditional white Arab garb. What emerges from the colliding images of the spacious hotel room and the crowded ship is the Marxist nightmare of a world wherein people—as opposed to things—are transported in bulk by cargo ships; Menkes's camera thus simulates the perception of Western eyes, which see profit rather than people, as the ship travels through one of the most essential trade routes in the world. Moreover, this scene, in which Africans are conflated with cargo, is clearly reminiscent of the once-lucrative slave trade.

At this juncture, "fair," while associated with "whiteness," acquires a more sinister, double meaning that refers to policy rather than pigment; in other words, fair is foul indeed, as the fairest of all countries are also the most unfair. This statement is reinforced by the do not enter signs that are featured between the interrogation scenes and the African footage. Ostensibly alerting desert rovers that they are about to trespass on U.S. military-owned property, the sign also suggests a comment on U.S. and, indeed, an increasingly global immigration policy among "white," Anglo-European countries. Writing in 2002, Slavoj Žižek describes "an ominous decision of the European Union," which "passed almost unnoticed: the plan to establish an all-European border police force to secure the isolation of the Union territory and thus to prevent the influx of immigrants." "*This*," he continues, "is the truth of globalization: the construction of new walls, safeguarding prosperous Europe from the immigrant flood" (emphasis original; 149).

Mission Impossible?

Unquestionably, Menkes fulfills her stated objective of connecting "the West's destructive relationship with Africa to that which the dominant Western culture has to women" through the film's inherently violent juxtapositions of the murder scenario with the African material, a technique that reads cinematographically as a tale of parallel political lives—one personal, the other collective ("Active Dreams"). But Menkes constructs a more complex sociopolitical calculus, as the film triangulates away from its equation of women and the Arab (land) toward an equation of the marines themselves with Arabs—and, more specifically—with (dead) Iraqis. What *The Bloody Child* ultimately accomplishes is a far more subtle meditation on race within the New World Order. Exploiting a process that Richard Dyer identifies as "making whiteness strange" (1997, 4), Menkes explores strategies for defamiliarizing "Africa" and "blackening" the marines, as she works toward refining what it means to be "white."

The Bloody Child must be contextualized by the film's companion piece, titled *The Dead Iraqi*, in which Menkes interviews a group of inebriated, enlisted veterans of the first Gulf War. In this brief, exclusively audible piece, Menkes protects the men's identities by "showing" only their voices superimposed over a still shot of marines enjoying drinks in a bar. The piece starts off by testifying to the fact that "*Semper Fi*" means little when war conditions make it impossible to distinguish enemy from ally. Indeed, in this respect, the marines' status as "Iraqis" and, simultaneously, *homo sacer*, is based on their shared status as "bare life," or a category of being that is, quite literally, "presumed dead." Sent into a forced retreat during Operation Desert Storm, the Iraqi ground forces were the target of gratuitous air and ground strikes, which were intended to annihilate Saddam Hussein's elite Republican Guard. His men walked right into the arms of death. So too the marines—always the first to the front—were sent on a mission from which they were not expected to return. As one of the vets explains, "None of us were supposed to be here today; they'd already formed units to replace us, 'cause we were supposed to get wiped out." "But," he adds, with a slight chuckle, "at the time I didn't know that." This marine, without using such terms, acknowledges himself as *homo sacer*—a life that can be killed but not sacrificed, as an impersonal throng of replacements had already been called up to fill his place.

However, it is the figurative meaning of "presumed dead" that most interests Menkes, since it could be argued that marines, above all other branches of the military, wholly consent to their status as *homo sacer*. This conversation is introduced by the most inebriated man, who, having explained earlier, "I qualify for fucking food stamps," fails to recognize the fact that, in the perceptual schemes of late capitalism, he is—if not "black"—then not "white" either. This marine proceeds to boast about the particularly horrific behavior he participated in while on an expedition in the first Gulf War; goaded on by the others, he exclaims, "We took a picture with some dead Iraqis. Oh my God it was great, you know, uh, it was terrific. You know why? Because the dead Iraqis didn't give a shit." When the other guys start to get excited about this kind of "off-the-record" talk, several marines initiate discussions of some of the crazier things they saw (or did) while in Saudi Arabia.

At this juncture, several of them start muttering "1/9," "yeah, 1/9," to which Menkes, confused, asks, "What does 'one nine' mean?" Becoming colder and distinctly curt with the director, the marines define 1/9 for the still confused Menkes in two words: "walking dead." Suddenly nervous, the marines are quick to distinguish themselves from this cadre, even though they bragged moments ago about being part of it. A phone interview with U.S. Army helicopter test pilot Scott Hetrick confirms that although "1/9" literally refers to the marine battalion (9th Division) that, during the Vietnam War, sustained the highest casualty rate of any battalion in marine history, the more generalized, figurative interpretation of 1/9 means "someone without a soul"—a human being, in other words, who has lost the right to claim this privileged category of life. In the eyes of the marines, the Iraqis were little more than beasts, for as one of the men explains, "the people of that country were simply too weak-minded" to do anything other than retreat into suicide—an event that the marines interviewed failed to interpret as a mass *homicide*. Instead, the Iraqis embodied the literal definition of *homo sacer*: a population "produced" to die. But the reverse is also true since, on the basis of the abominations committed by the 1/9ers, the marines prove not only "weak minded" but also "soulless" (acts that would be grossly revisited in the offenses committed at Abu Ghraib, for example). "Blackened" in ways they would never come to acknowledge, the marines interviewed by Menkes only experience their own status as *homo sacer* in terms of getting caught in "friendly fire." As the most vocal member of the group testifies in a statement reminiscent of *The Merchant of Venice*'s "which is the merchant here, and which the Jew?": "You couldn't even see three feet in front of you, 'cause of the oil fires.... [Then] everybody just fuckin' opened up on us... 'cause they thought we were the enemy, *'cause it was so dark you couldn't tell*." Proof of the complete absence of difference between the marines and Iraqis, this phenomenon occurs when the phantasmatic projection of the enemy reveals that "fair is foul and foul is fair"—and the distinction between "us" and "them" is only nominal; ontologically, the marines and the Iraqis are no different.

THE END?

By way of conclusion, I argue that this incarnation of *homo sacer* as the "walking dead" is precisely what it means to be "white," a designation that does not pertain to skin color but, rather—as the setting of Menkes's film in the Mojave and the Sahara suggests—the strategic *desertion* of reality. Specifically, in the figure of the female captain who looks on, passively gazing at her own reflection as though it were not part of her, we see an entirely new incarnation of the "walking dead," even as she bears witness to her own embodiment of the "new whiteness." Indeed, when *this woman looks*, she invokes the numb citizenry featured in films like *The Matrix*, in which an entire society has been duped into believing that comfort is what constitutes life. Like this "walking dead" citizenry, the captain detaches herself so efficiently from the obscene material reality she witnesses precisely because her relationship to the real is profoundly *unreal*. Always framed by Menkes within other frames—car windows, mirrors, motel rooms, and doorways—the captain lives a life that is hypermediated and, as such, is systematically deprived of direct engagement with reality. As Žižek might explain regarding the captain's predicament, "it is not only that Hollywood stages a semblance of real life deprived of the weight and inertia of materiality—in late capitalist consumer society, 'real social life' itself somehow acquires the features of a staged fake.... Again, the ultimate truth of the capitalist utilitarian despiritualized universe is the dematerialization of 'real life' itself, its reversal into a spectral show" (13–14). As a consequence of this "hypnotic consumerist state" (9), we are all, in effect, reduced to 1/9, the "walking dead," whose only cravings are for those things that synecdochize this relationship between the real and the spectral—things such as caffeine-free coffee and sugar-free ice cream—which, in turn, thematize a bland, white(bred) life that is bereft of any "flavor." This relationship between consenting adults and the unreal is nothing less than the homogenizing process—not unlike that which results in white milk—of "the basic sameness of global capitalism" (7). And is this "basic sameness," in a sense, what it means to be a marine, but without the risk—the "hurly burly"—that comes with the commitment "*Semper Fi*"? Like the wicked relativism of the witches in *Macbeth*, under such conditions does it matter whether "the battle's lost [or] won"? Indeed, put reductively, both sides of the present war on terror boil down to less or more "toil and trouble," depending on *which* nihilism one respectively chooses, passive or active: if "We in the West" are associated with the former, "immersed in stupid daily pleasures" while "real life" passes us by, then the "Muslim radicals are...engaged in the struggle" of living "even up to their own self-destruction" (40).

An admonitory vision, Menkes's *The Bloody Child* hereby becomes an act of bearing witness to the choice of nihilism that defines the new "whiteness." This is a nihilism similar to Baudrillard's "hell of the same," born of a reduction to a bare life from which there is no obvious exit strategy other than to ex-ist, that is, to cease to remain in the trance induced by global capitalism (122). Hence, in Menkes's film, the proverbial last word—rather than being awarded to the witches ("when shall we three meet again")—might be better stated by the bloody child himself, who never appears in the *mise-en-scene*. In the play, the apparition known as the "bloody child" hails Macbeth, advising him: "Be bloody, bold, and resolute. Laugh to scorn / The power of man, for none of woman born / Shall harm Macbeth" (4.1.95–97). Translated into the rhetoric of contemporary culture, this is precisely the attitude that constitutes the Bush doctrine, embodied by a man who fully believes he leads a charmed life, despite *all evidence to the contrary*. Particularly telling in terms of what I am calling the "new whiteness" are the comments from a Congolese teacher: "'Our misfortune,' he explains, 'is that we have gold, diamonds and precious wood, but, unfortunately, no white farmers'" (quoted in Žižek 125). The fact that Menkes shoots her entire film in desert landscapes underscores the association of the "new whiteness" with the *desertion* of our own humanity, which, in George Kennan's words, means being "always faithful" (*semper fi*) to one vow alone: "to dispense with all sentimentality [and] cease thinking about human rights, the raising of living standards and democratization." The "bloody child" in *Macbeth* may be no more than a spectral apparition, a mere illusion conjured with the help of the witches' sophistry, but for Menkes, the bloody child is a specter that returns—as Derrida would argue—from the *future*, to remind us that such ongoing, opportunistic triangulations between gender, race, and geopolitical exigency will, in all *reality*, return to haunt us in the form of the righteous whiteness that is the sequel to Menkes's film: the second Gulf War.

22

SHADES OF SHAKESPEARE:
COLORBLIND CASTING AND INTERRACIAL
COUPLES IN *MACBETH IN MANHATTAN*,
GREY'S ANATOMY, AND *PRISON MACBETH*

Amy Scott-Douglass

INTRODUCTION

When we talk about Shakespeare and race; when we talk about actors' ethnicities; when we talk about casting policies that result in diverse casts; when we talk about how audiences, critics, producers, playwrights, and actors respond to integrated-cast productions, often (but not always) what we are really talking about is how people feel about interracial relationships; and often (but not always) we are talking in particular about how interracial couples are treated today: not interracial couples in Shakespeare, not interracial couples 400 years ago, interracial couples right here and right now.

In Fall 2002, when I was teaching Shakespeare as part of the California State University's study abroad program in London, I took my students to see a Royal Shakespeare Company Academy Production of *King Lear* at the Young Vic, starring the twenty-three-year old Nonso Anozie, the sole black actor in an otherwise white cast. During the intermission, as I chatted with a group of my students, another came over and interrupted "the girls," as he put it. In an effort to get us to laugh, and evidently thinking that we would be sympathetic to his point-of-view, this young, white male jokingly tried to assume the voice of one of Lear's daughters. "Dad, I've got some bad news. It looks like Mom must have cheated on you. At least three times." I did not respond. My student continued talking, in an attempt to clarify his story, as if we did not understand what he meant. "Why did they cast a black guy when all the women are white?" I answered, "The RSC has a longstanding colorblind casting policy which says that they will cast actors based upon their talent, not upon their skin color." "Well," he said, "the RSC may be colorblind, but I'm not. You know a white man is the real father of those daughters, not a black dude."

My student's comment was instructive. The first part of his response—
"The RSC may be colorblind, but I'm not"—is a point that has been made by
many Shakespeare-in-performance scholars. Ralph Berry, for instance, writes
about what he calls the "audience difficulties" with integrated casting, particularly when it comes to "certain problems" like "kinship." Berry argues:

> A black Cordelia engages the mind in torturous moves. (Did Lear remarry?)...For the audience, part of the dramatic energy is dissipated in such cumbersome explanations. Or it is a burden of convention that the audience has to shoulder. Even a non-naturalistic play needs to be rooted in some kind of social reality, "to hold the mirror up to nature." (36)

According to Berry, audience difficulties with integrated casting and kinship result in anxieties that are in some way amplified by what Berry sees as the theatre critics' self-censorship on the topic.

Coming at the topic from another angle, other scholars, like many of the contributors in Ayanna Thompson's collection *Colorblind Shakespeare*, argue against colorblindness on the basis that, in practice, "colorblindness ultimately signifies assimilation" (L. Anderson 91). "When white students claim that they 'don't see color,' or that a racialized person is 'just a person,' they are reading the racialized-other as a being like them: 'white.'...It is a part of our cultural ontology to see race and to assess people according to race," Lisa Anderson argues (91). Or, as Thompson herself puts it, a colorblind "approach is dangerous because it requires that viewers ignore 300 years of cultural, social, and political history" (19).

But I would like to go back to the second part of my student's response. There is some anxiety about black and white couples and their children underlying my student's comment. "You know a white man is the real father of those daughters, not a black dude," he said, which might suggest a certain gratefulness for the visual evidence that would render illegitimate such a relationship. I believe that the anxiety about casting choices that result in interracial couples works both ways. In this essay, I consider how casting actors of different skin colors as the Macbeths plays out in four *Macbeth*-based shows: the 1995 independent film *Macbeth in Manhattan*, an episode from the popular television series *Grey's Anatomy*, and two contemporary prison productions.

MACBETH IN MANHATTAN

Macbeth in Manhattan seems to assert the association between darkness and badness, all the while appearing to embrace a colorblind approach to casting. Released in 1999, *Macbeth in Manhattan* is an independent film about a group of actors who are rehearsing an off-off-Broadway performance of *Macbeth*. Gloria Reuben (the biracial Jamaican and white Canadian-born actress of television's *ER* fame) stars as Claudia, the actress playing Lady Macbeth. Her live-in fiancé, Max (played by the white actor David Lansbury), is slotted to play Macbeth, but he is usurped when a Hollywood actor named

William (played by the white actor Nick Gregory), flies in from California and takes the role away. In very short time, William steals not only Max's leading role but also his leading lady. But when William abandons Claudia, Max cannot contain his anger and kills William the night before the show's opening. In the final scene, Max and Claudia, reunited, are in the wings preparing to go on stage. Claudia seems unaffected by William's death. She coos at Max, "This is what you wanted, isn't it?" William's bloody ghost appears in the audience, pointing his finger at Max, and Claudia smiles and chuckles wickedly to herself, suggesting that she was the one who put Max up to killing William, driven by her ambition for her fiancé *and* by her own desire for revenge.

One of the remarkable scenes in *Macbeth in Manhattan* is a sex scene. It begins with Claudia and William having sex standing up while rehearsing *Macbeth* in her apartment, and then it cuts to Claudia and William having sex standing up as Lady Macbeth and Macbeth. The remarkable aspect of this scene, however, is that as soon as Claudia and William/Lady Macbeth and Macbeth start having sex, another actor pops into the foreground of the frame. That actor is Harold Perrineau, the African-American actor whom Shakespeareans will recognize as Mercutio from Baz Luhrmann's *Romeo + Juliet*. As the Chorus in *Macbeth in Manhattan*, Perrineau provides occasional narration and plays several different characters: a street drummer, the theatre's effeminate West Indian costume designer, the guy who hangs the lights, and the camera person who films an interview with the play's director. In the sex scenes, Perrineau addresses the camera directly and says, "Lady Macbeth is no fool. She knows what a man wants, and she knows how to give it to him. Being bad never felt so good," and he claws the walls with his fingers and lets out a laugh of pleasure (see Figure S.6).

What in the world is Harold Perrineau doing in the foreground of the sex scenes? Of course, what his character represents depends upon the viewpoint of the audience member watching the film. Is Perrineau upstaging the coupling? Is he disrupting the coupling? Is he legitimizing it? Is he making a joke of it? Is he facilitating it? Given that Perrineau's lines link him with Lady Macbeth, is his character supposed to be articulating Lady Macbeth's forbidden and "dark" desires? Is he yet another black, albeit male, representation of the witches?

The three white men who provide the commentary on the DVD—who never directly mention their races, the races of the actors, or the casting process and/or history—take a seemingly colorblind approach to their narration. Interestingly, as they watch the sex scene, their attention is entirely on Perrineau. The director of the film says, "There Harold is again, egging things on." The director of music says, "Yeah, Harold is so cool." The third man agrees, "Harold is the coolest. Harold was born cool." Yet the three men neither indicate why there has been a deliberate choice to have Perrineau appear on screen at this time, nor how the continuum of skin colors on the screen might suggest the racist stereotype of the hypersexual black man and woman. Again, the filmmakers' rhetoric appears to embrace a colorblind approach, but the film eerily links racial blackness with the metaphorical darkness of ambition, sexuality, and bloodlust.

GREY'S ANATOMY

Colorblind casting has been the guiding mantra of the dramatic television show *Grey's Anatomy* from its debut in 2005, and much of the action has focused on the romance of heart surgeon Preston Burke (played by African-American actor Isaiah Washington) and intern Christina Yang (played by Korean-Canadian actor Sandra Oh). Although the show featured this particular interracial couple for its first three seasons, the writing stressed the differences between Drs. Yang and Burke in just about every way possible *except* race: he was an attending, and she was an intern; he was raised Christian, and she was raised Jewish; he was deeply spiritual, and she had become an atheist; he was a concert-quality trumpet player, and she was tone deaf; he was emotionally expressive, and she was cold; he wanted to get married, and she did not want to live together. They fought about everything, and they listed their differences in every single episode. But they never mentioned their different racial and ethnic backgrounds. This was purposeful on the part of the show's creator, producer, and head writer Shonda Rimes, who told one news reporter, "It's incredibly encouraging that our viewers haven't gotten hung up on the race thing. It's not about the fact that she's Asian and he's black. It's about the fact that she's a slob, and he's a neat freak. That's what the whole relationship is all about" (quoted in Barney).

Yang and Burke are the focus of one of the best episodes of the show, "From a Whisper to a Scream" (November 23, 2006). The plot and writing of this episode borrow heavily from Shakespeare's *Macbeth*. Rimes makes all of the writers post online blogs; for this episode, guest script writer Kip Koenig points out that Shakespeare's play inspired his writing, joking that he felt like he was writing a school essay for which the prompt was "compare and contrast tonight's episode with Lady Macbeth" (Koenig).

As the episode begins, Dr. Burke has always aspired to be named Chief of Surgery. Following an incident in the hospital, in which he was shot in the hand, he has been having hand tremors that make it difficult and dangerous for him to operate. Rather than report this to the Chief, Burke has been continuing to operate, aided by his partner Yang, herself an extremely talented and ambitious intern. Through this process, she climbs up her own career ladder, even replacing Dr. Bailey, her supervisor, who used to scrub in on Burke's surgeries. Just as the Chief decides to step down and appoint Burke as his successor, people begin to figure out Burke and Yang's secret, most importantly an intern named George. When George's father is admitted to the hospital and Burke is scheduled to perform his heart surgery, George asks for Burke's rival heart surgeon Dr. Hahn to perform the surgery. After Burke shames Yang during a surgery in which he has to remove a shard of glass from a woman's body, Yang responds to Burke's betrayal by informing the Chief herself.

Shades of Shakespeare

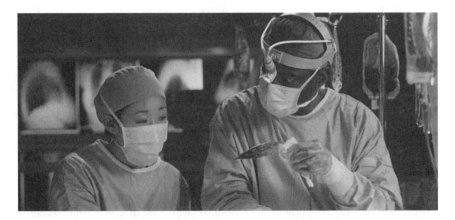

Figure 22.1 Sandra Oh and Isaiah Washington, "From a Whisper to a Scream," *Grey's Anatomy*, November 23, 2006.

Writer Kip Koenig's borrowings from *Macbeth* are subtly underscored by many of the choices of the episode's director Julie Ann Robinson. In one scene between the Chief and Burke in the Chief's top-story office there are trees in the background, reflected from the window, showing how, with clever cinematography, a director can "impress the forest [to] bid the tree / Unfix his earth-bound root" for a screen version of Great Birnam wood coming to Dunsinane hill (4.1.111–12). Likewise, the scene in which Yang walks out of surgery and down the hallway covered in blood is written as a dream sequence that it is reminiscent of Lady Macbeth's sleepwalking scene. And finally, in Burke and Yang's final surgery together, he removes a dagger-shaped, blood-covered shard of glass from a patient's heart, and holds it up, putting one in mind of Macbeth's line, "Is this a dagger which I see before me, / The handle toward my hand" (2.1.33–34; see Figure 22.1).

But the scene that is most similar to Shakespeare's comes in the middle of the episode when Burke tells Yang that he has been named the new Chief and that his conscience is pained: a scene reminiscent of *Macbeth* 1.7 in which Lady Macbeth encourages her husband to "screw [his] courage to the sticking-place / And we'll not fail" (1.7.60–61).

> BURKE: I'm going to be Chief.
> YANG: And you didn't tell him anything about your tremor, right?
> BURKE: I'm going to be Chief of Surgery at Seattle Grace Hospital.
> YANG: Why do you sound like that?
> BURKE: Because I'm going to be Chief of Surgery at Seattle Grace Hospital!
> YANG: This is what you worked for. This is what you wanted. I can work harder. I can learn more procedures, whatever you need. [...]. And once you're Chief, it won't matter, cause you'll be Chief. This is good. Nobody will ever know.
> BURKE: I'll know. I can't be Chief. Not now; not like this. Do you know how long I've wanted this? My entire career. And when I finally get it, there's blood

on it. I had a tremor, and I didn't say anything. It's unimaginable. It's unethical. I crossed the line.

YANG: We crossed the line. Together. I crossed the line with you.

BURKE: You dragged me across the line. You made us a team. [...]

YANG: I did what you needed me to do. You were standing there looking at me telling me your whole life was your hands, if you couldn't operate, if you couldn't be Preston Burke. [...] When you got shot, I walked away. And you just can't let go of that, can you? Well I'm sticking now; I'm sticking.

As far as the show's producers are concerned, if there is an ideological message underlying this *Grey's Anatomy* plotline, it has to do with gender rather than race. In one of the final scenes, Dr. Hahn, the other heart surgeon, tells Yang, "You're lucky to study under Burke. If your little intern brain can retain any of what he teaches you, maybe you can become half the surgeon that he is." Meanwhile Yang has been doing Burke's most complicated surgeries successfully without anyone guessing, so in that sense she has been passing for Burke, she has been passing for male. Yang is just as ambitious as Burke, just as smart, just as talented, and yet her talents are not acknowledged.

In interviews, the show's creator and actors repeatedly avoided commenting on the interracial component of the relationship. When *USA Today* did an article on television portrayals of interracial relationships, for instance, Isaiah Washington "didn't want to talk about his character's romance, saying through his publicist that drawing attention to the races takes away from the fact that it's quietly and happily existing without being an issue" (Oldenburg). Similarly, the *USA Today* article quotes Rimes as saying, "The pairing stems from 'casting whoever we thought was best for the part'" (quoted in Oldenburg). Does this philosophy of casting sound familiar?

But when word got out, just a bit before the *Macbeth* spin-off episode aired, that Isaiah Washington had started a physical fight with fellow castmember Patrick Dempsey, Washington thought it was time to "draw attention to the races" (to use his publicist's language) and point out "what he saw as existing prejudices" on the set of *Grey's Anatomy* (Samuels):

> I had a person in human resources tell me after this thing played out that "some people" were afraid of me around the studio. I asked her why, because I'm a 6-foot-1, black man with dark skin and who doesn't go around saying "Yessah, massa sir" and "No sir, massa" to everyone? (quoted in Samuels)

Just after the story about Washington broke in the news, the "dagger" photo from the *Macbeth* episode was posted on the *Grey's Anatomy Insider* website as part of their weekly caption contest. While all of the contributors are anonymous, most of them assume female pseudonyms, and the captions they write are troubling. One imagines that Burke is saying, "Let this be a lesson, Cristina...see what happened to her! If you don't keep your mouth shut you may accidently [*sic*] end up like this!" Another writes, "I will kill you Christina....Remember I will kill you." Yet another writes, "Come on

Christina...just one time...I'll stitch you back together as soon as I'm done."[1] Given the fact that there is nothing on the show to suggest that there is even the threat of physical abuse in this relationship, the captions suggest that audiences are writing a different narrative; and it is *not* a colorblind narrative. Instead, it is a narrative that is hyper-attentive to race and to the Burke/Yang relationship as interracial. That narrative is not the *Macbeth* story; it is, of course, *Othello*. These fans' reactions are another example of what Celia R. Daileader calls "Othellofication" made manifest—a sort of racist cultural tic whereby audiences interpret virtually *any* story about an interracial relationship through the lens of the Othello/Desdemona story (2000, 199), often, as in the case of the *Grey's Anatomy* fans, subscribing to the fallacy that when a black man has a relationship with a woman of another race it may start romantically but can only end tragically.

Prison *Macbeth*

This begs the question: are there theatrical spaces in which colorblind Shakespeare exists? Can Shakespearean, or even Shakespeare-inflected, productions with interracial couples exist without race being at the forefront of the actors' and audiences' minds? For that space, we might have to turn away from commercial theatre, for the time being. As a Shakespeare-in-performance scholar who has, for the past several years, been focusing on secure-setting Shakespeare, I have had what are probably very unique experiences with integrated cast productions in prisons. Surely prison must seem an unlikely place for colorblind casting to thrive. How could race not matter in prison Shakespeare? Race, performances of race, and the injustices of racism certainly matter when it comes to prison: parole board rooms, courthouses, and virtually every section of the prisons themselves—where allies or enemies are made—are affected by skin color. In addition, there is the fact that currently all extant Shakespeare prison programs are overseen by white theatre directors or professors, while inmates of color frequently comprise *at least* 50 percent of the casts. In many cases Shakespeare himself is presented to the inmates by these directors as someone to laud and revere, someone above critique, someone who can save them. Representing Shakespeare as a (white) savior to populations with a high percentage of African-American inmates has the potential to be a problematic enterprise. Despite the good intentions of the directors, one might worry that the phenomenon of race-blind casting in prison Shakespeare is a result of white dominance and "the 'dominant group...dismiss[ing]' race as an irrelevance" (Iyengar 2006, 61).

But then there is the point that the inmates themselves insist that race does not factor into the discussions when they are casting their plays. In *Shakespeare Inside*, I relate a discussion between me and the three black lead actors of the Shakespeare Behind Bars program in Kentucky. The conversation begins with me saying, "You [say] that when it comes to casting decisions race doesn't matter," and with them responding, "When it comes to blacks and whites, Shakespeare Behind Bars is a unique group in the sense that we don't see." "I don't think we think of it in terms of race when we're casting or anything," they say (Scott-Douglass 58). What I did not include in the book is that our conversation went on for hours, with me resisting the idea that colorblind casting was feasible, particularly in prison Shakespeare, and the men refusing to budge on the point. And this is because race, ethnicity, and skin color truly are not important factors in the Shakespeare Behind Bars casting process.

What matters more than anything else, they will tell you, is who is the best actor for the part. This, of course, is the mantra of colorblind casting in commercial theatre, but the unspoken and subjective criteria for what makes an actor the "best" actor in commercial theatre is hard to measure. In prison theatre, the criteria are very clear. The best actor for the part is the one whose history is the most similar to the character he or she is playing. So, when the inmates at Vandalia's Women's Eastern Reception, Diagnostic and Correctional Center in Vandalia, Missouri were casting *Macbeth: Act Five*, Macbeth went to the actress who could say during the talkback after the show, "I identify with Macbeth because he has power, and is greedy, and look where it got him. I connect that to my situation. You make poor choices, there will be consequences" (*Macbeth: Act V*). That actress happens to be a black woman. Lady Macbeth went to the actress who has a fifty-year sentence for killing her husband, a white woman ("Yet who would have thought the old man to have had so much blood in him?" [5.1.33–34]). Perhaps the part hardest to cast was Lady Macduff, because almost every one of the women at Eastern identifies with having lost her children.

Similarly, the lived experience of brotherhood and not the physical approximation of kinship is often the rationale guiding the casting talks. Recently the stars of the 2009 Shakespeare Behind Bars' production of *Macbeth*, a black man (Macbeth) and a white man (Lady Macbeth), spent a good thirty minutes telling me about the thoughts that guided them throughout the self-casting process: first and foremost, their own relationship had been stormy in the past, so they thought it would be good for them to play the Macbeths; secondly, they are both in for murder, so they identify with the Macbeths and feel as though they have a lot to learn from playing them. Never once did they mention race as a factor either for or against their casting choices.[2]

CONCLUSION: MY WISH, OR WHY I WROTE THIS ESSAY

Ralph Berry says:

> In *A Midsummer Night's Dream*, the tall blonde gets to play Helena, the short brunette plays Hermia. They might like it the other way round, but they have to submit to the imperatives of text and casting.... However you cast Othello, Iago has to be played by a white actor, else the play is destroyed. (35, 36)

A while ago, I might have agreed with his claims. But now I have *seen* Iago played by a black man. For that matter, I have seen Desdemona played by a white man; I have seen Tamora, the Scandinavian Queen in *Titus Andronicus*, played by a black man; I have seen Petruchio played by a black woman; I have seen Macbeth played by a black woman. All of these productions were prison productions, and after having witnessed them, it is harder for me to "get" some of the arguments being made about cross-race and cross-gender casting.

I do not know whether colorblindness is an attainable goal or even if it should be one. At a roundtable titled "Questions of Color and the Current State of Casting" at the 2008 Association for Theatre in Higher Education (ATHE) conference, one African-American theatre professor remarked that in response to her student's claim "I don't see color onstage," she replied, "Well, you'd *better* see color onstage." An audience member, an African-American man, spoke up and said, "Before we throw the baby out with the bath water, we need to remember that colorblind casting and nontraditional casting came from a place of love and inclusion, and we needed it at the time. I couldn't get work before colorblind casting." Another audience member, an African-American woman, replied, "But we're still only getting token roles. It's 2008, and this is still happening." Speaking for myself, I acknowledge that multicultural casting is an important, complicated topic. Yet my own ideas about the possibilities of casting in Shakespeare have broadened in the past few years, largely because of the unique perspective that prison Shakespeare acting companies have provided me. I value what Claire Conceison calls "total casting," that is, casting that is "welcoming of all possibilities—of all races, bodies, ages, genders—and that asserts that any cast that does not allow for these possibilities is incomplete" (Conceison). But I also recognize that changing terms without changing practices and mindsets is counterproductive. When a philosophy of casting—whether it is called "colorblind casting" or "total casting"—results in a Shakespeare production that reinscribes racist stereotypes, then it is important for audience members to question the production companies. But we must also be open to the possibility of true "total casting," like that practiced in many prison programs.

Finally, my ideas about nontraditional casting choices that result in interracial couples onstage come from being part of an interracial couple myself. People in interracial relationships have a keen sense of what it feels like to navigate between experiences of other people's colorblindness and color-consciousness, between our own colorblindness and color-consciousness, between spaces of racism and pain, and spaces of restfulness and peace. My hope is that a generation from now, even a presidential term from now, there will be a number of people who read my observations in this essay as an anomaly. I wish for sympathetic audiences for both theatrical and real interracial couples and their children, whether the members of those audiences are colorblind or color-conscious.

Notes

1. All quotes are taken from "Grey's Anatomy Caption Contest XXXIII."
2. In 2003 the series finale of the HBO television series *Oz*, about a fictional super-maximum prison, ended with a depiction of a multiracial prison production of

Macbeth that took a less optimistic approach to the subject of race and casting in prison Shakespeare. In the *Oz Macbeth*, all of the leads, including Macbeth, Lady Macbeth, and Macduff, are white. The only African-American men in the production play the witches. When Macbeth's stage knife is replaced with a real knife, and the actor ends up killing his real-life archnemesis, a black inmate in the audience stands up and shouts at the stage, "That motherfucker's dead!"("Exeunt Omnes"). This moment in *Oz* inverts Stendhal's anecdote of the 1822 performance of *Othello* in Baltimore, in which a white soldier in the audience stood up and shot the white actor playing Othello, exclaiming, "It will never be said that in my presence a confounded Negro has killed a white woman!" (Bate 222).

7
SHAKESPEAREAN (A)VERSIONS

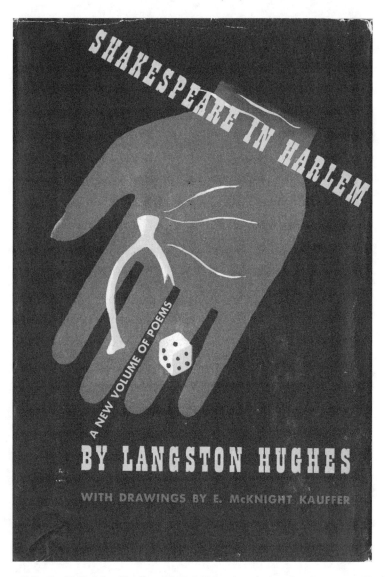

Figure S.7 E. McKnight Kauffer (1890–1954), dust jacket for Langston Hughes's *Shakespeare in Harlem* (1942). Copyright © Simon Rendall, Estate of E. McKnight Kauffer. Image courtesy of the Emory University Manuscript, Archives, & Rare Book Library.

23

THREE WEYWARD SISTERS: AFRICAN-AMERICAN FEMALE POETS CONJURE WITH MACBETH

Charita Gainey-O'Toole and Elizabeth Alexander

Because Shakespeare sits at the center of a canon with which American writers must engage and even master, many African-American writers, not surprisingly, make dynamic allusions to his works, often, in the words of Peter Erickson, "encompassing jauntiness, improvisation, and trickery" (2007, 1).[1] *Macbeth* seems to open a particularly inviting door of intertextuality in the work of black poets. The examination of moral weakness, unflinching ferocity, and ambivalently-constructed characters becomes excellent fodder in a tradition that, through figurative language, has contemplated human nature at its most craven as well as at its most improbably hopeful. In the hands of African-American poets, characters like Banquo's ghost can stand in for the "swarthy-hued procession" of justice-seeking African Americans (for nineteenth-century poet Josephine Delphine Henderson Heard[2]); Macbeth can become Richard Nixon (for June Jordan and Gil Scott-Heron); a national "eagle" can try to wash its Watergate stains "like Lady Macbeth" (for Amiri Baraka). Citations abound: Gerald Barrax alludes to the phrase "murders sleep" (l. 31); May Miller recalls Lady Macbeth's "Damned spot" (l. 5), Gwendolyn Brooks her "milkofhumankindness" (l. 15). Claudia Rankine draws on the voice of Lady Macbeth herself.

In addition to signifying on the play's characters, African-American poets have situated their poems in tension with the performance of the very play itself. Most famous is Langston Hughes's "Note on Commercial Theatre," which Thompson and Kolin discuss in their respective essays. More recently than Hughes, Al Young's poem "Identities" unfolds a dreamlike vision of: "So youre playing / Macbeth in Singapore / 1937." The date seems intended to flag the Welles production, which would have closed the previous year. In the midst of this performance, an "old seacoast drunk"

>   stomps up one-legged
>   onto the stage to

> tell you youve got no
> business playing this
> bloody Macbeth,
> not a lonely black boy
> like you [...]
> (ll. 1–2, 9, 11–17)

Young, like Hughes, meditates upon the almost hallucinatory incongruity of the "bloody / business" (see *Macbeth* 2.1.48) of being "put in *Macbeth*," to paraphrase Hughes's words. Although one may initially read the drunk as someone who questions the young black actor's right to play Macbeth, a Scottish King, it seems that he in fact begs the actor to ponder the implications of embodying the murderous character, suggesting that Shakespeare's lexicon ("whom / thou knowest" [20–21]) may not be worth the damage arising from temporarily escaping his identity ("who you know" [22]). The poem concludes in the midst of this hiatus, awaiting the outcome of a decision in which "Shakespeare" is now used as a verb:

> Well, do you go on
> & Shakespeare anyway
> or reach for the sky
> for the 500th time?
> (ll. 34–37)

Beyond these more familiar invocations of characters, lines, and performances of *Macbeth*, we find that African-American female poets have been peculiarly drawn to the figures of the three weyward/weird sisters. The performative power of incantation that these witches wield disturbs the seemingly intractable, bringing us closer to revelation through the transformative power of language, both sturdy tropes of African-American literature and especially black women's poetics. More specifically, it is likely the *conjuring* power of the weyward sisters that particularly appeals to this poetic tradition, which itself has been characterized by Houston A. Baker, Jr. as

> a phenomenology of conjure ... the poetry of *conjure* as an image resides in the secrecy and mysteriousness of its sources of power, in its connection to ancient African sources syncretized by a community of diasporic believers with Christian scriptures (66, 89)[3]

In this essay we briefly survey three disparate allusions to *Macbeth* by Rita Dove, Julia Fields, and Lucille Clifton, and explore the ways in which the figures of the weyward sisters fuse with (and confuse) a parallel tradition of conjuring within the African-American literary tradition.

I Conjure You by That Which You Profess (4.1.66)

At the moment of *Macbeth*'s creation, the word "conjure" had many active and overlapping senses in Shakespearean England, as Arthur Kinney details:

the word *conjure* was an especially powerful one in 1606, carrying a number of meanings: to swear together by a private agreement or contract; to plan a conspiracy; to constrain by oath or appeal to something sacred, to abjure; or to entreat, plead, beseech, or implore. Especially in this second meaning, it carried political overtones—and overtones of resistance or rebellion. It also had strong religious meaning. Since the Puritans often used the word to denounce recusant priests (whose acts were conjurations, i.e. black art), *conjurors* became a synonym for both papists and Jesuits. (Kinney 2001, 245)

When many today continue to identify "conjure" as "black art," they not only employ the signifier "black" to indict such belief systems as dealing in supernatural arts whose goal is to harm, but also associate such arts with people of African descent. Some aspects of the early modern connotations linger in the African-American tradition of conjuring, often seen as a conspiratorially taboo and covert alternative to "pure" Christianity. For example, in Charles Chesnutt's tales, the conjure woman was often a figure who professed religion in public, but assisted both African Americans and whites through conjure under clandestine circumstances.

In his seminal prose poem "Neo-HooDoo Manifesto" from his book *Conjure* (1972), Ishmael Reed articulates the fundamental connection between poetic and religious expression: "Neo-HooDoo believes that every man is an artist and every artist a priest." Reed suggests that artists and priests alike serve as intercessors between spiritual and earthly realms, and urges us to consider that both have comparable interpretive authority (21). For Reed, the conjuring of revision is a core principle, as "Neo-HooDoo is a litany seeking its texts"—texts that are the vehicles through which various identities speak and are thereby inherently intertextual. Reed's Neo-HooDoo aesthetic informs playwright Suzan-Lori Parks's essay "Possession" from *The America Play and Other Works* (1995). Whereas Reed seems to envision more of a priestly class of artists engaging with their sources, Parks wants to revise "The definition of possession" to the point where it "cancels itself out. The relationship between possessor and possessed is, like ownership is, multidirectional" (3). In other words, Parks seeks to envision a more reciprocal relationship between the artist and her art: the writer shapes her texts, but the texts themselves also take on lives of their own. Parks's concepts of possession and ownership deconstruct the conflict between originality and appropriation that critics often wage against artists who revise canonical texts. She levels literary stalwarts by insisting that interpretation and reinterpretation are essential to making meaning of texts. While Orson Welles depicts voodoo (or conjure) as a dark force that possesses Macbeth and robs him of agency, African-American artists such as Ishmael Reed and Suzan-Lori Parks adopt conjure, haunting, and possession as a means of *enacting agency* in their own writing processes, while also giving voice to those who truly have been robbed not only of agency, but often of life and limb.

Rita Dove enacts her enmeshment with and mastery of Shakespeare's texts in her poem "In the Old Neighborhood" by placing her reading of his texts among the everyday activities of life. She boasts of her familiarity with Shakespeare and his place in her childhood home when she recalls, "I've read every book in this house, / I know which shelf to go to / to taste crumbling saltines / ... and the gritty slick of sardines" (Dove 1993, xxiii). Dove comes to associate each Shakespearean play not with the language, characters, or plots—though she is thoroughly familiar with those aspects—but with the foods that she ate while reading them. The foods, along with her saliva and fingerprints, literally become a part of the pages of the books. Here she removes Shakespeare's tomes from the hallowed halls of rare manuscript libraries and puts them in the possession a child's greasy hands. Dove specifically mentions *Macbeth*—the "very first Shakespeare play I read"[4]—when she says, "But Macbeth demanded dry bread, / crumbs brushed from a lap" (xxiii). We might hear in this phrase an echo of the First Witch's report of "A sailor's wife [who] had chestnuts in her lap" (1.3.3). But more strikingly, in these lines Dove plays with the notion of Macbeth as both the title of the play *and* the name of the main character. Because she does not italicize Macbeth as she does *Romeo and Juliet* and *King Lear*, Dove evokes a double meaning: she herself requires dry bread (often prescribed to remedy nausea) to read the tragedy; yet the eponymous character requires it, too. *Both* need this bland meal to stomach the drama's gratuitous violence. Dove takes possession of Shakespeare's Macbeth by inserting herself as an actor in a play notoriously associated with banqueting. She simultaneously provides the characters with sustenance while feeding herself, putting Dove on equal footing with the iconic playwright and his equally iconic characters.

While Dove communes with Macbeth, another African-American poet, Julia Fields, carves up Lady Macbeth's language to underscore the violence prostitution perpetrates against women. In her poem "Vigil," Fields evokes Lady Macbeth's last lines from the play, which read "To bed, to bed. There's knocking at the gate. Come, come, come, come, give me your hand. What's done cannot be undone. To bed, to bed, to bed" (5.1.57–59). While the "sorely charged" (5.1.44) Lady Macbeth claims that "What's done cannot be undone," Fields's lines 30–35 read:

> Quick, quick, quick
>   Instant
>   Condensed, fleet
> It must be done
> And done and done
> Until it is undone.
>   (96–97)

Lady Macbeth's lines themselves are a feverish amalgamation of her prior statements from the play, such as "that which rather thou dost fear to do / Than wishest should be *undone*" (1.5.22–23). Fields conjures such lines even further, by remarkably repossessing not only allusions to the Lady's famous parting speech but also her earlier "The future in the *instant*" as well as Macbeth's own "If it were *done* when 'tis *done*, then 'twere well / It were done *quickly*" and "for't *must be done* to-night" (emphases added; 1.5.56, 1.7.1–2, 3.1.132). The prostitutes in Fields's poem follow Lady Macbeth's instructions and go to bed (or rather to the car) and they do so at the quick pace that mimics the Lady's compulsive speech: "Come, come, come, come," says Lady Macbeth; "Quick, quick, quick," Fields insists. While Lady Macbeth intones that the murders cannot be undone, Fields's speaker suggests that the possibility of undoing damage exists. Just as Dove makes Shakespeare's plays commonplace by leaving the vestiges of meals in their pages, Fields accomplishes a similar task by taking his words spoken by a "lady" and applying them to prostitutes. With allusive dexterity, Fields challenges the boundary that marks these ladies of the night as marginal, outsider figures.

Lucille Clifton likewise draws upon Shakespearean characters to address marginalized women, in this case those accused during the Salem witch trials. In the "in salem" section of her long poem "sisters," Clifton, like Welles, racializes the weird sisters, thereby taking possession of Shakespeare's characters (and, secondarily, Welles's interpretation of them) as a means to critique the fanaticism and xenophobia that fanned the flames at Salem.[5] Clifton opens this section of her poem with the lines:

> weird sister
> the black witches know that
> the terror is not in the moon
> choreographing the dance of the wereladies
> (ll. 1–4)

Calling her addressee "weird sister" and aligning her with "the black witches" marks both the addressee and Shakespeare's witches as black women. The poet's use of lower case letters also suggests a lack of formality between the addressee and her "sister." Furthermore, she distinguishes the weird sister and the black witches from the "wereladies," a neologism that Clifton coins in the poem. According to the *Oxford English Dictionary*, the prefix "were-" indicates:

> The first element of werewolf used in combination, chiefly with names of animals, to indicate a human being imagined to be transformed into a beast; as were-animal, -ass, -bear, etc.; also were-man.

Clifton's term denotes figures that are part beast and part lady—which might serve as one interpretation of Lady Macbeth. Clifton asserts that it is not supernatural forces (in this case, of the moon) that dictate the behaviors of "beastly" characters.

This contrasts with Welles's production, which seems to attribute Macbeth's murderous deeds to the power of the black magic voodoo priestesses thrust upon him. Contrariwise, Clifton insists that supernatural entities are not responsible for humans' revolting behavior. When Clifton writes "and the terror is not in the broom / swinging around to the hum of cat music / nor the wild clock face grinning from the wall," she catalogs items that are often linked to witches in superstitious lore (5–7). However, the poet defangs the false notion that these representations of witchcraft are where true evil resides. Instead, she identifies a more valid source of horror when she asserts:

> the terror is in the plain pink
> at the window
> and the hedges moral as fire
> and the plain face of the white woman watching us
> as she beats her ordinary bread,
> (ll. 8–12)

Clifton contrasts "black witches" and "plain pink," deliberately marking not only differences in their skin colors, but also their intentions. The poet rejects the notion that the blackness indicates evil and reflects upon the capacity for whites to evoke terror. However, it is important to note that the poet specifies the source of the terror by utilizing the definite pronoun: "*the* plain pink at the window" and "*the* plain face of the white woman." If she were to couch her critique in more general terms, she would run the risk of simply reversing, rather than suspending, racial essentialism. We should seek to complicate such essentialism by opening up intertextual possession with our own interpretive conjuring: conspiring to bring texts together to resist singular meanings.

Notes

1. For a survey of the travels of the phrase "The forms of things unknown" (*A Midsummer Night's Dream* 5.1.15) in African-American critical discourse, see Carpenter.
2. Around the same period as Heard, W. E. B. Du Bois likewise deployed the word "swarthy" in connection with Banquo's ghost; see the Moschovakis essay. Heard follows Robert Southey in imagining a ghastly slave pleading for judgment; Southey slightly modifies Macbeth's vision of Duncan's virtues "trumpet-tongued against / The deep damnation of his taking off" (1.7.19–20) in the stirring close of Sonnet 6 of his "Poems of the Slave Trade": "the Slave / Before the Eternal thunder-tongued shall plead / 'Against the deep damnation of your deed'" (Wood 218).
3. For a critique of Baker's use of the "conjure" trope in black poetics, see Murray; for further contextualization of the history of conjuring in African-American culture more generally, see Chireau.
4. In her interview with Peter Erickson, Dove notes that this was her first play "mainly because I heard my mother reciting from it" (2007, 26).
5. For more on the debates surrounding the ethnicity of Tituba (one of the accused women at Salem) as of either Indian or African descent, see both Hansen and V. Tucker.

24

"BLACK UP AGAIN": COMBATING *MACBETH* IN CONTEMPORARY AFRICAN-AMERICAN PLAYS

Philip C. Kolin

In the 1940 poem "Note on Commercial Theatre," Langston Hughes protests white culture's appropriation of the blues and the marginalization of African Americans, stating:

> You also took my spiritual and gone.
> You put me in *Macbeth* and *Carmen Jones*.
> And all kinds of *Swing Mikados*.
> And in everything but what's about me
> (ll. 8–11)

Continuing the poem, Hughes yearns for somebody to "talk," "write," "sing," "And put on plays about me," whom he describes emphatically as, "Black and beautiful" (ll. 13, 14, 16, 17, and 15). While Hughes's criticism of the erasure of a black presence rings painfully true in American culture, his diatribe tells only part of the story of *Macbeth*'s impact on African-American lives. More often than not, black actors were taken "out" of a play like *Macbeth*, not "put" into it. Consigned to play stereotypical roles on stage or in film, many African-American actors saw *Macbeth* as a fortress of Eurocentric casting, a shibboleth of a white theatre culture. Moreover, Hughes's assumption that *Macbeth* could not sing "about me" does not reflect the complex ways contemporary African-American playwrights have voiced their demands for identity through Shakespeare's play.

For example, *Macbeth* occupies a central place in Lonne Elder, III's *Splendid Mummer* (1986), a one-character play about the (highly fictionalized) life of Ira Frederick Aldridge (see the Lindfors essay). In the play, Aldridge recalls that he began as a janitor at the Chase Playhouse in London where "I was so mesmerized by Charles Wallack's spellbinding characterization of the tragic Scotsman [that] while tidying up his dressing room the day after the play's opening, I slipped into one of Macbeth's costumes and commenced to imitate his 'He's here in double trust' monologue" (9). But when the distinguished white actor walked in, "I scrambled out of the costume and dropped to my knees" and begged forgiveness for "adorning my unlicensed person in [his] costume." Commending Aldridge's "audacious thrust," Wallack, surprisingly, agrees to tutor him, adding this keen advice about playing Macbeth: "An actor does not simply read Macbeth, boy! You do battle with him! You arm and shield yourself as any good soldier would. Mount your steed, your lance, and you charge with alacrity! If not, Macbeth will slaughter you no sooner than first curtain rise" (9).

This fictional incident sums up both the curse and the promise *Macbeth* poses in contemporary African-American theatre. That Aldridge had to slip into Macbeth's costume secretly and then apologize for his "unlicensed person" shows that a white theatre regarded him and all black actors as cultural criminals forbidden to wear Macbeth's robes or speak his lines. Wallack's advice about playing Macbeth contextualizes the relationships between *Macbeth* and contemporary African-American drama. According to Wallack in *Mummer*, the script is an enemy determined to defeat Aldridge. Physicalized, *Macbeth* becomes an instrument of black oppression and a battleground between the races. In the allegory of racism fashioned in *Splendid Mummer*, however, *Macbeth* was in reality taken over by a black man who interpreted the role with acclaim. Taking Wallack's criticism to heart, Elder's Aldridge further recuperated/recruited Shakespeare's scripts to counter bigotry. He relished playing Shylock in one of the most anti-Semitic countries in nineteenth-century Europe—Poland—and reveled in performing Aaron, the invidious black villain of *Titus*, boasting that it enabled him to play an even more powerful Othello. Contrary to Langston Hughes's view, Elder envisioned and constructed an Aldridge who successfully used the role of Macbeth to be "about me."

Entering *Macbeth*'s (con)textual spaces, several black playwrights have attacked its colonizing legacy even as they have displaced or recast ritualized behaviors troped as white in *Macbeth*. Through allusion, adaptation, or racialization, they have incorporated *Macbeth* and resisted or renegotiated Shakespeare's authorial exclusivity. In this essay, I explore how *Macbeth* has, paradoxically, both repressed and represented an African-American experience in plays by Adrienne Kennedy, Ntozake Shange, August Wilson, and Suzan-Lori Parks.

Adrienne Kennedy, whose surrealistic plays influenced a generation of African-American women dramatists, has a combative, torturous relationship with Shakespeare and *Macbeth*. In her *Owl Answers* (1965), Shakespeare denounces Kennedy's heroine, an aspiring writer, yellow-skinned Clara Passmore who travels to England to claim her white father's heritage. Symbolizing the white patriarchy of authors who refuse to recognize a black woman, Shakespeare bristles at Clara: "Why is it you are a Negro if you are his ancestor?" (31) and imprisons her in the Tower of London. As Aldridge found out in *Mummer*, Clara's body is not licensed to claim a white role.

Even so Kennedy incorporates and racializes the iconic characters, imagery, and sounds of *Macbeth* in another play, *Funnyhouse of a Negro* (1964; see also Kolin). Chronicling events from the early 1960s, including her divorce, Kennedy confesses, "Often I thought of Lady Macbeth; she could not get the blood from her hands" (Kennedy 1990, 103). Lady Macbeth is trapped inside Kennedy and her characters in *Funnyhouse* just as they are trapped inside a white society that punishes them for being black. Looking at Kennedy through Shakespeare's outcast sister Judith, fictionalized by Virginia Woolf, Claudia Barnett argues that Kennedy's plays are "permeated with figures groping futilely for any sense of identity, attempting to create identities for themselves... her characters are frequently motherless women furtively seeking matriarchs among royalty and movie stars—clutching at socially acceptable, though inaccessible, ancestry" (48). Of Negro Sarah's four distinct identities/selves in *Funnyhouse*—Queen Victoria, Duchess of Hapsburg, Jesus, and Patrice Lumumba—two are indeed royal matriarchs.

Yet emblematic of a mad queen, Lady Macbeth is also a submerged voice for the schizophrenic Sarah and her selves. Sarah thus emerges as a textual outcast/offspring of Shakespeare's creation. In the process, Kennedy's black heroines in *Funnyhouse* are both ennobled by their relationship to *Macbeth* and alienated/subjugated through their performance of it. Evidencing Lady Macbeth's tragic white madness, Sarah shares with Shakespeare's queen a host of tortures, including being haunted by the bloody cries of children, suffering overwhelming guilt, and performing harrowing purgation rituals. Evoking 5.1 of *Macbeth*, the first stage direction in *Funnyhouse* reads: "*Before the closed Curtain a Woman dressed in a white nightgown walks across the Stage carrying before her a bald head. She moves as one in a trance... Her hair is wild, straight and black and falls to her waist*" (11). This is Sarah's narcoleptic mother who leaves a Lady Macbeth legacy of nightmares and tainted blood to her hysterical daughter and her daughter's selves. They cry of foul deeds (rape, miscegenation, murder) in language overflowing from *Macbeth*. Inscribing Lady Macbeth's "perturbation" on Sarah's body reifies how Kennedy's black protagonists desire to participate in the white world even to the point of owning white-colored guilt. Moreover, like Lady Macbeth in 5.1, Sarah writes in her somnambulistic nightmare. Attempting to purge the guilt of race from her conscience, Sarah is "preoccupied with the... positions of words on paper. [She] write[s] poetry filling white page after white page with imitations of Edith Sitwell" (14). Recruiting white scripts, Shakespeare's or Sitwell's, Sarah tries to gain access to the white world of letters and culture that also denied Clara in *The Owl Answers*. In confronting and suitably blackening the white fortress of *Macbeth*, then, Kennedy represents the plight of African-American women victimized by colonization. Shakespeare, for Kennedy, symbolizes this colonization.

Unlike Kennedy's combative relationship with Shakespeare, Ntozake Shange re-inscribes *Macbeth* in her 1979 play, *Spell #7*. While the play includes only one overt reference to *Macbeth*, it is, nonetheless, a major aperture into the larger incorporation of Shakespeare's play. As in her earlier choreopoems,

Shange deconstructs white archetypes and conventions by breaking models and "deny[ing] the power of language which...she wishes to 'attack, deform, maim'" (Bigsby 303). In *Spell #7*, she assaults the colonizing representation of African-American life, art, and language with the aggressiveness that Wallack urged Aldridge to adopt when he played *Macbeth*.

In *Spell #7*, a group of off-work black actors gather at a bar to bemoan the prejudice of a white-dominated theatre. Lily, "an unemployed actress working as a barmaid" (3), declares, "i wish i cd get just one decent part" (13) to which the proprietor Lou responds, "say as lady macbeth or mother courage" (14). But Eli, the poet-bartender, asks, "how the hell is she gonna play lady macbeth and macbeth's a white dude" (14). To challenge/battle this exclusionary white world, Shange parodically structures *Spell #7* as a minstrelsy, or a Jim Crow musical, portraying the life of "Negroes" in the Old South. The set is dominated by a "huge black-face mask hanging from the ceiling" (7), and characters wear blackface masks and dress like field hands, dancing and performing racist acrobatics. Minstrelsy, which became the "white man's joke," is, unquestionably, at the other end of the spectrum from *Macbeth* (Cronacher 180).

Yet in the space between the high culture of *Macbeth* and the coarse, disdainful comedy of minstrelsy, Shange erects a platform for her "unlicensed" black actors to represent themselves. For Shange, *Macbeth* symbolizes not only the world from which her black minstrelsy-linked actors are excluded but also the world they can reclaim on a new stage free from a white hegemony. As Lily argues before a bigoted director, "so let me play a white girl/ i'm a classically trained actress & i need the work & i can do it/ he said that wdnt be very ethical of him. can you imagine that shit/ not ethical" (47). In *Spell #7*, Shange gives her the chance to play a black version/aversion of *Macbeth*. As one of the actors in *Spell #7* admits, black actors often need to "Black up again" (46). Shange enters *Macbeth*, subtextually and imagistically, even as she subverts Shakespeare's signifying symbology. Weaving *Macbeth*-like images of blood, hair, children, clothes, madness, curses, nightmares, and trees throughout *Spell #7*, she fragmentizes, eroticizes, and racializes *Macbeth* to privilege a black presence resisting a dominant white culture.

The central figure in Shange's juxtaposing dramaturgy is Lou, "a practicing magician" (3), who is the master of ceremonies for the ensuing performances. Walking "thru the black-faced figures in their kneeling poses, arms outstretched as if they were going to sing 'mammy,'" Lou at first calls for stereotypical music and tap dancers, but his ancestry resonates with Shakespearean signifiers as well (9). Recognizing Lou's African roots as well as his Western ones, Jane Splawn argues that he reminds audiences of Prospero (391), but, more to the point, Lou has the magical powers of the "weird [weyward] sisters." Yet unlike the witches and their mistress Hecate who deal in evil threes so that "magic sleights / Shall raise such artificial sprites / As by the strength of their illusion / Shall draw him on to his confusion" (3.5.26–29), Lou is a benevolent magician. He "come[s] from a family of retired sorcerers/ active houngans & pennyante fortune tellers," and, "Yes, yes, yes 3 wishes is all you get" (8). Though he cannot turn a black child white, he can make being black enjoyable. Battling *Macbeth* and the restraining cultural codes it propagates, Shange from the start of *Spell #7* alludes to and then challenges Shakespeare's play.

On his magical set, Lou performs a destabilized version of the witches' theatre, empowering his black actors to "shed" their masks and the "lies" of white repression (11, 27). Unlike Macbeth and Banquo, enslaved by and impaled on the weird sisters' words, the actors in *Spell #7* come under the influence of a sorcerer/director who wants them to be free and not "sell yr skin" (27).

> you have t come with me/ to this place where magic is/
> to hear my song
> ...
> in this place where magic is involved in
> undoin our masks/ i am able to smile & answer that.
> in this place where magic always ask for me
> i discovered a lot of other people who talk without mouths
> who listen to what you say
> ...
> & in this place where magic stays
> you can let yrself in or out
>
> (27)

His actors will not, like a white Macbeth, be undone. His invitation to the actors to unmask situates *Spell #7* as a racialized *Macbeth*, a black *Macbeth* fantasy countering the evil forces in art or politics that force blacks to keep their masks on as minstrels. Revealing how the vestiges of a white play/culture/identity can be shed and transformed by following Lou's black magic, the characters in *Spell #7*, all "lookin' for parts" (14), receive their chance to perform in a black version of *Macbeth*. They escape into a world where black actors/artists are fashioners of their own dreams, their own aesthetics. After all, as one actor proclaims, "I am theatre" (24). As Neal Lester argues, *Spell #7* "demonstrates the need for individual fantasy as a tool for survival in a bigoted society and shows the definitive need for racially segregated experience to dispel oppressive and internalized myths about black cultures and celebrates...black identity" (7). Shange's fantastical *Macbeth* divests white theatre of its imperious control.

The most *Macbeth*-like performance in *Spell #7* is Natalie's in Act 1 through her persona of Sue-Jean, with commentary by Alec. Evoking the tragic signifiers of mothers and children in *Macbeth*—sleep, dreams, power, spells, cries, blood, baby's milk—Sue-Jean/Natalie, with her mask removed, presents a black Lady Macbeth, decolonizing the role in her own words, time, and place. As Sue-Jean, Natalie had "always wanted a baby/ who wd suckle & sleep / a baby boy who wd wet/ & cry/ & smile" (28). But "the men in the bar never imagined her as someone's mother" (28). Like Lady

Macbeth, Sue-Jean plays a power game with her offspring, committing murder for ambitious ends. Sue-Jean confesses, "i hadda a bad omen on me/ from the very womb/ i waz bewitched is what the ol women usedta say" (30). She felt "an earthquake" in her womb and when the child was born, his "name was myself" (30, 31). Alec adds, "the nites yr dreams were disturbed by his cryin" (31). Things were going well for Sue-Jean until "myself" started to crawl "& discover a world of his own" (31). As Alec explains to her, "you became despondent/ & yr tits began to dry & you lost the fullness of yr womb/ where myself/ had lived" (31). But because Sue-Jean "wanted back the milk" "& the tight gourd of a stomach," she "slit his wrists" while he was sleeping and "sucked the blood back into [herself] & waited" while her infant "shriveled up in his crib" (30). Sue-Jean concludes the horrific tale by explaining that, upon realizing what she had done, she "screamed in [her] bed" (32).

As Natalie tells her story through the eyes of Sue-Jean, and as Alec describes her birthing experience, we hear echoes of Lady Macbeth's horrific imprecations: "make thick my blood" and "Come to my woman's breasts, / And take my milk for gall, you murd'ring ministers" (1.5.41; 45–46). Welcoming infanticide for royal rule, she vows to Macbeth:

> I have given suck, and know
> How tender 'tis to love the babe that milks me.
> I would, while it was smiling in my face,
> Have plucked my nipple from his boneless gums
> And dashed the brains out, had I so sworn
> As you have done to this.
> (1.7.54–59)

Incorporating Lady Macbeth's role and language, Sue-Jean creates her own black tragedy. Natalie's narrative of "myself" eerily recalls Lady Macbeth, whose dreams of motherhood are replaced with nightmares over her guilt (28). In the end, too, Sue-Jean's fate is not unlike Lady Macbeth's. "She is left psychologically traumatized and paralyzed by the incident. Not only does Sue-Jean lose power over her environment, but she loses control of her own rational faculties" (Lester 103). Paradoxically, she has to kill "myself" in order to perform as a black, fantasized protagonist in *Macbeth* (28). Unmasking the marginalized black woman, Natalie plays a role that historically excluded her, racially and politically. Shange thus appropriates a white iconic script that outlawed black voices.

But on a deeper level, Shange incorporates *Macbeth* not solely as theatrical archive but as an icon to indict white culture. The color of evil in *Spell #7* is not black but white, and *Macbeth*-like action and imagery demonize and exorcise Shakespeare's Scottish play and the white culture in which it is enshrined. Describing the white race, Lily thinks of the violence of Madison Square Garden and concludes, "oh how they love blood" (15). In having Natalie enact a role steeped in the metaphors of the quintessential white play,

Shange thus combats Shakespeare's play, as Wallack instructed Aldridge to do, and reinterprets the racial signifiers *Macbeth* privileged. *Spell #7* signals an earthquake-size shift in the way actors, directors, and audiences think about race and *Macbeth*. *Spell #7* turns upside down the bigoted theatre conventions that forced actors like Ira Aldridge to appear in symbolic whiteface, hiding while revealing their black identities.[1] The title of Frantz Fanon's classic study of race—*Black Skin, White Mask*—ironically sums up Shange's message—it takes a black actor to get a white performance right.

August Wilson's *King Hedley II* (1999), the eighth work in his ten-play chronicle of African-American life in the twentieth century, does not resist the white identity and authority imprinted in *Macbeth* but, rather, appropriates and reconfigures the play's culturally-constructed tragic roles. Empowering a black character not just to battle but overcome Shakespeare, Wilson writes a mythic black *Macbeth* far different from Shange's adaptation of the play. Repeatedly advocating the establishment of a black cultural identity, Wilson nonetheless confirmed in a 1994 interview that "The life I know best is African American life...[but] I realized that you could arrive at the universal through the specific" (1994, 71). While Wilson resolutely situates his plays in an African-American tradition, he also admits that "In one guise, the ground I stand on has been pioneered by the Greek dramatists—by Euripides, Aeschylus, and Sophocles—and by William Shakespeare, O'Neill, Miller, and Tennessee Williams. In another guise, the ground I stand on has been pioneered by my grandfather, Nat Turner,...Marcus Garvey and the Honorable Elijah Muhammad" (1996, 11). These two worlds—one based on white canonical authors such as Shakespeare and the other on black historic black figures—complement each other in *King Hedley II*.

At first Wilson's hero King, a scar-faced ex-con robbing and murdering in Pittsburgh's drug-infested Hill District in the mid-1980s, seems far removed from Shakespearean tragedy. Reviewing a production of Wilson's play, however, Les Gutman asserted that in King Hedley "we have a later day Hamlet." But Hedley is closer to Macbeth than he is to Hamlet. Wilson's play, like *Macbeth*, is filled with epic-like quests, robberies, bloody knives, deaths of saintly spiritual/civic leaders, gory murders, fatal rivalries, superstitions, ghosts, fatal games of chance, cursed seeds, stark landscapes, dead cats, prophecies, and doomed lineages. Even more important, intersections between *Macbeth* and *King Hedley II* emerge in the ways each play foregrounds ideas of manhood and valor. Unlike Shange's play, Wilson's black tragic protagonist acquires heroic, even mythic status, displaying the tragic characteristics that audiences valorize in *Macbeth*. Speaking of King, Wilson called him "a man of high integrity and purpose" (quoted in Fluker 59).

Articulating the topic sentence of Wilson's, and Shakespeare's tragedy, too, King's best friend Mister warns, "He got to look in the mirror and see what kind of man he gonna be" (98). Like Macbeth, King is filled with self-defeating "vaulting" ambition and tragic overweening faith in his own manhood. King boasts: "Ain't nothing I can't do. I could build a railroad if I had the steel and a gang of men to drive the spikes. I ain't limited to nothing" (55).

"I want everyone to know King Hedley II is here," he proclaims (58). But his sense of manhood and honor are rooted in crime and violence, catching him in a tragedy of relativism. "If you try and take King's honor he'll kill you. Whether he right or wrong," attests Mister (98). With Macbeth-like resolve, King challenges the institutional powers that be: "I don't care what the law say. The law don't understand this" (56). Identifying King's hamartia, Stool Pigeon, the neighborhood prophet, warns: "King want to be like the eagle. He want to go to the top of the mountain. He wanna sit on top of the world. Only he ain't got no wings" (60). Similarly, Macbeth, too, is compared to an eagle in the Sergeant's description of his valor (1.2.35). Like King's, Macbeth's unbridled quest for control is connected to his ideology of manhood, referred to over thirty times in the play but best summed up in his declaration that "I dare do all that may become a man; / Who dares do more is none" (1.7.46–47). Macbeth and King become the "secret'st m[e]n of blood" (3.4.125), asking audiences to judge their proud masculine ideals. Macbeth and King, thus, espouse a personal code of honor that propels them into tragedy. "I got honor and dignity even though some people don't think that. I was born with it," asserts King (57). Their honor is rooted in the battles/causes they wage and the successive robberies/murders they execute. Wilson empowers King, as Shakespeare did Macbeth, with unassailable bravery.

Paralleling *Macbeth* further, King's manhood/fate is tied up with names, games of chance, and soul-crushing secrets that come back from the grave to destroy him. He is overly proud of the blood heritage of his name, being the son of the manly, fearless Hedley, who ruled the neighborhood. Thus, he styles himself King Hedley II. Throughout the play, he plants seeds (his lineage) and tries to protect his threatened plot by putting barbed wire around it, but others constantly threaten that lineage. Tonya, King's wife, tries to "get an abortion without telling" him (38). When Stool Pigeon gives him a bloody machete that Hedley used decades ago to kill a man, King learns he "has to find a way to wash that blood off...[to] sit on top of the mountain" (62). The machete is analogous to the fatal dagger Macbeth confronts, a tragic reminder that murder is in his blood. King's familial pride extends to biblical proportions. His father, just as King imagines himself, is hailed as "the Conquering Lion of Judea" (61). So proud of his name is Hedley that he kills a man for calling him "Champ" (73). "He don't know my daddy killed a man for calling him out of his name" (73). His name is his title, the assurance that he will be victorious, no matter what, a conviction analogous to Macbeth's belief that the witches speak an unconquerable futurity in hailing him as Thane of Cawdor and, ultimately, King of Scotland. At the end of the play, Elmore, the "consummate hustler," tells King the truth about his name (29). His father was not King Hedley but Leroy Slater, the man whom Elmore killed because he refused to pay a gambling debt. Hearing that, King "picks up the machete, turns and walks out of the yard" (97). Symbolically, like Macbeth, King is twice accursed, once by Elmore's tragic revelation (one's fate is based on names) and again by being linked/doomed to a bloody weapon.

Filled with revenge, King challenges Elmore: "the way I see it...Leroy owed you the fifty dollars...I'm gonna pay my daddy's debt...Now we straight on that. But see...my name ain't Leroy Slater, Jr. My name is King Hedley II and we got some unfinished business to take care of" (99). Contemporizing *Macbeth*, Wilson weaves games of chance/fortune throughout *King Hedley II*. The dark, foreboding landscape where the witches play Macbeth for a fool becomes the seedless yards and decaying houses of Wilson's own Hill District. Nothing can stop King from playing this fatal game of dice with Elmore: "We gonna play man to man" (99). As in *Macbeth*, King bases his life on the idea of a winning manhood and willingly gambles his future on the "fantastical" secrets, the verbal sorcery, that Elmore brings. Elmore's words are for King "strange intelligence" (1.3.74). An "imperfect speaker" who can "palter...in a double sense," like "juggling fiends" (1.3.68; 5.10.20, 19), Elmore sells guns (protection) that do not work, steals wives, and tricks men out of their money. As Ruby, King's mother, warns her son, "Elmore got a way with gambling...You'll lose all your money if you gamble with him" (17). Money for King, as becoming king for Macbeth, is power, control, keeping his family, and staying out of jail, protecting and preserving his turf. It is the historicized currency of *Macbeth*. Like Macbeth, King thinks he is supernaturally blessed, poised to win. Earlier, he told everyone he had a halo, but Mister humorously undercuts King's assertion: "You ain't got no halo. The devil is looking for you," an ironic prefigurement of King's encounter with Elmore (14). Just as the witches play their deceptive games of fate with Macbeth, who is not willing to see the tragedy on the other end of their secrets, King learns too late that names and games of chance sign his destruction.

Like Macbeth, too, King tries to follow his destiny amid a dizzying bombardment of prophecies and curses. Stool Pigeon repeatedly quotes Scripture to warn of God's impending scourge and calls for sacrificial blood to be spilled on the grave of a dead cat that belonged to the saintly and prophetic Aunt Ester who died at the age of 366 and whose death "stand[s] as the greatest evidence of the present danger for the community" (Herrington 171): just as Duncan's murder does in Scotland. King, too, summarizes the fate that stands against him, "They got everything stacked up against you as it is" (53). In King's house, as in Macbeth's, "One minute [it is] full of life. The next minute it's full of death" (39). Even though despair and death mark Wilson's play, his protagonist is a larger than life warrior whose manly feats bring him the tragic dignity invested in Macbeth. As Joan Herrington argues, "Despite King's less admirable qualities, we care for him, root for him, and find him heroic" (179). Wilson entered the world of *Macbeth* and victoriously captured and rewrote its conventions.

More blatantly than any other contemporary black dramatist, Suzan-Lori Parks savagely combats *Macbeth* through her witty spoof "Project Macbeth" included in her monumental *365 Plays/365 Days* (2006). Stretching to barely 250 words, "Project Macbeth" seizes on the major themes of *Macbeth*—fate, prophecies, rivalry, power/patriotism, murder, blood, and war—but Parks's sardonic incarnation of the play is closer to an episode in *Sanford and Son* or *The Jeffersons* than to a Shakespearean tragedy. Parks transforms Shakespeare's Scottish heath into a street in the projects where Macbeth and Banquo are rechristened MacSmith and MacJones, "2 Army Guys walk[ing] along," boasting about who killed more of the enemy "For God. And country" (133, 134). While MacJones accuses MacSmith of "zero" hits, MacSmith rebuts with "I hit 20" and "See my gold star," eliciting MacJones's less than eloquent rejoinder—"Bullshit" (134). Amid their humorous rancor, they decide to swear "eternal brotherhood," "cut their hands," and "trad[e] blood," at which point "a trio of women" (one even identifies them as "me and my 2 weird sisters") slink on stage, "perhaps dressed like the Supremes," and prophesize to MacSmith that he will kill MacJones and "build a kingdom on his bones," mocking Macbeth's fears that his rule would not "stand in thy posterity" (3.1.4). Paralyzed in fear (mocking Banquo's fright), MacJones screams, "I'm outa here," but he cannot move. When the Supremes/witches order MacSmith to "Draw yr. knife," he wails like a coward, "Do I have choice?" The third sister smartly bristles back, "We are the Fates, brother" (134–135). Parks's skit ends with MacJones "suddenly regain[ing] his ability to run" as MacSmith chases him.

"Project Macbeth" undercuts the eerie prophecies of the witches through the flip responses of the Supremes look-a-like "trio" and the comic and irreverent performances of MacJones and MacSmith. After trading blood with MacJones, MacSmith declares, "I've changed my fate line, cutting my hand like that" as if Macbeth's fate/destiny were transferable and hence not "ordained" or "immutable." Parks's two-minute play about these two brothers in the "Projects" totally drains the dignity and supernatural horror out of Shakespeare's tragedy, reducing Macbeth/MacSmith's portentous doubts to the monosyllabic "Shit" (135). Moving far beyond Wallack's advice to Aldridge on how to play Macbeth, or Hughes's lament that Shakespeare's work did not "talk about me," Parks shows that African Americans can indeed do so in her mini-version of/assault on the play. In the process, Parks gives the Bard the bird.

Macbeth has haunted the African-American presence and in a sense has helped to define it. Determined to penetrate its murky landscape of ghosts, superstitions, and witches, African-American actors and theatre companies since Ira Aldridge have left their mark on the play. Landmark black productions extend from Orson Welles's 1936 "Voodoo" *Macbeth* in Harlem to the 2002 hip-hop *Macbeth* produced by the African-American Shakespeare Company to the National Spirit Project's 2001 *Vo-Du Macbeth* to the planned African-diasporic film in 2010 (see the Rippy, Sloan, and Lennix essays). More subtly, though no less important, contemporary African-American playwrights have also tackled this classic white script and its profound impact on racial memory and identity. As we have seen, these playwrights have battled *Macbeth* for suppressing or displacing a black presence while, ironically, embracing and orchestrating a black agon. *Macbeth* enters their plays, directly or subtextually, in a variety of ways, from Lonne Elder's racial allegory of a defiant Aldridge to Adrienne Kennedy's combative "Shakespeare" to Ntozake Shange's choreopoetic anti-minstrelsy *Macbeth* to August Wilson's creation of a heroic black Macbeth to Suzan-Lori Parks's transformation of Shakespeare's play into a black sitcom. Through this broad range of African-American drama, *Macbeth* emerges as a consummate script on the politics of race in contemporary America. Ultimately, *Macbeth* has become "about [Black and beautiful] me."

Note

1. It was a custom for white actors in blackface to leave their hands uncovered. Like Krystyna Courtney, however, I prefer to believe that for Aldridge, leaving his hands unpainted, was a sign of racial pride (114–116).

25

BLACK CHARACTERS IN SEARCH OF AN AUTHOR: BLACK PLAYS ON BLACK PERFORMERS OF SHAKESPEARE

Peter Erickson

I begin with two passages by African-American writers. First are the reflections of an author who, at age 12 or 13, witnessed the black *Macbeth* in 1936:

> Bill takes me to see my first play, the Orson Welles production of *Macbeth*, with an all-black cast, at the Lafayette Theatre, on 132nd Street and Seventh Avenue, in Harlem.... [B]efore the curtain rose, I knew the play by heart.... I knew enough to know that the actress (the colored lady!) who played Lady Macbeth might very well be a janitor, or a janitor's wife, when the play closed, or when the curtain came down. Macbeth was a nigger, just like me, and I saw the witches in church, every Sunday, and all up and down the block, all week long, and Banquo's face was a familiar face. At the same time, the majesty and torment on that stage were real: indeed they revealed the play, *Macbeth*. They *were* those people and that torment was a torment. For, they were themselves, these actors—these people were themselves. They could *be* Macbeth only because they were themselves.

The second passage records the opposite experience, finding that black identification with Shakespeare is impossible:

> [T]he most crucial time in my own development came when I was forced to recognize that I was a kind of bastard of the West; when I followed the line of my past I did not find myself in Europe but in Africa. And this meant that in some subtle way, in a really profound way, I brought to Shakespeare, Bach, Rembrandt, to the stones of Paris, to the cathedral at Chartres, and to the Empire State Building, a special attitude. These were not really my creations, they did not contain my history; I might search in them in vain forever for any reflection of myself. I was an interloper; this was not my heritage.

Both speakers are James Baldwin (1976, 499 & 504; 1955, 7–8). In order to coordinate Baldwin's two attitudes, we must try to separate the black performer and the classic text: the potential strength to which Baldwin responds is in the actor, not the text. The distinction resides in the thin line between valuing the black actor and rejecting the canonical white author. While I want to honor Baldwin's tribute to Orson Welles's black *Macbeth*, I also want to activate here, in the strongest form, Baldwin's adversarial, critical voice. I support Baldwin's edgy ambivalence because it tells us that our job is not merely to celebrate, but rigorously to analyze and to evaluate black actors' access to Shakespearean roles.

The positive gains for black actors in the 1936 Harlem *Macbeth* are achieved at a cost. Two elements in particular dovetail and reinforce the problem of black stereotyping: Welles's relocation of *Macbeth* to Haiti and his evocation of Eugene O'Neill's *The Emperor Jones* (1920). The Haitian connection refers us not to the inspiring moment of the successful revolution led by Toussaint L'Ouverture against French colonial power in 1804 and the establishment of an independent black republic, but rather points to the later moment of collapse under Henri Christophe. Instead of a symbol of black freedom and hope, Haiti becomes an emblem of political failure. The opportunity to play Shakespeare is thus crossed, through the association with Haiti, with the enactment of black defeat; at the same time that black actors are elevated by entry into the Shakespearean realm, they are diminished by an image of inability to handle freedom that keeps them in their racial place. This dimension is compounded by the exoticism derived from *Emperor Jones*, including the use of drumming, that creates "a sort of Brutus Jones Macbeth" and makes Orson Welles the self-proclaimed "King of Harlem" (Callow 1995, 240; 244; see the Rippy essay). The net effect is encoded in the inadvertently demeaning description of Welles's production as "Voodoo" *Macbeth*.

Moving from Welles's landmark production, I turn to three contemporary plays in which black dramatists focus on black actors who perform Shakespeare. The three—Phillip Hayes Dean's *Paul Robeson* (1978), Derek Walcott's *A Branch of the Blue Nile* (1983), and Carlyle Brown's *The African Company Presents Richard III* (1988)—address a range of Shakespeare's works: *Othello*, *Antony and Cleopatra*, and *Richard III*.[1] In their various ways, the contemporary plays probe what it means to be a black actor playing Shakespeare. Each asks implicitly the further question: what is the relationship between Shakespeare as the starting point and the individual black dramatist in the present, and how does this doubled authorship play out over the course of the drama?

"I'VE ALWAYS THOUGHT OF THAT LADY WITH THE TORCH AS DESDEMONA"

Phillip Hayes Dean's *Paul Robeson* has two thematic links with Orson Welles's black *Macbeth*. First, Dean's presentation of Robeson displays a similar conjunction of Shakespeare with O'Neill's *Emperor Jones*. Second, Dean's dramatization of Robeson's career also uses Haiti as an emblem of black revolution, with the difference—in keeping with Robeson's vision of world liberation for non-white peoples—that Robeson's connection is to Toussaint L'Ouverture as the symbol of liberty instead of Henri Christophe as a failed dictator (11–12). It is the tension between these two themes in Dean's play that I highlight here.[2]

Paul Robeson's performance career unfolds in multiple media—theatre, film, music, and visual art—all of which involve intensely physical displays of Robeson's black male body as a work of art. However, it is crucial not to blend these professional venues but to examine each separately. In the retrospective survey of Robeson's life dramatized in Dean's play, we note a distinct difference between theatre and singing with regard to changing a given text. In the second act, while witnessing the fight against Franco during the Spanish civil war, the Robeson character calls attention to the changes he has made in the song "Ol' Man River": "I don't sing it that way any more" (61). Especially resonant is the shift from passive to assertive stance registered in the change from "I'm tired of livin' and scared of dyin'" to "I must keep fightin' until I'm dying" (62).[3]

These new fighting words are more congruent with the action of the anti-fascist fighters whom Robeson wishes to support. At the very end of the play, the music of "Ol' Man River" returns to remind us of Robeson's transformation over the course of his career (81). However, in the same concluding moment, Robeson recites Othello's final speech unchanged. This juxtaposition confronts us with two different attitudes toward textual revision: evidently Robeson feels a flexibility with respect to song lyrics that he cannot apply to the Shakespearean script.

The combination of Shakespeare and O'Neill in Welles's *Macbeth* had already occurred in Robeson's compressed trajectory from his 1924 performance in *Emperor Jones* to his 1930 performance of Othello in London. Dean's play acknowledges that Robeson received his opportunity for rapid advancement at the expense of the black actor, Charles Gilpin, who originated the role of Brutus Jones when O'Neill's drama opened in 1920. In recalling the brief but moving encounter between Gilpin and Robeson that records the transition, the Robeson character reflects:

> He mumbled his congratulations, and left. And it's with me still—all the things I didn't say to him.
>
> Sir, I am sorry. I never wanted to take your place. I never did.
>
> Why couldn't you leave that stuff alone? Look at yourself. You had no right to do this to yourself. We needed you. The Negro people need the great ones like you. We owe a place to the ones who come after us....
>
> Will I be the next ex-nigger of the year? No! They'll not do that to me. I'll not be found staggering through Harlem—chasing a ghost of what I might have been. (41)

The "stuff" and the "staggering" that Robeson condemns refer to Gilpin's alcoholism.

Yet in considering the difference between failure and success for a black male actor in the 1920s, we need to take into account the other element in O'Neill's dismissal of Gilpin: "he said Mr. Gilpin has started to drink a lot and was changing the dialogue" (40). The changes Gilpin made in the language were his responses to the racial stereotyping in O'Neill's text.[4] By contrast with Robeson's changing lyrics, it is as though he learns from Gilpin's fate that play texts should be left untouched. This perspective gives Robeson's conception of the need for black generational continuity a different meaning because, in the specific cultural arena of theatre, what Robeson passes on may not be salutary. The Robeson character's words come back to haunt him because the role of Othello leaves him "chasing a ghost of what he might have been" (41).

In terms of dramatic structure, the exchange with Gilpin occurs at the very end of act one, while Robeson's recitation of Othello forms a key component of the play's conclusion. When connected, the two culminations speak to each other in a way that makes Othello's speech seem flat and inadequate as a summary of Robeson's life. "Speak of me as I am" does not allow the play to speak with the emotional depth that is called for. "Nothing extenuate" short-circuits the process because the play does not tell what needs to be extenuated. The sense of circularity and entrapment is poignantly compounded by the fact that the Robeson character was performed by James Earl Jones, Robeson's most prominent successor in the sequence of black actors adopting the role of Othello.

In addition to Robeson as singer and actor, Dean's play stages a third form of iconicity for Robeson as the subject of sculpture and photography. The motif of sculpture is present from the outset in two works. The more well-known is the full-size nude of Robeson by the sculptor Antonio Salemmé. A second example immediately visible as an onstage prop—"the sculptured bust of Paul Robeson sits on a pedestal. It is lighted dramatically" (4)—all but announces that Robeson's self-fashioning as an icon will be a major issue. Yet the play does not engage this problem in a sustained and critical way.

According to Martin Duberman's biography, "After Robeson became more political and more outspoken on racial questions, he let his friendship with Salemmé dissolve" (590n11). Yet the play does not drop Salemmé, but instead begins with Robeson genially reminiscing about his collaboration with the sculptor. This expansive gesture of fusion makes it difficult to come to terms with the sharp split between the two phases in Robeson's career between the initial artistic orientation and the subsequent political emphasis, a split reflected in the overarching division in the play between act one and act two.

The sculpture metaphor reaches its pinnacle when an eloquent evocation of the Middle Passage is followed by the implausible equation of the Statue of Liberty with the putative freedom represented by Desdemona: "I've always thought of that Lady with the Torch as Desdemona" (67). Equally unconvincing is the interpretation of *Othello* as a liberal icon. Robeson's claim—"By playing Othello, I could help Americans look inside themselves—look into the deepest fabric of American society and see that its tragic flaw was racism" (67–68)—is asserted but not demonstrated. America's tragic flaw is racism, but it is far from clear that *Othello* is the means to address it (see Quarshie). Dean's recourse to Othello's speech at the end of the play is incongruous in view of the political stand the Robeson character takes in the second act.

"Can't Talk Shakespeare, Though"

Walcott's *A Branch of the Blue Nile* creates a fictional, play-within-a-play scenario by having black actors rehearse a performance of Shakespeare's *Antony and Cleopatra*, while reflecting on the tense relationship between their acting roles and their lives.[5] Because it stages the Africa connection through Cleopatra's location in Egypt, the choice of this particular Shakespeare play for Caribbean actors produces an automatic conflict between English and African legacies that evokes Walcott's early formulation—"how choose / Between this Africa and the English tongue I love?"—in the bitterly punning title of the poem "A Far Cry from Africa."

I approach Walcott's play through the perspective of a later poem. The line—"But the blacks can't do Shakespeare, they have no experience"—from poem 23 in the volume *Midsummer* voices the key issue at stake in the drama (1984). In similar terms, two of the black actors, Sheila and Chris, act out a ritual satire on the stereotype of the limitations of black actors: "We got rhythm." / "Can't talk Shakespeare, though" (1983, 228–29). This exchange is a microcosm of the play because its bitterness cuts two ways that are so evenly counterbalanced that they preserve the conflict in unresolvable stasis. On one hand, the ironic bite of this moment mocks the stereotype and thereby asserts that blacks can do Shakespeare. On the other hand, the moment also conveys an underlying anguish of exclusion. This division expresses uncertainty, which is dramatized especially through the self-doubt experienced by Sheila, the only member of the acting company who remains on the island at the end of the play.

An analogous stalemate occurs in Walcott's poem 23 because the poet represents himself as trapped and divided by issues of "color" that remain seemingly incapable of solution except in the explosive, apocalyptic imagery of the Brixton riots in the poem's last three lines. But this leaves the question of Shakespeare not fully processed, and Walcott's position as poet highly ambiguous. "I was there to add some color to the British theatre," Walcott sardonically observes. But his color becomes whitened: "Praise had bled my lines white of any more anger, / and snow had inducted me into white fellowships." The lines movingly disclose the pain of the poet's anomalous position. Walcott is caught and isolated in the middle: he can identify neither with the white riot police nor with the Brixton blacks. Yet the potential third option of a fully articulated critical perspective remains elusive and hence a loose end.

A Branch of the Blue Nile stages an equivalent three-part negotiation because Walcott's drama contains three different plays. In addition to Shakespeare's *Antony and Cleopatra* with a white British director, there is the folk play written by Chris, one of the black actors. The rehearsal of this second play precipitates a complete artistic breakdown. As Marylin remarks: "All you making us schizophrenic. We talking Shakespeare like 'rangatang people...And vice versa. [*Points to the two directors*] Over there you see our split personality" (244). In a parody of James Baldwin, Chris's effort explicitly opposes Shakespeare—"I ain't care who the arse it is, Shakespeare, Racine, Chekhov, nutten in there had to do with my life" (246)—while also immediately compromising his argument by casting his ideal audience in Shakespearean terms—"Phil is my Lear, my Mad Tom out in the rain" (247). By Chris's own self-deprecating admission, his play amounts to "a damned stupid West Indian back-yard comedy" (235) and is therefore incapable of providing a convincing alternative.

The possibility of synthesis rather than split is deferred until the third play, which is Chris's second effort: "It's not finished....It's about us....Everything we tried to do is in here" (306). This new project is, in effect, the actual play of multiplicity that we see unfolding before us and hence serves as a surrogate for *A Branch of the Blue Nile*. But this interpretive move contains a throwaway gesture because Walcott's play itself does not achieve unity, but remains divided and precarious. The all-inclusive mixture stages the problem, but does not offer a solution.

The ongoing conflict is expressed in the counterpoint between the two major speeches that the defeated derelict Phil gives at the end of the play. In the first, Phil delivers a message of failure, underscored by the story of the demise of his own singing group (297–99). When Phil rises to the level of eloquence, the conditional force of his "if" sharply qualifies his statement: "But if it was in my power to sprinkle benediction on your own kind, to ask heaven to drizzle the light of grace on the work you trying to do here...you knows I would" (300). The ecstatic language—"benediction," "the light of grace"—is recognizably Walcott's, but Phil uses it to express the futility of this artistic ideal.

Phil modifies the sense of impossibility when, speaking the final words of the play, he makes a plea that defies his previous despair. This time the elevated rhetoric, now addressed specifically to Sheila, sketches a more encouraging view of performance:

> They clap, and is like waves breaking. They laugh, and what they saying is thanks. And it does lift your heart up like a wave, so high you does feel the salt prickling in your eyes. Oh, God, a actor is a holy thing. A sacred thing....And you know the gift ain't yours but something God lend you for a lickle while. Even in this country....It touch me once, that light. It fill me full....For all our sakes, I beg you. Please. Continue. Do your work. Lift up your hand, girl. (311–12)

Yet the tentative hope that this conclusion envisions is conspicuously unrealized. Phil is a key figure through whom Walcott explores his own implied position as dramatist. The open-endedness of his exploration is engaging, but it is also painfully inconclusive. The result is that, within the context of this drama, the status of Shakespeare remains deeply uncertain. By contrast, Carlyle Brown's play directly confronts the ambivalence and moves toward a more decisive resolution.

"I Rather Have Me Own Play"

Like Dean's *Paul Robeson*, Carlyle Brown's *The African Company Presents Richard III* involves a specific event in American theatre history. While Dean focuses on Robeson's historic breakthrough as the first black actor to play Othello in a major American venue, Brown deals with an earlier point, over a century before in 1821, concerning the African Company, an all-black group managed by William Henry Brown. Their production of *Richard III*, starring James Hewlett, is forced to close at Mercer Street and at City Hotel in New York so that it cannot compete with the same play at the white theatre, Stephen Price's Park Theatre (see E. Hill 11–14; and Hill and Hatch 24–40). The sharp contrast between the African Company's blocked access to Shakespeare and Robeson's success on Broadway suggests an overall narrative of progress, leading to Robeson as the symbol of the triumphant vindication of black performers of Shakespeare.

Yet Carlyle Brown's play reverses this expectation by rejecting the idea of Shakespeare as the ultimate goal of professional fulfillment. What drives this surprising dramatic twist, however, is not the external opposition by the white director but rather the internal dissension within the black company. Brown presents an extended rehearsal over several scenes of the exchange in 1.2 of *Richard III* in which Lady Anne is wooed by Richard Duke of Gloucester. The two characters in Brown's play who perform the two Shakespearean roles, are Ann Johnson and James Hewlett, and as their relationship emerges, Ann makes an implicitly feminist criticism of Shakespeare's characterization of women: "And it makes me mad every one a them nights, actin' out such a huck 'em so woman as she is" (21). In her objection to the part, she refuses to treat Shakespeare's words as sacred. When Hewlett tells her "That's not the next line," she responds, "If I were Lady Ann[e], it would be the next line" (25). Refusing to be bound by the limits of the script, she moves outside the Shakespeare framework: "You mean to tell me that the only thing we have to talk about is in this play?" (25).

Ann's insistence on the difference between herself and the role—"This ain't no Lady Ann[e], this Ann. This is me" (26)—becomes a means of questioning Shakespeare's value that signals wider possibilities of cultural change. When changing the text proves impossible—"it's not in the play" (25), Ann opts out: "I'm not gonna be in this play.... I'm out" (30–31). The rejection of Shakespeare thus occurs well before the African Company's performance is shut down by outside forces. Ann's individual action prefigures the inner dramatic action as a whole.

Another strand that points beyond Shakespeare is the play's invention of his double, Papa Shakespeare, whose name begins by mocking him and ends by mocking Shakespeare. The shift is expressed in Papa Shakespeare's two-part comment at the start of act two: "My master 'dere he call me Shakespeare so to mock me, 'cause I don't speak the way he do. He laugh.... He no hear me thinkin' in me own way" (32). In terms of Paul Laurence Dunbar's "We Wear the Mask," the poem Carlyle Brown uses as his epigraph, Shakespeare is "the mask" that conceals a more complex inner life. When Brown's play turns away from Shakespeare, the redirection has been anticipated by Papa Shakespeare's summary: "I rather have me own play" (36).

In the play's final scene, the company's mission is suddenly redefined as the desire for alternative theatrical resources. The rival white theatre director stipulates, with a Lear-like flourish, that the African Company "swear never, never, never to play Shakespeare again" (49). But, in the play's delicious irony, his emphatic prohibition unwittingly endorses the company's new sense of opportunity. Abandonment of the struggle for access to Shakespeare, instead of being portrayed as a defeat, is presented as a positive outcome.

The alternative to Shakespeare takes the literal form of a play, written by William Brown himself, about "an insurrection of the Caribs on St. Vincent Island" (53). *The Drama of King Shotaway* is not fictional but an actual work by Brown (Hill and Hatch 32–35). Particularly astonishing is the overt political focus on a "Black-Carib King" who "bleed his blood for freedom" (Brown 53–54). The climactic final speech, given by James Hewlett, quotes not Shakespeare but Brown's play. By contrast, Dean's *Paul Robeson* makes us feel the unbridgeable division between Robeson's embodiment of Othello and Robeson's reading of the abolitionist Wendell Phillips's 1861 speech on Toussaint L'Ouverture (11–12). From the comparative perspective of *The African Company Presents Richard III*, where the new drama supports revolution, Robeson's unwavering Shakespearean commitment appears to be a dependency that hinders his political views.

"Shakespeare Is Not the Only One with a Tongue"

In conclusion, I return to my starting point in Baldwin's "Autobiographical Notes" at the beginning of *Notes of a Native Son*. After announcing his discovery that Shakespeare, as an epitome of European culture, is "not my heritage," Baldwin goes on to add: "At the same time I had no other heritage which I could possibly hope to use—I had certainly been unfitted for the jungle or the tribe. I would have to appropriate these white centuries, I would have to make them mine—I would have to accept my special attitude, my special place in this scheme—otherwise I would have no place in *any* scheme" (1955, 8). Baldwin thus raises the theme of appropriation, which has been so crucial to my generation of Shakespeare scholars.

Writing in 1974–75 in *The Devil Finds Work*, however, Baldwin indicates how constrained and fraught the goal of appropriation can be when he observes:

> It is scarcely possible to think of a black American actor who has not been misused: not one has ever been seriously challenged to deliver the best that is in him. The most powerful examples of this cowardice and waste are the careers of Paul Robeson and Ethel Waters. If they had ever been allowed really to hit their stride, they might have raised the level of cinema and theater in this country.
>
> What the black actor has managed to give are moments—incredible moments, created, miraculously, beyond the confines of the script: hints of reality, smuggled like contraband into a maudlin tale, and with enough force, if unleashed, to shatter the tale to fragments. (1976, 554)

This shattering of the tale to fragments, especially when applied to Shakespeare, is hard to come by. I think Baldwin's remarks on "waste" help us to address the tragic depth of distress we sometimes feel in relation to appropriative efforts to reformulate Shakespearean themes. We like to believe that classic texts can always be transformed through contemporary acts of revision, but the three plays under discussion enact versions of failed appropriation of Shakespeare. In each case, the Shakespearean material is shown to be too resistant to permit revisionary triumph.

A review of the three plays suggests a progression. Dean's *Paul Robeson* stages a stance of uncritical celebration of black performance of Shakespeare, in which both the character Paul Robeson and the contemporary author are completely enmeshed and from which neither can be extricated. Walcott's *Branch of the Blue Nile* reveals an intermediary position of unresolved ambivalence. The action of the play can neither translate nor convert the intractable Shakespearean material. The play's ending leaves us in an emotional state both of deep loss and of deep uncertainty. At the end of *Branch of the Blue Nile*, what we feel most strongly is this insufficiency—the need for something else, a longing that Shakespeare cannot satisfy. Finally, Brown's *The African Company Presents Richard III* concludes that access to Shakespeare is no longer the be-all and end-all. The path to freedom may go through Shakespeare, but it may also go around Shakespeare.

Brown's drama opens up the route of bypass, as August Wilson's summary describes: "There is a moment in Carlyle Brown's *The African Company Presents Richard III* when the company is jailed for performing William Shakespeare's plays, and as they wrestle with the demand that they stop, one of the characters agrees, saying, in essence, that Shakespeare is not the only one with a tongue" (2004, ix–x). Wilson refers to the turning point in the play's final scene expressed by the company's manager, William Henry Brown: "What, we got to do Shakespeare like his mouth the only mouth what speak? Cat have tongue too ya know. Taste just as good sayin' words comin' from one who knows who you are, than one who don't know ya a' tall" (53). In Brown's alternative drama, we hear Shotaway make a direct appeal for liberation: "Restore yourselves, your wives and your children to the inheritance of your ancestors, who inspire your fury and who show you the way. Those marvelous, struggling spirits who suffered to you the air you breathe; who knit time for you to walk on; who give you stars to cover your body" (54). The ecstatic lyricism of this last phrase connects with August Wilson's view:

> When we, as black Americans, sit down to create in the formal language of theater, we are describing our ideas about the world. We are making an aesthetic statement that makes use of ourselves in total. Our history, our experience, our culture, our ideas about pleasure and pain, our concepts of beauty and justice, our ideas of story and song. It doesn't leave any part of ourselves out of the equation. (2004, x)

The line of development from Baldwin to Wilson supports this new direction by giving us confidence that there is another place to go, that we can live in more than one world. This is not to dismiss renewed efforts to revise Shakespeare but rather to insist that we do not have to put all our eggs into one Shakespearean basket.

Notes

1. Dates given in the main text are the dates of first performance.
2. See Dyer 2004 (64–136); Carby (45–83 and 91–101); and Erickson 2007 (77–101).
3. See Duberman for a summary of the evolution of Robeson's intervention in the lyric (604–605n14).
4. Krasner describes how Gilpin's limited attempts to change the language were met with the author's anger: "his drinking was probably related to the tight rope he walked as a 'race man' trying to avoid any appearance of betrayal, and a 'Negro actor' in a role that indeed stressed pejorative racial characteristics" (200).
5. See MacDonald 2005; Juneja (specifically 243–45); B. King (319–21 and 331–36); and Dasenbrock.

Epilogue

Figure S.8 Obamicon, Paste Media Group, 2009.

26

OBAMACBETH: NATIONAL TRANSITION AS NATIONAL TRAUMISSION

Richard Burt

I TELL YOU YET AGAIN, BANQUO'S BURIED (5.1.53)

During Barack Hussein Obama's presidential inauguration ceremony, several of the featured speakers and broadcasting pundits noted the importance of the United States' publicly-performed and peacefully-enacted transition of power. The ceremony performs both the sign and the signifier of a democracy in which old leaders gracefully recede and new ones graciously ascend *without violence*. The "vaulting ambition" (1.7.27) and the "horrible imaginings" (1.3.137) of both the outgoing and incoming leaders—which obviously played crucial roles in their respective rises to power—are suppressed in the performance of an irenic hand-off of authority. What could be further from *Macbeth*, in which power is seized through unbridled corruption, violence, and greed? What could be further from *Macbeth*, in which performances of power are forgone for the physicality of force? And yet the election of Barack Obama haunts this volume, recurring like Banquo's ghost in several of the essays. Even outside the pages of this collection, the "analogy to the start of an Obama presidency is made complete" in the Chicago Shakespeare Theatre's production of *Macbeth* (running at the time of the inauguration!) with a Malcolm (Phillip James Brannon) who distinctly resembled Obama (Polkow).

Although the inauguration ceremonies never explicitly invoked *Macbeth* (or Shakespeare, for that matter), *Newsweek* ran a story on the day before the inauguration entitled "Will Act for Food," which featured the "Voodoo" *Macbeth* directed by Orson Welles (McCarter). In the print edition, a small reproduction of the red and black poster for the Federal Theatre Project *Macbeth* (with a caricatured Macbeth) appears in the upper left corner of the first page of article, facing a full-page photo of the African-American actor Jack Carter in costume as Macbeth. The author invites Obama to include a New Deal for culture in his agenda, to remake the NEA (National Endowment of the Arts) as the WPA (Works Progress Administration).

That Shakespeare should be recalled in the context of a U.S. presidency should come as no surprise; for many decades presidential candidates and presidents of both parties have been compared to Shakespearean characters (Newstok). The U.S. presidential primaries and election campaigns of 2008 were no exception: both Senators McCain and Obama were compared by pundits and bloggers to Macbeth, and their wives to Lady Macbeth, as was Senator Clinton. These often tortured comparisons constitute a form of *caricature criticism* akin to political cartoons, shorthand for both celebration and vilification. The *Newsweek* article stands out as an exemplary "traumission" of transition: that is, a transmission traumatized by awkward pauses, false starts, restarts, and repetitions. To go forward, the author of the *Newsweek* article implies, Obama should go *back*. Furthermore, Obama should go back to unspeakable Shakespeare, to a modernized production of a play whose title dare not be named. The caption across the photo in the magazine reads: "THE SCOTTISH PLAY: A scene from 'Voodoo Macbeth' which one critic said 'overwhelms you with its fury and phantom splendor'" (McCarter, citing Atkinson).

In addition to producing a temporal oddity—by calling for change by returning to the past—the author seems unaware that *Macbeth* itself poses a famous staging problem in the transition of power from Macbeth to Malcolm: Macbeth dies on stage when Macduff kills him, yet his severed head is brought onstage without indication when the corpse should be removed (see Swander). After the line "Lay on, Macduff, / And damned be him that first cries 'Hold, enough!'" (5.10.33–34), the stage directions in the Folio read:

> *Exeunt fighting. Alarums*
> *Enter Fighting, and Macbeth slaine.*
> *Retreat, and Flourish. Enter with Drumme and Colours,*
> *Malcolm, Seyward, Rosse, Thanes, & Soldiers.*
> (Hinman 2477–79)

The stage direction "[*Exit Macduff with Macbeth's body*]" has been added between "*slain*" and "*Retreat*" by modern editors. The only stage direction for Macduff in the Folio reads: "*Enter Macduffe, with Macbeth*s *head*" (Hinman 2504).

Moreover, the *Newsweek* author assumes that what he calls Welles's "trailblazing, voodoo-inflected, all-black staging of *Macbeth*" marks social progress, forgetting that the production was in fact quite controversial because of the Haitian setting and voodoo drums (see the Rippy essay). The *Newsweek* author also overlooks the fact that the Bush administration's NEA, led by Dana Gioia and with the advocacy of First Lady Laura Bush, embraced Shakespeare for a reactionary brand of multiculturalism. In short, the history of race-based casting in Shakespeare in the United States has been anything but a linear narrative of progress, as I have previously argued (see Burt). The assumption that the first black president of the United States would necessarily create a progressive WPA with Shakespeare at its heart is *traumissive* at its core.

Perhaps most notably, the *Newsweek* article calls up a "spooky" *Macbeth* by quoting a critic on the production's "phantom splendor." What remains of the "Voodoo" *Macbeth* production is footage at the end of in the WPA documentary film *We Work Again* (1937), devoted to employment for African Americans. Ironically, none of the actors in Welles's "Voodoo" *Macbeth* performance is credited in *We Work Again*: another instance of the traumatic underpinnings of supposedly progressive politics. The end of the play appears as the film's capstone, the last shot being the curtain call of the cast. The film has its own *traumissive* aspect, disrupting a scene of political transition: Jack Carter does not appear as Macbeth; instead, Maurice Ellis, who played Macduff in the original Harlem production, plays Macbeth in Carter's place.

Faith, Here's an Equivocator (2.3.7–8)

I introduce the neologism "Epit(gr)aph" at the end of this brief epilogue to mark a problem of writing *on* the end, and *to* the end. This problem is linked to problems of narrating transitions of power in linear, chronological terms—narrations structured, that is, by a clearly demarcated *before* and *after*, *beginning* and *end*. As historicism and cultural studies have taken a so-called "materialist" turn, they have regressed, in my view, into an empiricism and positivism. These approaches seem to imagine that spectrality has been exorcised from materiality, and that history is reducible to linear, causal chains—chains without trauma. As *Weyward Macbeth* demonstrates, however, progression and regression often occur simultaneously in the same instance, like Welles's "Voodoo" *Macbeth*. The double troubled ObaMacbeth traumission of power suggests that we remain haunted by ghosts of Shakespeare and racist violence that continue to be a deeply disturbing part of American history, precisely because the history continues to be so difficult to tell by white and black Americans. President Obama provided perhaps the most uncanny example of national traumission when he visited "Ford's Theater, site of President Lincoln's assassination," as a CNN reporter put it, to pay "tribute to the 16th president's ability to recall passages of Shakespeare's *Hamlet* and *Macbeth*" ("In Ford's Theater").[1] By referring to *Macbeth* by name rather than as "the Scottish play," Obama, the reporter continued, "unwittingly ventured into what many theater hands believe to be dangerous territory: any mention of [*Macbeth*] will result in a cursed production—including a greater possibility of injury, bankruptcy, even death." We may perform peaceful transitions of power, but they are as haunted as Macbeth's banquet, and as equivocal as the Porter.

Note

1. For a discussion of *Macbeth* as "Lincoln's favorite play," see J. C. Briggs, and Beran.

Appendix

Figure S.9 Playbill from the Los Angeles FTP production of *Macbeth* (1937). Library of Congress, Music Division, Finding Aid Box 1095.

27

SELECTED PRODUCTIONS OF *MACBETH* FEATURING NON-TRADITIONAL CASTING

Brent Butgereit and Scott L. Newstok

This is by necessity an incomplete list. While we concentrated primarily on professional productions within the United States, we make occasional exceptions for innovative community and school productions, as well as for a few non-U.S. productions (those that include U.S. cast or crew, influence U.S. productions, or tour/are screened in the United States). To our regret, this excludes many suggestive international entries in which race and performance provocatively intersect. Nonetheless, in an attempt at coherence, we maintain an American focus, in spite of how diasporic influences inevitably complicate such a focus.

For earlier materials, we often gratefully followed the scholarship of Errol G. Hill and James V. Hatch.

1821–24
From September 1821 to the summer of 1824, the African Theatre in New York City stages at least ten major productions, including two composed of excerpted scenes from *Julius Caesar*, *Macbeth*, and possibly *Romeo and Juliet* (Hill and Hatch 26–27).

1827–67
Ira Aldridge plays Macbeth a number of times in England and Europe (see the Lindfors essay).

1866–69
Philadelphia-born black actor Morgan Smith performs excerpts from Shakespeare at various English venues; Macbeth is included among his roles (E. Hill 34; Hill and Hatch 64–65).

1876
March 28: the Amateur Dramatic Company of San Francisco stages Sheridan Knowles's *The Wife; or, A Tale of Mantua* in conjunction with scenes from Shakespeare's *Macbeth* at Platt's Hall. Cecelia V. Williams takes the female leads in both plays, performing in heavy whiteface for the sleepwalking scene (Hill and Hatch 69–70).

1884
January: Henrietta Vinton Davis performs selections from *Macbeth* alongside Powhatan Beaty in Cincinnati, OH; on May 7, Davis and Beaty perform at Ford's Opera House in Washington, DC (where Lincoln was assassinated) and the program includes three scenes from *Macbeth* (E. Hill 62–68).

1887–88
April 29: John A. Arneaux holds a benefit performance to mark his retirement; he performs scenes from *Macbeth* (as Macbeth) alongside Henrietta Vinton Davis (as Lady Macbeth; see 1884) and Thomas Symmons (doubling as Macduff and Duncan). On January 3, 1888 at Baltimore's Wilson Post Hall, Arneaux appears again in scenes from *Othello* and *Macbeth* with assistance from Hurle Bavardo and Alice M. Franklin (Hill and Hatch 81).

1888
The Booth-Barrett Amateurs advertise a Louisville, KY, production that includes selections from *Richard III*, *Othello*, *Macbeth*, and Bulwer-Lytton's *Richelieu*, "but no record of performance, presuming one did occur, has survived" (Hill and Hatch 86).

1880s–1910s
R. Henri Strange includes the character of Macbeth in his repertoire of roles (Hill and Hatch 89).

1892
Affluent black Seattleites present an evening of Shakespeare at the Opera House, including excerpts from *Richard III* and *Macbeth* (Hine 507).

1890s–1920s
Richard B. Harrison makes a living doing poetic recitations on tour, including a full one-man *Macbeth* (Nouryeh).

1898
The Lend-a-Hand Club of Allegheny City, PA, performs *Macbeth* (Glasco 298).

1912
Under the direction of Charles Burroughs, a twenty-four player company in New York produces a pageant of "pantomime of scenes" from *Macbeth*, *The Merchant of Venice*, and *Othello*. With A.L. Halsey as Macbeth and Minnee Brown as Lady Macbeth, the pageant benefits the Detention Home for Colored Girls at Young's Casino (E. Hill 86).

1913
At commencement, a production of *Macbeth* is presented by the HBCU Lincoln University Stagecrafters, Jefferson City, MO (Peterson 1997, 127).

1928
Ralph Chessé produces marionette versions of *Hamlet*, *The Emperor Jones*, and *Macbeth* in San Francisco; in 1979–80 he creates a new production of *Macbeth* with his sons Bruce and Dion at the University of San Francisco. While Chessé's ancestors in New Orleans chose to pass for white in the late

nineteenth century, they fought in black regiments during the Civil War, and were listed on the census as black.

1931
Anne Cooke directs *Macbeth* with the University Players at HBCU Spelman College, Atlanta, GA (E. Hill 84).

September 15–19: Forbes Randolph directs the New York musical revue *Fast and Furious*, which closes with a number by John Wells called "Macbet." Performers include Jackie (later "Moms") Mabley as Lady Macbeth and, as Young Siward, Maurice Ellis, who would subsequently play Macduff (and, on tour, Macbeth) in the FTP production (see 1936; Peterson 1985, 128–29).

1934
March 8: Owen Dodson (see 1959) plays *Macbeth*'s "Old Man" with the Bates College (Maine) 4A Players (Hatch 335).

1935
September–October: Ralf Coleman directs an all-black Boston production of *Macbeth* (see the Simmons essay).

1936
February 23: Lillian Voorhees directs Thomas Wood Stevens's abridged version of *Macbeth* at HBCU Talladega College, Talladega, AL.

April 14: The Harlem Negro Theatre Unit of the WPA Federal Theatre Project launches what comes to be known as the "Voodoo" *Macbeth*. John Houseman produces and Orson Welles directs an all-black cast, first at the New Lafayette Theatre in Harlem (through June 20), on Broadway at the Adelphi Theatre (July 6–17) and eventually on a national tour (July 21–October 4) to Bridgeport, Hartford, Dallas, Indianapolis, Chicago, Detroit, Cleveland, and finally returning to New York at the Majestic Theatre in Brooklyn (October 4–17). Performers include Jack Carter (Macbeth), Edna Thomas (Lady Macbeth), Maurice Ellis (Macduff), Canada Lee (Banquo); Sierra Leone drummer Asadata Dafora Horton coordinates the percussion (see the Rippy essay). At the close of its U.S. tour. Welles declines an invitation from Charles Blake Cochran to bring the production to London.

1937
July 14: *Macbeth* premieres at the Maya Theatre, Los Angeles, CA. Despite the intention to remount the FTP production with different crews, Los Angeles is the only Negro Theatre Unit that actually does, with Max Pollock directing and Jess Lee Brooks (Macbeth) and Mae Turner (Lady Macbeth) starring (see Figure S.9 and the Newstok essay).

1950
HBCU Atlanta University Summer Theatre (AUST) produces *Macbeth*, starring Georgia Allen as Lady Macbeth (Peterson 1997, 23).

1955
British actor John Ainsworth stages a racially integrated production of *Macbeth* in Trinidad, Barbados, and Guyana, including Trinidadian Freddie

Kissoon as Malcolm and future theatre scholar Errol Hill as associate director. Kissoon would himself later write a play called *Shakespeare's Dream* in which characters from *Macbeth*, among many plays, speak (Banham 1125).

1956
Negro Actors Associated advertises an "interracial cast" for *Macbeth*, directed by Chestyn Everett. A white Lady Macbeth (Joanna Merlin) plays across a black Macbeth (Robert Decoy) at the Dance & Drama Playhouse, Los Angeles, CA (Harvard Theatre Collection).

1957
Akira Kurosawa films an adaptation set in medieval Japan, known in English as *Throne of Blood*; many subsequent directors cite this movie as an influential precedent (e.g. Zaslove, 1990; Bully, 1998; Chappelle, 2010).

1959
February 17–21: Owen Dodson (see 1934) directs *Macbeth* with the HBCU Howard Players at Spaulding Hall, Howard University, Washington, DC; James W. Butcher (see 1982) plays Macbeth and serves as the production manager.

1961
December: American-born, Harvard-educated volunteer teacher Nathaniel Frothingham directs Makerere College students in *Macbeth*, the first all-African production at the National Theatre in Kampala, Uganda (Smalls 249).

1963
Puerto Rican actor Raul Julia (see 1966, 1989) stars as Macbeth at the Tapia Theatre, San Juan.

1963–64
Earle Hyman (see 1968) and Vinie Burrows tour in a revue, *The Worlds of Shakespeare*. The program, which includes excerpts from *Macbeth*, is conceived and adapted by Marchette Gaylord Chute and Ernestine Perrie, who "insisted on Negro actors" since "universality is aimed at" (Calta).

1966
Gladys Vaughan directs New York Shakespeare Festival's Mobile Theatre's all-black *Macbeth*, starring James Earl Jones and Michael McGuire (alternate Macbeth/Macduff), Ellen Holly (Lady Macbeth), and Cleavon Little in his first NYSF role (see 1967). A slightly modified cast tours NYC schools in the fall (Kuner).

June 28–August 20: Osvaldo Riofrancos directs (and plays Duncan) in a Spanish-language production (translated by Leon Filipe) with the New York Shakespeare Festival's Spanish Mobile Theatre, which tours the boroughs. Igancio Lopez Tarso (Macbeth); Maria Brenes (Lady Macbeth); Raul Julia (Macduff; see 1963, 1989); Osvaldos (Duncan).

1967
February–March: Black actress Diana Sands is the lead in a production at HBCU Spelman College, Atlanta, GA.

February 22: Barbara Garson's play *MacBird!*, a satire of President Johnson's administration, premieres at the Village Gate in New York, closing January 21, 1968. Cleavon Little (see 1966) plays the second Witch, specified in the stage directions as "a Negro with the impeccable grooming and attire of a *Muhammad Speaks* salesman."

1968

October: Richard Dorso and Robert Sweeney, producers of the Doris Day TV series, announce a contemporary screen adaptation to be performed by "all-Negro cast," with a script by David Karp, and (unrealized) plans to film in Spring 1969 in Watts. According to Dorso, "the conflicts exploding in Negro ghettoes parallel those in the Scotland of Shakespeare's *Macbeth*" (Weiler).

Barry Boys directs Earle Hyman in an otherwise white cast for the Actors Company filmed production, which is released on November 17, 1968 on WNET (Channel 13 New York).

1969

Bernard C. Jackson directs Yaphet Kotto and Beah Richards at the Inner City Cultural Center of Los Angeles (Lane).

1970–2001

South African Welcome Msomi's *Umabatha*, which comes to be known as the "Zulu *Macbeth*," premieres at the University of Natal, Durban, and tours extensively (London, the United States, Israel, and Italy) in the 1970s. In 1995 it is revived in South Africa at the request of Nelson Mandela before it tours internationally again in 1997, including performances at the rebuilt London Globe (1997, and again in 2001); nearly a dozen cities in the United States; and Mississauga, Canada (Wright).

1972

June 13–26: Peter Coe's *The Black Macbeth* appears at The Roundhouse, London, featuring Oscar James (Mbeth), Jeffrey Kissoon (Meru), and Mona Hammond (Wife of Mbeth; see 1991–95); set in Barotseland. According to Hammond, this is the first all-black Shakespearean play performed in the UK (Lambert).

1973

Trinidad-born Sullivan Walker performs a one-man show called *Black Macbeth* in his Caribbean Experience Theatre in Brooklyn (Hill and Hatch 296).

Dr. Carroll Dawes and Len Anthony Smith co-direct at the Creative Arts Center on the Mona campus of the University of the West Indies, Jamaica; Rastafarians (including spiritual leader Mortimer Planno) act in the production, set within a Caribbean dictatorship. Dawes would herself perform in a 1974 Jamaican television production directed by Wycliffe Bennett.

1977

Yoland Bavan plays Lady Macbeth, directed by John Dillon at the Cleveland Playhouse (E. Hill 164).

May 10–29: Edmund Cambridge, Jr. directs an all-black version of *Macbeth* based on Welles's recently rediscovered production notes for Woodie King Jr.'s New Federal Theatre (New York); Esther Rolle and Lex Monson star (see the Newstok essay).

May: Lithuanian exile Jonas Jurasas's American directorial debut is *Macbeth* for La Mama Experimental Theatre Company (New York) with Barbara Montgomery as Lady Macbeth.

July 6–August 27: E. Keith Gaylord directs *Macbeth* for the Champlain Shakespeare Festival in Burlington, VT, with two black actors in the leads (Jennifer A. Cover and Ray Aranha [see 1979]) in an otherwise white cast (Jorgens and Levie).

1979

Ray Aranha (see 1977) plays Duncan in a Folger Theatre Group production in Washington, DC, directed by Mikel Lambert (Roberts).

1981

April 8–May 31: Danny Glover (see 2010) stars in a post-apocalyptic version of the play directed by William Bushnell, Jr., at the Los Angeles Actors' Theatre (Wilds).

1982

January 28–30: James W. Butcher (see 1959) directs an all-black version of the play at HBCU Howard University in Washington, DC, billed as "An Adaptation Set In West Africa" and starring William Marshall and Carol Foster Sidney; unfortunately, the Folger Shakespeare Library video recording is damaged.

Winter: the Inner City Cultural Center of Los Angeles mounts Felton Perry's comic musical adaptation called *Sleep No More*; Perry stars in a nearly all-black cast (E. Hill 181).

1985

Shakespeare Festival of Dallas premieres John R. Briggs's adaptation *Shogun Macbeth*; Pan Asian Repertory Theatre in New York performs the play in 1988 and revives it in 2008 (see the Huang essay).

November 14–17: David Richard Jones directs *Macbeth: A Modern Mestizo Story Set in Central America* at the KiMo Theatre in Albuquerque, NM, for La Compañía de Teatro de Alburquerque; sets include Mayan designs, with the witches as ancient Mayan spirits, and monarchical Scotland is transferred to "caudillo" contemporary Latin America. The cast includes Marcos Martínez (Macbeth), Laura Worthen (Lady Macbeth), and witches Mónica Sanchez, Elena Parres, and future film actress Angie Torres.

1986

Directors Clinton Turner Davis and Harry Newman, as part of the Actors' Equity Association (AEA) project, launch a "nontraditional casting" festival at the Shubert Theatre in New York. They invite five-hundred actors, producers, and directors to reanimate eighteen scenes, including ones from *Macbeth* (Davis and Newman).

1987
April 14: Peter Gill and John Burgess co-direct an all-black "work in progress" production at the Studio of the Royal National Theatre, London; Wilbert Johnson and Decima Francis MBE (see 1989) star.

June 11–October 2: Demetra Pittman portrays Lady Macbeth in the Oregon Shakespeare Festival's *Macbeth* directed by James Edmondson.

1989
Nick Ward directs an all-black *Macbeth* for London's Royal National Theatre, which tours New York, Chicago, St. Louis, and other United States locations; Herman Jones serves as associate producer. Amelda Brown (Witch/Lady Macbeth), Ken Drury (Macbeth), and Cyril Nri (Banquo) star.

Richard Jordan directs Raul Julia (see 1963, 1966) as the lead in the New York Shakespeare Festival's production at the Public/Anspacher Theatre; *New York Times* reviewer Frank Rich characterizes it as an "irrelevant" and "bloodless" production, finding confusing "the periodic brandishment of what looks like a Navajo blanket" (Rich).

The Roxbury [MA] Outreach Shakespeare Experience (R.O.S.E.) presents as its first play "an all-black, full-length production of *Macbeth* that fall at the University of Massachusetts, Lowell and at the Strand Theatre in Dorchester." Decima Francis MBE directs (see 1987); Nefertiti Burton alternates scenes as Lady Macbeth with Christina R. Chan (Engle, Londré, and Watermeier 185).

1990
May 13–July 1: G. Valmont Thomas (see 2002) plays the lead in a production set in the 1870s American frontier for the Bathhouse Theatre in Seattle, WA. Director Arne Zaslove notes, "It turns out that a lot of blacks did head for the high plains after the Civil War.... And then, of course, there's that line about 'black Macbeth'" (Adcock).

1991
Andre Braugher plays the lead in the Philadelphia Drama Guild's production of the play, which Mary B. Robinson directs.

1991–95
Haworth [NJ] Shakespeare Festival, founded by Cindy Kaplan, produces *Macbeth* in collaboration with African-American producer Voza Rivers, using leading black actors from the Royal Shakespeare Company and the Royal National Theatre, including Mona Hammond (see 1972) as one of the Witches; South African-born Stephen Rayne directs. The play is set in a contemporary, vaguely African locale; venues include festivals and universities in New York, Ann Arbor, Lowell, Memphis, Nashville, Atlanta, Miami, Raleigh, Charlotte, Milwaukee, and London (Hampton; Engle, Londré, and Watermeier 208–09).

1994
Lee Breuer, founder of Mabou Mines, considers a revival of the FTP production (see 1936), which was to have starred Carl Lumbly (see 1997; Winn).

1995

September 12–November 5: Joe Dowling directs the play at the Shakespeare Theatre at the Lansburgh, Washington, DC; a reviewer notes that "Popular television actor Stacey Keach played the lead, with other key parts, including Macduff [Chris McKinney], taken by African Americans. Curiously, the effect of this casting was less visual than vocal" (quoted in Maley).

1996

Nina Menkes's film *The Bloody Child* is released (see the Lehmann essay).

1997

February 5–7: Luther D. Wells directs an all-black production at HBCU Florida A&M University Essential Theatre, Tallahassee, FL; Wells draws upon the production notes from the 1936 FTP play.

February 14–April 4: Berkeley Repertory (CA) features Carl Lumbly (see 1994) and L. Scott Caldwell as the Macbeths, whom Tony Taccone directs (Winn).

1998

February 27–March 22: Rich Crooks adapts *Macbeth: The Black Side* at the House of Candles (NY), with a primarily African-American cast, although both Macbeths are white actors (Jared Voss and Jennie Israel).

Earl Warner stages Jamaican playwright Alwin Bully's adaptation, *McB*, at the Phillip Sherlock Centre at the University of the West Indies (UWI), Jamaica. The play was originally written in 1986, and had a staged reading in 1990. Bully notes both Kurosawa (1957) and Dawes and Smith (1973) as influences.

George C. Wolfe directs Alec Baldwin and Angela Bassett at the Joseph Papp Public Theatre, New York City, NY as part of the New York Shakespeare Festival.

Clyde Ruffin adapts a Caribbean-inflected version in St. Lucia under the auspices of Derek Walcott; this one-act version provides the inspiration (as well as a Trinidadian actor) for Ruffin's April 15–25, 1999, production at the University of Missouri-Columbia's Black Theatre Workshop.

1998–99

British actor Ken Campbell stages *Makbed*, a version in the pidgin English of the South Pacific islands of Vanuatu, as an attempt to promote the potential for a universal language. After performances in London, the show tours Papua New Guinea, New Zealand, Australia, the Solomon Islands, and Vanuatu (see the Carroll essay).

1999

Javon Johnson performs Macbeth with Shakespeare-in-the-Schools (SITS), a touring company associated with the University of Pittsburgh; Greg Longenhagen directs.

Greg Lombardo's film *Macbeth In Manhattan* includes a black actress, who plays Lady Macbeth (Gloria Reuben), and a black actor, who plays a Chorus figure (Harold Perrineau) (see the Scott-Douglass essay).

English-born Alexander Abela produces and directs *Makibefo*, using local actors in Madagascar (Burnett).

1999–2001
Polish writer/director Agnieszka Holland, with American Zev Braun Entertainment, makes (unrealized) plans for a feature film retelling *Macbeth* in the environment of a contemporary African country.

1999–2006
July 12–August 3: Alfred Preisser opens the inaugural season of the Classical Theatre of Harlem (CTH) with *Macbeth*, starring Ty Jones and April Yvette Thompson; the show is revived in three subsequent seasons, July 9–August 5, 2003; June 12–20 (Biennale Bonn, Germany), and July 14–August 8, 2004; and July 28–August 5, 2006, making it the most frequently produced drama in their repertory. According to CTH, the play "combines the European cadences of Shakespeare's English with the rhythms and imagery of Africa and the Caribbean" (see the Newstok essay).

2000
July 29–August 20: The Queen's Company, New York, presents an all-female production; Rebecca Patterson directs. Dialogue from the "Wyrd Sisters" is enhanced, and there is borrowing from *Richard II*, *Coriolanus*, and *King Lear*, among other Shakespearean plays.

2001
Frank Cwiklik creates and directs *Bitch Macbeth* for Dans Macabre Theatrics (NY), a stylized reworking with S&M motifs; the original cast stars Rasheed Hinds; the production is remounted in 2003 and again in 2008.

2001–05
Lenwood Sloan's *Vo-du Macbeth* revives the FTP production (see 1936) with local theatre companies across the nation with an all-black production crew (see the Sloan essay).

2002
February 26–November 3: G. Valmont Thomas (see 1990) and B. W. Gonzalez play the Macbeths in Libby Appel's Oregon Shakespeare Festival production.

June 28–August 24: Sean Kelley directs an adaptation with "an African flavor" for the Colorado Shakespeare Festival at the University Theatre, Boulder; the lead is originally to be Jeffrey Nickelson, who withdraws and then performs in another version at the Shadow Theatre the following year (see 2003); in his place, Macbeth is played by Kyle Haden, with Candace Taylor as Lady Macbeth. According to their press release, the production involves "the creation of a tribal setting ... based on West African and ancient Mayan cultures," with two Ghanian drummers performing in the show.

Victoria Evans Erville's *MacB: The Macbeth Project* adapts *Macbeth* to a hip-hop record-industry setting for the African-American Shakespeare Company, San Francisco, CA; the production is revived in 2008 with support from the National Endowment for the Arts' Shakespeare in American Communities

Program; it is performed at the Colorado Shakespeare Festival and the Buriel Clay Theatre, San Francisco (Hurwitt; see the Barnes essay).

2003
March: Teatro Tapia (San Juan, Puerto Rico) performs Adriana Pantoja's new Spanish translation *El Nuevo Macbeth*, which sets the play in a gypsy context.

October–November: Buddy Butler directs *Macbeth* for the African American Shadow Theatre Company in Aurora, CO, starring Jeffrey Nickelson (Macbeth) (see 2002).

November 18–23: Kimberley Lamarque's adaptation at HBCU Tennessee State University is reimagined as the 2100 world of *The Matrix*.

2004
Vishal Bharadwaj adapts the play as the film *Maqbool*, set in present-day Bombay.

Greg Salman's independent film *Mad Dawg* (a.k.a. *Fury*) is released, a "gangster" modernization of *Macbeth*.

Swedish Bo Landin and Norwegian Alex Scherpf release their film *Sámi Macbeth*, shot in Sweden. Their adaptation wins the first European Minority Film Festival for stateless European minority languages (November 18–19, 2007).

2004–05
Max Stafford-Clark's Out of Joint Theatre Company produces a *Macbeth* set in a contemporary "African" locale. The play opens West Yorkshire in 2004, "followed by its world tour in 2005 to London, the Czech Republic, Minneapolis, Massachusetts, Edinburgh, Bury St. Edmunds, Mexico, the Netherlands, and Nigeria" (Scheil).

2004–07
Anita Maynard-Losh directs the Perseverance Theatre production of "Tlingit *Macbeth*," first in Alaska but then touring nationally, eventually performing at the National Museum of the American Indian as part of the 2007 Shakespeare in Washington Festival (DC) (Stodard; see the Maynard-Losh essay).

2005
January 30–February 13: Justin Emeka sets *Macbeth* in the post-Civil War/Reconstruction era at the Meany Studio Theatre of the University of Washington (Seattle). Macbeth is a Sherman-esque general who falls in love with ex-plantation owner, Lady Macbeth; Duncan and Banquo are black; and the witches are freed slaves who still work for Lady Macbeth (Wick).

March–April: Fontaine Syer directors a predominantly African-American cast for the Saint Louis Black Repertory Company's production, set in Sudan, at the Grandel Theatre, St. Louis.

July 8–17: Peres Owino and Marc Ewing adapt and star in *Mocbet*, a production that recalls the regime of Idi Amin in Uganda. Coleman Sacks Artistic Company, the Pilot Light Theatre, Los Angeles.

October 20–22: The Cutting Ball New Play Festival, San Francisco, stages African-American playwright Robert Alexander's sci-fi spin-off, *Alien Motel 29: The Secret Outtakes of the Ebony Lady Macbeth*.

2006

September 30–October 9: Ayodele Nzinga adapts the play as *Mack: A Gangsta's Tale* for the Prescott-Joseph Center for Community Enhancement and the Lower Bottom Playaz. It is produced by Shakespeare in the Yard at the Sister Thea Bowman Memorial Theatre, Oakland, CA.

The film *The Last King of Scotland* opens, starring Forest Whitaker as Ugandan dictator Idi Amin. Giles Foden, British author of the novel (1998) upon which the film is based, acknowledges Shakespeare's play as a model for his own narrative, going so far as to call *Macbeth* "The African Play" in a review of the Stafford-Clark production (see 2004).

2007

January 4–27: Ato Essandoh leads an otherwise white cast in the adaptation *Macbeth: A Walking Shadow* for the Manhattan Theatre (NY); Andrew Frank directs this adaptation by Frank and Doug Silver.

January 5–February 4: Steve Marvel directs an all-black version at the Lillian Theatre, Los Angeles, starring Patrice Quinn and Harry J. Lennix (see 2010; see the Lennix essay).

February 2–4: Law Professor Marc Fajer directs at the University of Miami School of Law; according to their press release, "volatile politics of the Caribbean of the mid-20th century are the backdrop.... Music of Latin America and the British West Indies, the colorful costumes of the witches, and a multi-racial cast help create the Caribbean atmosphere."

April 4–6: Jacqui Carroll adapts and directs *Voodoo Macbeth* for OzFrank Theatre, Brisbane, Queensland, Australia. An all-white cast performs the story in the context of "French colonial slave trade: by day the Princes parade, by night the drums of the Haitian jungle raise the hair of the neck" (Forrest).

May 16–19: Richard J. Thomas directs the Kelvin Players (UK) in an adaptation "set in the jungles of Central America in the late 1970s / early 1980s" with Latin American juntas in mind as historical analogue.

2008

January 18–February 2: Julia Ewing directs *Macbeth* for Memphis's African-American repertory company, Hattiloo Theatre (TN).

March 20–22: Robin Jackson Boisseau directs a classical staging of *Macbeth* at HBCU Hampton University, Hampton, VA.

April 5–13: Glen Walford directs the play with a multiethnic cast in the Gale Theatre's continued collaboration between London and Barbados. While most of the cast are black Barbadians, Macbeth (Peter Temple), Duncan (Patrick Foster), and Ross (Ben Foster) are white British actors; Ghana-born Kawku Ankomah (Macduff) was Lennox in the Out of Joint production (see 2004).

May: Aaron Jafferis adapts and directs a "hip-hop opera based on Macbeth" called *Weird Sisters* at his arts high school alma mater, Educational Center for the Arts, New Haven, CT. Jafferis is also the co-author of *Shakespeare, The Remix* (see the Barnes essay).

May 15–June 1: José A. Esquea directs Teatro LA TEA's post-apocalyptic production *Macbeth 2029* with a multicultural cast at the Clemente Soto Velez Cultural Center, New York (see the Esquea essay).

October 16–November 8: Women's Will, an Oakland-based multiethnic all-female Shakespeare company, performs *Macbeth*. All roles beside Macbeth (Valerie Weak) and Malcolm (Desiray McFall) are played by three actors (Julia Mitchell, Leontyne Mbele-Mbong, and Treacy Corrigan), as if all parts emerged from the witches.

November 14–23: Paul Mitri directs a multicultural, multilingual version at the Kennedy Theatre, the University of Hawai'i-Mānoa (see the Carroll essay).

November 21–23: Ray Vrazel directs *Macbeth: "Something Wicked This Way Comes"* at HBCU Dillard University, New Orleans, LA; in an interview, Vrazel notes that they "re-set the play in a modern Third World country that's politically unstable" (Cuthbert).

2009

January 2–March 8: Barbara Gaines directs the Chicago Shakespeare Theatre's modern-dress version, with African American Karen Aldridge as Lady Macbeth performing across her white Macbeth (Ben Carlson); reviewers note that the black actor playing Malcolm (Phillip James Brannon) bears a distinct resemblance to Barack Obama.

January 23–24: At the Charles H. Wright Museum of African American History in Detroit, MI, 4Theatrsake performs "an up beat hip-hop twist" on the play, taking place in modern Brazil with Condomble practitioners in lieu of the witches.

February 13–November 1: Gale Edwards directs the Oregon Shakespeare Festival's racially mixed cast of *Macbeth*, at the Angus Bowmer Theatre, Ashland, OR. The African-American actors include Peter Macon (Macbeth), Kevin Kenerly (Macduff), Josiah Phillips (Duncan/Porter/Exorcist), and Perri Gaffney (Second Witch).

June 1–October 31: Colm Feore stars as Macbeth across black actors Yanna McIntosh (Lady Macbeth), Dion Johnstone (Macduff), and Timothy D. Stickney (Banquo) at the Stratford Shakespeare Festival, Canada; director Des McAnuff sets the play in "mythic 20th century Africa." Some reviewers perceive a resemblance between McIntosh and Michelle Obama.

2010

Director Aleta Chappelle plans an African-diasporic film version set in a Caribbean locale, tentatively starring Terence Howard, Sanaa Lathan, Harry J. Lennix (see 2007), Blair Underwood, and Danny Glover (see 1981).

References

Abdul, Raoul. "African American Advanced the Verdi Legacy." *The New York Amsterdam News* (May 3–9, 2001): 21.

Abuba, Ernest. "Director's Note." *Shogun Macbeth* playbill. Julia Miles Theatre, New York. November, 2008.

Abu-Lughod, Lila. "The Romance of Resistance: Tracing Transformations of Power though Bedouin Women." *American Ethnologist* 17:1 (February 1990): 41–55.

Accepted Addresses; or, Proemium Poetarum. To Which Are Added, Macbeth Travestie, in Three Acts, and Miscellanies, By Different Hands. London: Thomas Tegg, 1813.

"Active Dreams." Self-interview of Nina Menkes. Facets Video, 2007.

Adcock, Joe. "Deranged on the Range: 'Macbeth' Heads Out West." *Seattle Post-Intelligencer*. May 11, 1990.

Adjaye, Joseph K. "Popular Culture and the Black Experience." *Language, Rhythm & Sound: Black Popular Cultures into the Twenty-first Century*. Ed. Joseph K. Adjaye and Adrianne R. Andrews. Pittsburgh, PA: University of Pittsburgh Press, 1997. 1–22.

Agamben, Giorgio. *Homo Sacer: Sovereign Power and Bare Life*. Trans. Daniel Heller-Roazen. Stanford, CA: Stanford University Press, 1998.

Aidoo, Ama Ata. *Our Sister Killjoy: Or Reflections from a Black-Eyed Squint*. New York: N O K Publishers, International, 1977.

Alexander, Robert. *Alien Motel 29: Secret Outtakes of the Ebony Lady Macbeth*. Unpublished playscript, 2005.

Alker, Sharon, and Holly Nelson. "*Macbeth*, the Jacobean Scott, and the Politics of the Union." *SEL: Studies in English Literature, 1500–1900* 47 (2007): 379–401.

Allied Arts Players. Mission Statement. Program notes for *Dessalines, Black Emperor of Haiti*. Boston, MA: May 15, 1930.

"All Negro Cast Presents Shakespeare." *Boston Chronicle*. October 19, 1935.

Anbinder, Tyler. *Five Points: The Nineteenth-Century New York City Neighborhood That Invented Tap Dance, Stole Elections, and Became the World's Most Notorious Slum*. New York: The Free Press, 2001.

Anderegg, Michael. *Orson Welles, Shakespeare, and Popular Culture*. New York: Columbia University Press, 1999.

Anderson, Douglas. "The Textual Reproductions of Frederick Douglass." *Clio* 27 (1997): 57–88.

Anderson, Lisa M. "When Race Matters: Reading Race in *Richard III* and *Macbeth*." *Colorblind Shakespeare: New Perspectives on Race and Performance*. Ed. Ayanna Thompson. New York: Routledge, 2006. 89–102.

Anderson, Marilyn J. "The Image of the Indian in American Drama During the Jacksonian Era, 1829–1845," *Journal of American Culture* 1 (1978): 800–10.

André, Naomi. Email correspondence with Scott L. Newstok. January 4, 2008.

Aschenbrenner, Joyce. *Katherine Dunham: Dancing a Life*. Urbana: University of Illinois Press, 2002.

Asimov, Isaac. *Asimov's Guide to Shakespeare*. Garden City, NJ: Doubleday, 1970.

Atkinson, Brooks. "'Macbeth,' or Harlem Boy Goes Wrong, Under Auspices of Federal Theatre Project." *The New York Times*. April 15, 1936.

Babcock, Weston. "Macbeth's 'Cream-Fac'd Loone.'" *Shakespeare Quarterly* 4.2 (1953): 199–202.

Baker, Houston A., Jr. "Workings of the Spirit: Conjure and the Space of Black Women's Creativity." *Workings of the Spirit: The Poetics of Afro-American Women's Writing*. Chicago, IL: University of Chicago Press, 1991. 69–101.

Baldwin, James. *The Devil Finds Work*. 1976. *James Baldwin: Collected Essays*. Ed. Toni Morrison. New York: Library of America, 1998. 477–576.

Baldwin, James. *Notes of a Native Son*. 1955. *James Baldwin: Collected Essays*. Ed. Toni Morrison. New York: Library of America, 1998. 5–136.

Balfour, Lawrie. *The Evidence of Things Not Said: James Baldwin and the Promise of American Democracy*. Ithaca, NY: Cornell University Press, 2001.

Banham, Martin. *The Cambridge Guide to Theatre*. 2nd ed. Cambridge: Cambridge University Press, 1995.

Baraka, Imamu Amiri. "Watergate." *Selected Poetry*. New York: William Morrow and Company, 1979. 236.

Barnett, Claudia. "Adrienne Kennedy and Shakespeare's Sister." *American Drama* 5 (1966): 44–56.

Barney, Chuck. "Interracial TV Couples Old Hat: Relationships No Longer Considered a Tube Taboo." *The San Diego Union-Tribune*. February 15, 2006.

Barrax, Gerald W. "Gunsmoke: 1957." *Another Kind of Rain*. Pittsburgh, PA: University of Pittsburgh Press, 1970.

Bartlett, John, ed. *Bartlett's Familiar Quotations*. 10th ed. Rev. Nathan Haskell Doe. New York: Blue Ribbon, 1903.

Bate, Jonathan, ed. *The Romantics on Shakespeare*. New York: Penguin, 1992.

Baudrillard, Jean. *The Transparency of Evil: Essays on Extreme Phenomena*. Trans. James Benedict. London: Verso, 1993.

Beecher, Henry Ward. *Oratory: A Unique and Masterly Exposition of the Fundamental Principles of True Oratory*. 1892. Philadelphia, PA: Penn Publishing Company, 1908.

Beran, Michael Knox. "Lincoln, *Macbeth*, and the Moral Imagination." *Humanitas* 11.2 (1998): 4–21.

Bergeron, Arthur W., and Richard M. Rollins. *Black Southerners in Gray: Essays on Afro-Americans in Confederate Armies*. Redondo Beach, CA: Facts on File, 1994.

Berlin, Normand. *O'Neill's Shakespeare*. Ann Arbor: University of Michigan, 1993.

Bernardi, Daniel. "Introduction: Race and Hollywood Style." *Classic Hollywood, Classic Whiteness*. Ed. Daniel Bernardi. Minneapolis: University of Minnesota Press, 2001. xiii–xxvi.

Bernstein, Charles. "Objectivist Blues: Scoring Speech in Second-Wave Modernist Poetry and Lyrics." *American Literary History* 20.1–2 (Spring/Summer 2008): 346–68.

Berry, Ralph. "Shakespeare and Integrated Casting." *Contemporary Review* 285 (July 2004): 35–39.

Berthold, Dennis. "Class Acts: The Astor Place Riots and Melville's *The Two Temples.*" *American Literature* 71 (1999): 429–461.

Bigsby, C. W. E. *Modern American Drama, 1945–2000*. Cambridge: Cambridge University Press, 2000.

Blank, Paula. *Broken English: Dialects and the Politics of Language in Renaissance Writings*. London: Routledge, 1996.

Blassingame, John, et al., eds. *The Frederick Douglass Papers*. New Haven, CT: Yale University Press, 1979.

The Bloody Child: The Interior of Violence. Dir. Nina Menkes. Perf. Tinka Menkes. Nina Menkes, 1996.

"The Bloody Shirt." *The Republic: A Monthly Magazine Devoted to the Dissemination of Political Information*. 7.2 (Aug. 1876): 89–91.

Bontemps, Arna. *Drums at Dusk: A Novel*. New York: Macmillan, 1939.

Bontemps, Arna, and Langston Hughes. *Arna Bontemps—Langston Hughes Letters 1925–1967*. Ed. Charles H. Nichols. New York: Dodd, Mead, 1980.

Boose, Lynda. "Techno-Muscularity and the 'Boy Eternal': From the Quagmire to the Gulf." *Cultures of United States Imperialism*. Ed. Amy Kaplan and Donald E. Pease. Durham, NC: Duke University Press, 1993. 581–616.

Bourne, Kay. "*Macbeth* Performance Wins Strand Audience." *Bay State Banner*. January 4, 1990.

Boyd, Frank. *Records of the Dundee Stage from Earliest Times to the Present Day*. Dundee, Scotland: W. & D.C. Thomson, 1886.

Bradbury, David. *Duke Ellington*. London: Haus Publishing, 2005.

Brantley, Ben. "Something Wicked This Way Comes." Rev. of *Macbeth*, dir. Rupert Goold. *New York Times*. February 15, 2008.

Brantley, Ben. "Theater Review; So Steeped in Blood: A Couple on the Edge." *The New York Times*. March 16, 1998.

Briggs, John C. *Lincoln's Speeches Reconsidered*. Baltimore, MD: Johns Hopkins University Press, 2005.

Briggs, John R. Email interview with Alexander C. Y. Huang. January 17, 2009.

Briggs, John R. *Shogun Macbeth*. New York: Samuel French, 1988.

Brode, Douglas. *Shakespeare in the Movies: From the Silent Era to Shakespeare in Love*. Oxford: Oxford University Press, 2000.

Brooks, Gwendolyn. "Young Afrikans." *Blacks*. Chicago, IL: Third World Press, 1987. 494–95.

Brown, Carlyle. *The African Company Presents Richard III*. 1988. New York: Dramatists Play Service, 1994.

Brown, Lois. *Pauline Elizabeth Hopkins: Black Daughter of the Revolution*. Charlotte: University of North Carolina Press, 2008.

Burnett, Mark Thornton. "Madagascan Will: Cinematic Shakespeares / Transnational Exchanges." *Shakespeare Survey* 61 (2008): 239–55.

Burroughs, Norris. *Macbeth*. United Kingdom: Engine Comics, 2006.

Burt, Richard. "Civic ShakesPR: Middlebrow Multiculturalism, White Television, and the Color Bind." *Colorblind Shakespeare: New Perspectives on Race and Performance*. Ed. Ayanna Thompson. New York: Routledge, 2006. 157–85.

Callaghan, Dympna. "Othello was a White Man: Racial Impersonation on the Early Modern Stage." *Shakespeare Without Women*. New York: Routledge, 1999. 75–96.

Callaghan, Dympna, Lorraine Helms, and Jyotsna Singh. *The Weyward Sisters: Shakespeare and Feminist Politics*. Cambridge, MA: Blackwell, 1994.

Callow, Simon. *Orson Welles, Volume 1: The Road to Xanadu*. London: Jonathan Cape, 1995.
Callow, Simon. "Voodoo Macbeth." *Rhapsodies in Black: Art of the Harlem Renaissance*. Ed. Richard J. Powell and David A. Bailey. Los Angeles: University of California Press, 1997. 34–43.
Calta, Louis. "Two to Act Scenes of Shakespeare; Negroes as Stars Exemplify Universality of the Plays." *The New York Times*. September 13, 1963.
Carby, Hazel. *Race Men*. Cambridge, MA: Harvard University Press, 1998.
Carpenter, John. "'The Forms of Things Unknown': Richard Wright and Stephen Henderson's Quiet Appropriation." *Native Shakespeares: Indigenous Appropriations on a Global Stage*. Ed. Craid Dionne and Parmita Kapadia. Farnham, UK: Ashgate, 2008. 57–72.
Carter, Angela, ed. *Wayward Girls & Wicked Women: An Anthology of Subversive Stories*. New York: Virago, 1986.
Chamillionaire. "Ridin." *The Sound of Revenge*. Umvd Labels, 2005.
Cheatham, Wallace McClain. "Black Male Singers at the Metropolitan Opera." *The Black Perspective in Music* 16 (Spring 1988): 3–20.
Chesnutt, Charles W. *The Conjure Woman and Other Conjure Tales*. 1899. Ed. Richard Brodhead. Durham, NC: Duke University Press, 1993.
Child, Lydia Maria. *The Letters of Lydia Maria Child with a Biographical Introduction by John G. Whittier*. Boston, MA: Houghton, Mifflin, and Co., 1882.
Chireau, Yvonne P. *Black Magic: Religion and the African American Conjuring Tradition*. Berkeley: University of California Pres, 2003.
Christy's Nigga Songster. Containing Songs as Are Sung by Christy's, Pierce's, White's Sable Brothers, and Dumbleton's Band of Minstrels. New York and Boston, MA: T. W, Strong, G. W. Cottrell & Co., c. 1850.
Clark, J. C. "The Case of Marigny D'Auterive." January 7, 1828. *Register of Debates in Congress*. Vol. IV. Washington, DC: Gales and Seaton, 1828. 916–18.
Clark, VeVe A. "The Archaeology of Black Theatre." *The Black Scholar* 10.10 (1979): 34–50.
Cliff, Michelle. "Passing." *The Land of Look Behind*. Ithaca, NY: Firebrand Books, 1985. 19–23.
Clifton, Lucille. "sisters: in salem." *Good Woman: Poems and a Memoir, 1969–1980*. Rochester, NY: BOA Editions, 1987. 111.
Clinton, Catherine. *The Other Civil War: American Women in the Nineteenth Century*, revised edition. New York: Macmillan, 1999.
Cocteau, Jean. "Profile of Orson Welles." *Orson Welles: A Critical View*. André Bazin. Venice, CA: Acrobat Books, 1991. 28–32.
Cohen, Selma Jeanne, ed. *Dance as a Theatre Art: Source Readings in Dance History from 1581 to the Present*. New York: Dodd, Mead & Company, 1974.
Coleman, Ralf. Black History Week Speech. Reel-to-reel tape courtesy of Leona Coleman Flu via Lorraine E. Roses. Roxbury, MA, c. 1971.
Collier, John Lincoln. *Duke Ellington*. New York: Oxford University Press, 1987.
Collins, Charles. "Give 'Macbeth' Jungle Set for Colored Cast." *Chicago Daily Tribune*. September 1, 1936.
Coloma, Don Carlos. "Letter." August 10, 1624. *Thomas Middleton and Early Modern Textual Culture: A Companion to the Collective Works*. Ed. Gary Taylor and John Lavagnino. Oxford: Oxford University Press, 2007. 865–67.
Conceison, Claire. "Beyond Yellowface." Questions of Color and the Current State of Casting. Association of Theatre in Higher Education. 22nd Conference.

Difficult Dialogues: Theatre and the Art of Engagement. Denver, CO. July 31, 2008.

Copage, Eric V. "Spot in the Spotlight Again." Review of Classical Theatre of Harlem's *Macbeth*. *New York Times*. July 4, 1999.

Courage Under Fire. Dir. Edward Zwick. Perf. Meg Ryan and Denzel Washington. Twentieth Century Fox, 1996.

Courtney, Krystyna Kujawinska. "Ira Aldridge, Shakespeare, and Color-Conscious Performances in Nineteenth-Century Europe." *Colorblind Shakespeare: New Perspectives on Race and Performance*. Ed. Ayanna Thompson. New York: Routledge, 2006. 103–24.

Cronacher, Karen. "Unmasking the Minstrel Masks's *Black Magic* in Ntzokae Shange's *Spell #7*." *Theatre Journal* 44 (1996): 177–93.

Crouch, Stanley. "Stanley Crouch on *Such Sweet Thunder, Suite Thursday*, and *Anatomy of a Murder*." *The Duke Ellington Reader*. Ed. Mark Tucker. Oxford: Oxford University Press, 1993. 439–45.

Cullen, Countee. *My Soul's High Song: The Collected Writings of Countee Cullen, Voice of the Harlem Renaissance*. Ed. Gerald Early. New York: Anchor, 1991.

Cuney-Hare, Maud. *Negro Musicians and Their Music*. Washington, DC: The Associated Publishers, Inc., 1936.

Cuthbert, David. "Dillard's 'Macbeth' Staged Modern Militia-Style." *The Times-Picayune*. November 13, 2008.

Daileader, Celia. "Casting Black Actors: Beyond Othellophilia." *Shakespeare and Race*. Ed. Catherine Alexander and Stanley Wells. Cambridge: Cambridge University Press, 2000. 177–202.

Daileader, Celia. *Racism, Misogyny, and the Othello Myth: Inter-racial Couples from Shakespeare to Spike Lee*. Cambridge: Cambridge University Press, 2005.

Daileader, Celia. "Writing Rape, Raping Rites: Shakespeare's and Middleton's Lucrece Poems." *Violence, Politics, and Gender in Early Modern England*. Ed. Joseph Patrick Ward. Basingstoke, UK: Palgrave Macmillan, 2008. 67–89.

Dasenbrock, Reed Way. "Imitation Versus Contestation: Walcott's Postcolonial Shakespeare." *Callaloo* 28 (2005): 104–13.

Davis, Clinton Turner and Harry Newman, eds. *Beyond Tradition: Transcripts of the First National Symposium on Non-Traditional Casting*. New York: Non-Traditional Casting Project, Inc., 1988.

De Grazia, Margreta, and Peter Stallybrass. "The Materiality of the Shakespearean Text." *Shakespeare Quarterly* 44.3 (1993): 255–83.

Dean, Phillip Hayes. *Paul Robeson*. 1978. New York: Dramatists Play Service, 1997.

DeRego, Kris. "Don't Miss 'That Scottish Play.'" *Ka Leo*. November 17, 2008.

Desdunes, Rodolphe Lucien. *Our People and Our History: Fifty Creole Portraits*. 1911. Trans. Dorothea Olga McCants. Baton Rouge: Louisiana State University Press, 2001.

Douglass, Frederick. *Life and Times*. 1892. New York: Macmillan, 1962.

Dove, Rita. "Introduction." *"Harlem Gallery" and Other Poems of Melvin B. Tolson*. Ed. Raymond Nelson. Charlottesville: University Press of Virginia, 1999. xi–xxv.

Dove, Rita. "In the Old Neighborhood." *Selected Poems*. New York: Vintage Books, 1993. xxii–xxvi.

Dowden, Edward. *Shakspere*. New York: American Book Company, [1877].

Du Bois, W. E. B. "The Conservation of Races." 1897. In *The Souls of Black Folk*. Ed. Brent Hayes Edwards. Oxford: Oxford University Press, 2007. 179–88.

Du Bois, W. E. B. *The Souls of Black Folk*. 1903. Ed. Brent Hayes Edwards. Oxford: Oxford University Press, 2007.

Duberman, Martin. *Paul Robeson: A Biography*. New York: Knopf, 1988.

Dunham, Katherine. *Dances of Haiti*. 1947. Los Angeles: Center for Afro-American Studies at UCLA, 1983.

Dyer, Richard. "The Matter of Whiteness." *White Privilege: Essential Readings on the Other Side of Racism*. Ed. Paula S. Rothenberg. New York: Worth Publishers, 2005. 9–14.

Dyer, Richard. *Heavenly Bodies: Film Stars and Society*. 2nd ed. London: Routledge, 2004.

Dyer, Richard. *White: Essays on Race and Culture*. New York: Routledge, 1997.

Edwards, Brent Hayes. "The Literary Ellington." *Representations* 77 (Winter 2002): 1–29.

Effinger, Marta J. "Rose McClendon: First Lady of the Stage." *Footsteps* 3.3 (May/June 2001): 6–10.

Eidsheim, Nina Sun. "Voice as a Technology of Selfhood: Towards an Analysis of Racialized Timbre and Vocal Performance ." PhD thesis. University of California, San Diego, 2008.

Eisenstein, Sergei. *Film Form: Essays in Film Theory*. Trans. and Ed. Jay Leyda. New York: Harcourt, Brace, and Co., 1949; 1977.

Elder, Lonne, III. *Splendid Mummer*. Electronic Edition. Alexander Street Press, L. L. C., 2005. [PL000523].

Ellington, Duke. *A Drum is a Woman*. Columbia, CL 951, 1957; reissued, Jazz Tracks, JT933, 2008.

Ellington, Duke. *Music is My Mistress*. New York: Da Capo. 1973.

Ellington, Duke. *Such Sweet Thunder*. Columbia LP CL1033, 1957; reissued Columbia CD CK 65568, 1999.

Engle, Ron, Felicia Hardison Londré, and Daniel J. Watermeier. *Shakespeare Companies and Festivals: An International Guide*. Westport, CT: Greenwood Publishing Group, 1995.

Erickson, Peter. *Citing Shakespeare: The Reinterpretation of Race in Contemporary Literature and Art*. New York: Palgrave, 2007.

Erickson, Peter. "'God for Harry, England, and Saint George': British National Identity and the Emergence of White Self-Fashioning." *Early Modern Visual Culture: Representation, Race, and Empire in Renaissance England*. Ed. Peter Erickson and Clark Hulse. Philadelphia, PA: University of Pennsylvania Press, 2000. 315–345.

Ewbank, Inga-Stina. "Introduction to *The Tragedy of Macbeth: A Genetic Text*." *Thomas Middleton: The Collected Works*. Ed. Gary Taylor and John Lavagnino. Oxford: Oxford University Press, 2007. 1165–69.

"Exeunt Omnes." *Oz*. HBO. February 23, 2003.

Fanon, Frantz. *Black Skin, White Masks*. New York: Grove, 1967.

Feazell, Julius C. "Fit the Character Completely." *The Commercial Appeal*. February 4, 2008.

Fehrenbacher, Don E. *Slavery, Law, and Politcs: The Dred Scott Case in Historical Perspective*. New York: Oxford University Press, 1981.

Fernández-Vara, Clara. "Orson Welles's Intermedial Versions of Shakespeare in Theatre, Radio, and Film." Master's Thesis. Masschusetts Institute of Technology, 2004.

Fields, Julia. "Vigil." *Slow Coins: New Poems (& Some Old Ones)*. Washington, DC: Three Continents Press, 1981. 96–97.

Fisher, Burton D. *Macbeth*. Miami, FL: Opera Journeys, 2000.

Flanagan, Hallie. *Arena: The Story of the Federal Theatre*. New York: Duell, Sloan, and Pierce, 1940.

Fleming, Crystal M., and Lorraine E. Roses. "Black Cultural Capitalists: African-American Elites and the Organization of the Arts in Early Twentieth Century Boston." *Poetics* 35.6 (December 2007): 368–87.

Flocabulary [Blake Harrison and Alex Rappaport]. *Shakespeare is Hip Hop*. CD. Flocabulary, 2007.

Floyd-Wilson, Mary. *English Ethnicity and Race in Early Modern Drama*. Cambridge: Cambridge University Press, 2003.

Fluker, Elayne. "*King Hedley II*: Ruling the Stage." *Essence* 32 (2001): 58–59.

Foden, Giles. "The African Play." Review of Out of Joint's *Macbeth*. *The Guardian*. September 2, 2004.

Forrest, Florence. Dramaturgical Preview of Jacqui Carroll's *Voodoo Macbeth*. April 2007. http://designdramaturge.blogspot.com/2007/04/voodoo-macbeth-damaturgical-preview-1.html. Accessed January 6, 2009.

Fraden, Rena. *Blueprints for a Black Federal Theatre, 1935–1939*. Cambridge: Cambridge University Press, 1994.

France, Richard. *The Theatre of Orson Welles*. Lewisburg, PA: Bucknell University Press, 1977.

France, Richard. "The 'Voodoo' *Macbeth* of Orson Welles." *Yale Theatre* 5.3 (1974): 66–78.

Frederickson, George. *The Black Image in the White Mind: The Debate on Afro-American Character and Destiny, 1817–1914*. New York: Harper and Row, 1971.

Freire, Paulo. *Pedagogy of the Oppressed*. Trans. Myra Bergman Ramos. New York: Continuum, 1998.

French, Lawrence. Transcription of Orson Welles's remarks in *Filming Othello* (1978). http://www.wellesnet.com/filming_othello.htm. Accessed January 6, 2009

"From a Whisper to a Scream." *Grey's Anatomy*. ABC. November 23, 2006.

Furness, Horace Howard, ed. *A New Variorum Edition of Shakespeare*. New York: American Scholar, 1963.

G. I. Jane. Dir. Ridley Scott. Perf. Demi Moore and Viggo Mortensen. Caravan Pictures, 1997.

George, Don. *Sweet Man: The Real Duke Ellington*. New York: Putnam's, 1981.

Gill, James. *Lords of Misrule: Mardi Gras and the Politics of Race*. Jackson: University Press of Mississippi, 1997.

Glasco, Laurence, ed. *The WPA History of the Negro in Pittsburgh*. Pittsburgh, PA: University of Pittsburgh Press, 2004.

Godfrey, Esther. "'To Be Real': Drag, Minstrelsy, and Identity in the New Millennium." *Genders Online Journal* 41 (2005): 1–32.

Granville-Barker, Harley. *More Prefaces to Shakespeare: A Midsummer Night's Dream, A Winter's Tale, Twelfth Night, Macbeth*. Princeton: Princeton University Press, 1974.

Griboff, Debra. "*Shogun Macbeth*." *Encore: The Performing Arts Magazine* http://encoremag.com/?q=article&id=336. Accessed December 20, 2008.

Grimsted, David. *Melodrama Unveiled: American Theatre and Culture, 1800–1850.* Berkeley: University of California Press, 1968; 1987 reprint.

"*Grey's Anatomy* Caption Contest XXXIII." *Grey's Anatomy Insider.* November 28, 2006. http://www.greysanatomyinsider.com/2006/11/greys-anatomy-caption-contest-xxxiii.html. Accessed January 10, 2008.

Gruen, John. "Diva." *Opera News* 40 (January 1976): 8–11.

Guererro, Ed. *Framing Blackness: The African American Image in Film.* Philadelphia, PA: Temple University Press, 1993.

Gussow, Mel. Review of *Shogun Macbeth. New York Times.* November 21, 1986.

Gutman, Les. Review of *August Wilson's King Hedley II. CurtainUp.* May 1, 2001. http://www.curtainup.com/kinghedleyii.html. Accessed October 14, 2008.

Hadju, David. *Lush Life: A Biography of Billy Strayhorn.* New York: North Point Press, 1996.

Hall, Gwendolyn Midlo. "The Formation of Afro-Creole Culture." *Creole New Orleans: Race and Americanization.* Ed. Arnold R. Hirsch and Joseph Logsdon. New Orleans: Louisiana State University Press, 1992. 58–90.

Hall, Kim. *Things of Darkness: Economies of Race and Gender in Early Modern England.* Ithaca, NY: Cornell University Press, 1995.

Halpern, Richard. "Shakespeare in the Tropics: From High Modernism to New Historicism." *Shakespeare among the Moderns.* Ithaca, NY: Cornell University Press, 1997. 15–50.

Hampton, Wilborn. "A Contemporary 'Macbeth' Without Changing a Word." *New York Times.* June 18, 1991. B2.

Hansen, Chadwick. "The Metamorphosis of Tituba, or Why American Intellectuals Can't Tell an Indian Witch from a Negro." *The New England Quarterly* 47 (March 1974): 3–12.

Hapgood, Robert. "Shakespeare in New York and Boston." *Shakespeare Quarterly* 29.2 (Spring 1978): 230–32.

Harvard Theatre Collection. Clips, Shakespeare, *Macbeth.* Notice of Negro Actors Associated production of *Macbeth.* 1956.

Hatch, James V. *Sorrow Is the Only Faithful One: The Life of Owen Dodson.* Urbana: University of Illinois Press, 1995.

Haven, Gilbert. "The State Struck Down." June 11, 1854. *National Sermons: Sermons, Speeches and Letters on Slavery and its War: From the Passage of the Fugitive Slave Bill to the Election of President Grant.* Boston, MA: Lee and Shepard, 1869. 57–86.

Hendricks, Margo, and Patricia Parker, ed. *Women, "Race," and Writing in the Early Modern Period.* London and New York: Routledge, 1994.

Herrington, Joan. "King Hedley II: In the Midst of All This Death." *The Cambridge Companion to August Wilson.* Ed. C. W. E. Bigsby. Cambridge: Cambridge University Press, 2007. 169–82.

Hill, Errol. *Shakespeare in Sable: A History of Black Shakespearean Actors.* Amherst: University of Massachusetts Press, 1984.

Hill, Errol, and James V. Hatch. *A History of African American Theatre.* Cambridge: Cambridge University Press, 2003.

Hill, Leslie Pinckney. *Toussaint L'Ouverture: A Dramatic History.* Boston, MA: Christopher, 1928.

Hine, Darlene Clark. *Black Women in America.* 3 vols. Oxford: Oxford University Press, 2005.

Hinman, Charlton, ed. *The Norton Facsimile: The First Folio of Shakespeare.* New York: Norton, 1968.
hooks, bell. "The Oppositional Gaze." *Black Looks: Race and Representation.* Boston, MA: South End Press, 1992. 115–131.
The Hop of Fashion; or, The Bon-Ton Soiree. A Negro Farce, In Two Scenes. New York: Frederick A. Brady, 1856.
Houseman, John. *Unfinished Business: A Memoir.* London: Chatto & Windus, 1986.
Hsu, Hua. "The End of White America?" *The Atlantic* (January/February 2009): 46–55.
Huang, Alexander C. Y. "Asian Shakespeares in Europe: From the Unfamiliar to the Defamiliarised." *Shakespearean International Yearbook* 8 (2008): 51–70
Hudson, Theodore R. "Duke Ellington's Literary Sources." *American Music* 9.1 (Spring 1991): 20–42.
Hughes, Langston. *Emperor of Haiti.* 1936. *The Collected Works of Langston Hughes, Vol. 5: The Plays to 1942.* Ed. Leslie Catherine Sanders, with Nancy Johnston. Columbia, SC: University of Missouri Press, 2002. 278–332.
Hughes, Langston. "Note on Commercial Theatre." 1940. *The Collected Poems of Langston Hughes.* Ed. Arnold Rampersad and David Roessel. New York: Knopf/Random House, 1994.
Hurwitt, Robert. "Victoria Evans Erville on 'Macbeth'." *San Francisco Chronicle.* October 26, 2008. N–24.
"In Ford's Theater, Obama Invokes Forbidden Word." *CNN Political Ticker.* February 12, 2009. http://politicalticker.blogs.cnn.com/2009/02/12/in-ford's-theater-obama-invokes-forbidden-word. Accessed February 14, 2009.
Iyengar, Sujata. "Colorblind Casting in Single-Sex Shakespeare." *Colorblind Shakespeare: New Perspectives on Race and Performance.* Ed. Ayanna Thompson. New York: Routledge, 2006. 47–68.
Iyengar, Sujata. *Shades of Difference: Mythologies of Skin Color in Early Modern England.* Philadelphia, PA: University of Pennsylvania Press, 2005.
Jafferis, Aaron, and Gihieh Lee. *Shakespeare: The Remix.* Palo Alto, CA: Theatreworks, 2004.
Johnson, James Weldon, ed. *The Book of American Negro Poetry.* Rev. ed. 1931. San Diego, CA: Harcourt Brace, 1969.
Johnson, Jerah. *Congo Square in New Orleans.* New Orleans: Louisiana Landmark Society, 1995.
Johnston, David Claypoole. "A Proslavery Incantation Scene, or Shakespeare Improved." Cartoon. 1858.
Jonson, Ben. *The Masque of Blackness. Ben Jonson's Plays and Masques.* Ed. Richard Harp. New York: W.W. Norton, 2001. 314–24.
Jonson, Ben. *The Masque of Queens. Ben Jonson's Plays and Masques.* Ed. Robert M. Adams. New York: W.W. Norton, 1979. 321–40.
Jordan, June. "May 1, 1970." *Things That I Do in the Dark: Selected Poetry.* New York: Random House, 1977. 93.
Jorgens, Jack, and Jan Levie. "Champlain Shakespeare Festival." *Shakespeare Quarterly* 29.1 (Winter 1978): 228–30.
Jovicevich Tatomirovic, Aleksandra B. "The Theatre of Orson Welles, 1946–1960." PhD dissertation. New York University, 1990.
Juneja, Renu. "Derek Walcott." *Post-Colonial English Drama: Commonwealth Drama Since 1960.* Ed. Bruce King. New York: St. Martin's, 1992. 236–66.

Kalem, T. E. "Arc of Anguish." Review of *Julius Caesar*, dir. Joseph Papp. *Time.* February 5, 1979.
Kaplan, Michael. "New York City Tavern Violence and the Creation of a Working-Class Male Identity." *Journal of the Early Republic* 15 (1995): 591–617.
Katz, Wendy Jean. *Regionalism and Reform: Art and Class Formation in Antebellum Cincinnati.* Columbus: Ohio State University Press, 2002.
Katz, William Loren. *Black Indians: A Hidden Heritage.* New York: Atheneum, 1986.
Kennan, George F. U.S. State Department Policy Planning Study, #23, 1948.
Kennedy, Adrienne. *Deadly Triplets: A Theatre Mystery and Journal.* Minneapolis: University of Minnesota Press, 1990.
Kennedy, Adrienne. *Funnyhouse for a Negro* (1964). *The Adrienne Kennedy Reader.* Minneapolis: University of Minnesota Press, 2001. 11–26.
Kennedy, Adrienne. *The Owl Answers* (1965). *The Adrienne Kennedy Reader.* Minneapolis: University of Minnesota Press, 2001. 29–42.
King, Bruce. *Derek Walcott and West Indian Drama.* Oxford: Clarendon, 1995.
King, Woodie, Jr. "BTA Responds to Papp's Move on Black Theatre." *The Black Theatre Alliance Newsletter* 4.5 (May 5, 1979): 5.
King, Woodie, Jr. *Voices from the Federal Theatre.* Ed. Bonnie Nelson Schwartz and the Educational Film Center. Madison, WI: Terrace Books, 2003.
Kingston, Jeremy. "All-blacks tackle Wilde: *The Importance of Being Earnest* at the Tyne Theatre, Newcastle." *The Times* [London]. April 19, 1989.
Kinney, Arthur. *Lies Like Truth: Shakespeare,* Macbeth, *and the Cultural Moment.* Detroit, MI: Wayne State University Press, 2001.
Kinney, Arthur. "Shakespeare's *Macbeth* and the Question of Nationalism." *Literature and Nationalism.* Ed. Vincent Newey and Ann Thompson. Rowan and Littlefield Publishers, 1991. 56–75.
Kliman, Bernice. *Shakespeare in Performance: Macbeth.* 2nd ed. Manchester, UK: Manchester University Press, 2004.
Koenig, Kip. "Kip Koenig on 'From a Whisper to a Scream.'" *Grey Matter: From the Writers of Grey's Anatomy.* http://www.greyswriters.com/kip_koenig/index.html. Accessed January 14, 2009.
Kolin, Philip. "Shakespeare Knocks: *Macbeth* in Adrienne Kennedy's *Funnyhouse of a Negro.*" *Notes on Contemporary Literature* 34 (2004): 8–10.
Koolish, Lynda. "Maude Cuney-Hare." *The Concise Oxford Companion to African American Literature.* Ed. William Andrews, Francis Smith Foster, and Trudier Harris. New York: Oxford University Press, 2003. 95–96.
Krasner, David. *A Beautiful Pageant: African American Theatre, Drama, and Performance in the Harlem Renaissance, 1910–1927.* New York: Palgrave, 2002.
Kuner, Mildred C. "The New York Shakespeare Festival, 1966." *Shakespeare Quarterly* 17.4 (Autumn, 1966): 419–21.
Lachance, Paul F. "The Foreign French." *Creole New Orleans: Race and Americanization.* Ed. Arnold R. Hirsch and Joseph Logsdon. New Orleans: Louisiana State University Press, 1992. 101–30.
Lacher, Irene. "Taking a Different Road." *Los Angeles Times.* October 1, 2006.
Lambert, J. W. "Plays in Performance." *Drama: The Quarterly Theatre Review* 105 (Summer 1972): 15–28.
Lane, Bill. "Bill Lane from Hollywood." *Sacramento Observer.* January 30, 1969.
Lanier, Douglas. "Minstrelsy, Jazz, Rap: Shakespeare, African American Music, and Cultural Legitimation." *Borrowers and Lenders: The Journal of Shakespeare and*

Appropriation 1.1 (Spring/Summer 2005). http://www.borrowers.uga.edu/cocoon/borrowers/request?id=781409. Accessed November 10, 2008.

Lee, Esther Kim. *A History of Asian American Theatre.* Cambridge: Cambridge University Press, 2006.

Lee, Tonia. *Macbeth in Urban Slang.* New York: Urban Youth Press, 2008.

Leonard, William Torbert. *Masquerade in Black.* Metuchen, NJ: Scarecrow Press, 1986.

Lesher, Stephan. *George Wallace: American Populist.* New York: Da Capo, 1995.

Lessig, Lawrence. *Remix: Making Art and Commerce Thrive in the Hybrid Economy.* New York: Penguin Press, 2008.

Lester, Neal. *Ntozake Shange: A Critical Study of the Plays.* New York: Garland, 1995.

Levine, Bruce. "Conservatism, Nativism, and Slavery: Thomas R. Whitney and the Origins of the Know-Nothing Party." *Journal of American History* 88 (2001): 455–88.

Levine, Lawrence. *The Unpredictable Past: Explorations in American Cultural History.* New York: Oxford University Press, 1993.

Lewis, Emory. "*Macbeth* Splendid." *The Sunday Record* [Bergen County, N.J.]. May 13, 1977.

Lhamon, W. T., Jr. *Jump Jim Crow: Lost Plays, Lyrics, and Street Prose of the First Atlantic Popular Culture.* Cambridge, MA: Harvard University Press, 2003.

Lhamon, W. T., Jr. *Raising Cain: Blackface Performance from Jim Crow to Hip Hop.* Cambridge, MA: Harvard University Press, 1998.

Lindfors, Bernth. "Ira Aldridge's London Debut." *Theatre Notebook* 60 (2006): 30–44.

Lindfors, Bernth. "'Mislike Me Not for My Complexion': Ira Aldridge in Whiteface." *African American Review* 33 (1999): 347–54.

Lindfors, Bernth. "'No end to dramatic novelty': Ira Aldridge at the Royal Coburg Theatre." *Nineteenth Century Theatre & Film* 34 (2007): 15–34.

Linville, Susan. *History Films, Women, and Freud's Uncanny.* Austin: University of Texas Press, 2004.

Logsdon, Joseph, and Caryn Cossé Bell. "The Americanization of Black New Orleans, 1850–1900." *Creole New Orleans: Race and Americanization.* Ed. Arnold R. Hirsch and Joseph Logsdon. New Orleans: Louisiana State University Press, 1992. 201–61.

Loomba, Ania. *Gender, Race, Renaissance Drama.* Manchester and New York: Manchester University Press, 1989.

Loomba, Ania. "Introduction to *The Triumphs of Honour and Virtue.*" *Thomas Middleton: The Collected Works.* Ed. Gary Taylor and John Lavagnino. Oxford: Oxford University Press, 2007. 1714–18.

Loomba, Ania, and Jonathan Burton. *Race in Early Modern England: A Documentary Companion.* New York: Palgrave, 2007.

Lott, Eric. *Love and Theft: Blackface Minstrelsy and the American Working Class.* New York: Oxford University Press, 1993.

Ludlow, Noah. *Dramatic Life as I Found It.* St. Louis, MO: G.I. Jones, 1880.

"Macbeth." Review of Lafayette Theatre FTP production of *Macbeth*. *Variety*. April 22, 1936.

Macbeth. Dir. Roman Polanski. 1971. Colombia Pictures, 1971. DVD. Sony, 2002.

Macbeth: Act Five. Prison Performing Arts. Vandalia's Women's Eastern Reception, Diagnostic and Correctional Center. Vandalia, MO. January 24, 2006.

Macbeth Act V: Revenge of the Ghetto. Dir. Matt Dacek. YouTube.com. http://www.youtube.com/watch?v=oIVBdiwhAGw. Accessed December 1, 2008.

Macbeth in Manhattan. Dir. Greg Lombardo. DVD. First Rites, 1999.

MacDonald, Joyce Green. "Acting Black: *Othello, Othello* Burlesques, and the Performance of Blackness." *Theatre Journal* 46 (1994): 231–49.

MacDonald, Joyce Green. "Bodies, Race, and Performance in Derek Walcott's *A Branch of the Blue Nile.*" *Theatre Journal* 57 (2005): 191–203.

Maley, Willy. Review of Virginia Mason Vaughan's *Performing Blackness on English Stages, 1500–1800. Theatre Research International* 31.1 (Winter 2006): 102–03.

Mann, Horace. *Slavery: Letters and Speeches.* Boston, MA: Mussey and Co., 1853.

Marienstras, Richard. "Orson Welles: Shakespeare, Welles, and Moles." 1974. *Interviews with Orson Welles.* Ed. Mark W. Estrin. Jackson, MS: University of Mississippi Press, 2002. 146–172.

Marshall, Herbert. "Ira Aldridge as Macbeth and King Lear." *Ira Aldridge: The African Roscius.* Ed. Bernth Lindfors. Rochester: University of Rochester Press, 2007. 194–95.

The Matrix. Dir. Andy and Larry Wachowski. Perf. Keanu Reeves and Laurence Fishburne. Joel Silver, 1999.

Mazierska, Ewa. *Roman Polanski: The Cinema of a Cultural Traveller.* London: I. B. Tauris, 2007.

McAllister, Marvin. *White People Do Not Know How to Behave at Entertainments Designed for Ladies and Gentlemen of Color: William Brown's African and American Theater.* Chapel Hill: University of North Carolina Press, 2003.

McCarter, Jeremy. "Will Act for Food." *Newsweek* (January 19, 2009): 48–52.

McCarthy, Mary. "Wartime Omnibus." 1944. *Mary McCarthy's Theatre Chronicles, 1937–1962.* New York: Farrar, Straus, 1963. 65–72.

McDermott, William F. "Colored Actors Take *Macbeth* to Tropics." *Cleveland Plain Dealer.* September 30, 1936.

McDonald, Russ. *Shakespeare's Late Style.* Cambridge: Cambridge University Press, 2006.

McKay, Claude. *Complete Poems.* Ed. William Maxwell. Urbana: University of Illinois Press, 2004.

McLaren, Peter. *Revolutionary Multiculturalism: Pedagogies of Dissent for the New Millennium.* Boulder, CO: Westview Press, 1997.

Melville, Herman. "The Portent." 1859. *Heath Anthology of American Literature.* Ed. Paul Lauter, et al. Vol. 1. Boston, MA: Houghton, 1990. 2582.

Memoir and Theatrical Career of Ira Aldridge, the African Roscius. London: Onwhyn, 1848.

Mercer, Kobena. *Welcome to the Jungle: New Positions in Black Cultural Studies.* New York: Routledge, 1994.

Middleton, Thomas. *Thomas Middleton: The Collected Works.* Ed. Gary Taylor and John Lavagnino. Oxford: Oxford University Press, 2007.

Miller, May. "Come Morning." *Collected Poems.* Berkeley, CA: Creative Arts Books Company, 1989.

Mills, Gary B. *The Forgotten People: Cane River's Creoles of Color.* Baton Rouge: Louisiana State University Press, 1977.

Mire, Carissa D. "East Meets…Scotland? Woodlands Thespians Bringing Unique 'Macbeth' to Edinburgh." *Houston Chronicle.* July 31, 2003.

Mitri, Paul T. "Director's Notes." *Macbeth* Production Program, University of Hawai'i at Mānoa. 2008.

Mitri, Paul T. Grant Proposal to the "Special Fund for Innovative Scholarship and Creative Work," University of Hawai'i at Mānoa.
Mitri, Paul T. "Making Shakespeare Bi-Lingual." Paper delivered at International Association of Performing Language Conference, November 11, 2007.
Mitri, Paul T. Personal communication with William C. Carroll. January 12, 2009.
Mitri, Paul T. Working Script for *Macbeth*, University of Hawai'i at Mānoa. Provided by dramaturg Marie Charlson. 2008.
Monson, Lex. Production notes for New Federal Theatre's *Macbeth*. Box 1 / Folder 9, *Macbeth*, 1977. Lex Monson Collection, Schomburg Center for Research in Black Culture, The New York Public Library.
Moore, Sally. "It was a Gala for La Scala from Milan, but the Star was America's Shirley Verrett." *People Magazine* (September 27, 1976): 68–69.
Moschovakis, Nick. "Introduction: dualistic/problematic *Macbeth?*" Macbeth: *New Critical Essays*. Ed. Nick Moschovakis. New York: Routledge, 2008. 1–72.
Murray, David. "Black Arts: Conjure and Spirit." *Matter, Magic, and Spirit: Representing Indian and African American Belief.* Philadelphia, PA: University of Pennsylvania Press, 2007. 102–26.
Muto, Sheila. "Shakespeare Turns Japanese." *Asianweek* 13.32 (April 3, 1992): 19.
Myrdal, Gunnar. *An American Dilemma: The Negro Problem and Modern Democracy.* New York: Harper and Brothers, 1944.
Newstok, Scott L. "'Step aside, I'll show thee a president': George W as Henry V?" May 1, 2003. http://www.poppolitics.com/archives/2003/05/George-W-as-Henry-V. Accessed February 14, 2009.
Ng, David. "An Otello Who's Solidly Moored." *The Los Angeles Times.* February 10, 2008.
Nietzsche, Friedrich. *Daybreak: Thoughts on the Prejudices of Morality.* 1881. Ed. Maudemarie Clark and Brian Leiter. Tr. R.J. Hollingdale. Cambridge: Cambridge University Press, 1997.
Norrell, Robert J. *The House I Live In: Race in the American Century.* Oxford: Oxford University Press, 2005.
Northall, W. K. *Macbeth Travestie. In Two Acts.* New York: William Taylor & Co., 1847.
Nouryeh, Andrea J. "When the Lord Was a Black Man: A Fresh Look at the Life of Richard Berry Harrison." *Black American Literature Forum* 16.4 (Winter 1982): 142–46.
O'Connor, Marion. "Introduction to *The Witch*." *Thomas Middleton: The Collected Works.* Ed. Gary Taylor and John Lavagnino. Oxford: Oxford University Press, 2007. 1124–28.
Ogbar, Jeffery O. G. *Hip Hop Revolution: the Culture and Politics of Rap.* Lawrence: University Press of Kansas, 2007.
Oldenburg, Ann. "Love is No Longer Color-Coded on TV." *USA Today.* December 20, 2005.
Orman, Roscoe. Email correspondence with Scott L. Newstok. May 14, 2008.
Osborne, Charles. *The Complete Operas of Verdi.* New York, Alfred A. Knopf, 1970.
Osborne, Richard. Jacket notes for Verdi, *Macbeth.* Deutsche Grammophon 449733-2, 449734-2, 1976 CD.
Osumare, Halifu. "Global Breakdancing and the Intercultural Body." *Dance Research Journal* 34:2 (Winter 2002): 30–45.
Oxford Frederick Douglass Reader. Ed. William L. Andrews. New York: Oxford University Press, 1996.

Pan Asian Repertory Theatre. Company Web site. http://www.PanAsianRep.org. Accessed January 10, 2009.

Parker, Patricia. "Black *Hamlet*: Battening on the Moor." *Shakespeare Studies* 31 (2003): 127–64.

Parks, Suzan-Lori. "Possession." *The America Play and Other Works*. New York: Theatre Communications Group, 1995. 3–5.

Parks, Suzan-Lori. "Project Macbeth." *365 Days/365 Plays*. New York: Theatre Communications Group, 2006. 133–35.

People's Institute for Survival and Beyond. "Undoing Racism" workshop. New Orleans, 2003.

Peterson, Bernard L. *The African American Theatre Directory, 1816–1960: A Comprehensive Guide to Early Black Theatre Organizations, Companies, Theatres, and Performing Groups*. Westport, CT: Greenwood Press, 1997.

Peterson, Bernard L. *A Century of Musicals in Black and White: An Encyclopedia of Musical Stage Works By, About, or Involving African Americans*. Westport, CT: Greenwood Press, 1985.

Petrolle, Jean. "Allegory, Politics, and the Avant-Garde." *Women and Experimental Filmmaking*. Ed. Jean Petrolle and Virginia Wright Wexman. Urbana: University of Illinois Press, 2005. 93–104.

Pierce, Leonard. "Video of the Day: Orson Welles's Other *Macbeth*." *Screengrab: Nerve's Film Blog*. http://www.nerve.com/CS/blogs/screengrab/archive/2007/10/17/video-of-the-day-orson-welles-other-macbeth.aspx. Accessed January 6, 2009.

Polkow, Dennis. Review of *Macbeth*, Chicago Shakespeare Theatre. *Newcity*. January 13, 2009.

Pollock, Max. "Director's Report." *Macbeth* 1937 Los Angeles Production Bulletin. March 7, 1938. 8–9.

Porter, Cole. "You Do Something to Me." 1929. *The Complete Lyrics*. Ed. Roger Kimball. Cambridge: Da Capo, 1992. 117.

Pottlitzer, Joanne. *Hispanic Theatre in the United States and Puerto Rico*. New York: Ford Foundation, 1988.

Preisser, Alfred. Email correspondence with Scott L. Newstok. March 23, 2008.

Privett, Ray. "Secret Landscapes: A Conversation with Nina Menkes." *Senses of Cinema* 22 (2002). http://archive.sensesofcinema.com/contents/02/22/menkes.html. Accessed November 25, 2008.

Prodhomme, J. G. "Verdi's Letters to Léon Escudier." *Music & Letters* 4:1 (January 1923): 62–70.

Production Photographs for the New York Performance of *Macbeth*. Directed by Orson Welles, New Lafayette Theater, April 14–June 20, 1936—Finding Aid Box 1095. 1936. *The New Deal Stage: Federal Theatre Project, 1935–1939*. Library of Congress.

Pronko, Leonard C. "Approaching Shakespeare through Kabuki." *Shakespeare East and West*. Ed. Minoru Fujita and Leonard Pronko. New York: St. Martin's Press, 1996.

Quarshie, Hugh. "Conventional Folly: A Discussion of English Classical Theatre." *Black British Culture and Society*. Ed. Kwesi Owusu. London: Routledge, 2000. 289–94.

Rankine, Claudia. *The End of the Alphabet: Poems*. New York: Grove Press, 1998.

Rayam, Curtis. Email interview with Wallace McClain Cheatham. January 20, 2008.

Rayner, Alice. "To Do: 'I'll Do, and I'll Do, and I'll Do.'" *To Act, To Do, To Perform: Drama and the Phenomenology of Action.* Ann Arbor: University of Michigan Press, 1994. 59–82.

Rebhorn, Matthew. "Edwin Forrest's Redding Up: Elocution, Theater, and the Performance of the Frontier." *Comparative Drama* 40 (2006/2007): 455–481.

Redlich, Norman. "'Out Damned Spot; Out, I Say': The Persistence of Race in American Law." *Vermont Law Review* 25.2 (Winter 2001): 475–522.

Reed, Ishmael. "Neo-HooDoo Manifesto." *Conjure.* Amherst: University of Massachusetts Press, 1972. 20–25.

Remnick, David. "The Joshua Generation." *The New Yorker.* November 17, 2008. 68–83.

Reynolds, Bryan. "Untimely Ripped: Mediating Witchcraft in Polanski and Shakespeare." *The Reel Shakespeare: Alternative Cinema and Theory.* Ed. Lisa S. Starks and Courtney Lehmann. Madison, NJ: Fairleigh Dickinson University Press, 2002. 143–164.

Rich, Frank. "'Macbeth,' Its Lessons Ever Apt And Ever New, With Raul Julia." *The New York Times.* January 17, 1990.

Rippy, Marguerite. *Orson Welles and the Unfinished RKO Projects: A Postmodern Perspective.* Carbondale: Southern Illinois University Press, 2009.

Roberts, Jeanne Addison. "Shakespeare in Washington, DC." *Shakespeare Quarterly* 31.2 (Summer 1980): 206–11.

Roche, Theresa. "Latino Macbeth." *BBC Home.* May 22, 2007. http://www.bbc.co.uk/bristol/content/articles/2007/05/22/rev_kelvmacbeth_feature.shtml. Accessed November 10, 2008.

Roppolo, Joseph P. "Harriet Beecher Stowe and New Orleans: A Study in Hate." *The New England Quarterly* 30.3 (September 1957): 346–62.

Rosenfield, John Jr. "The Passing Show: Making Macbeth Haitian Emperor Not Without Precedent." *Dallas Morning News.* August 5, 1936.

Rothwell, Kenneth. *A History of Shakespeare on Screen: A Century of Film and Television.* New York: Cambridge University Press, 1999.

Rourke, Constance. *American Humor: A Study of the National Character.* 1931. New York: New York Review Books, 2004.

Royster, Francesca T. "White Limed Walls: Whiteness and Gothic Extremism in Shakespeare's *Titus Andronicus.*" *Shakespeare Quarterly* 51.4 (2000): 432–55.

Rux, Carl Hancock. "A Rage in Harlem: Is the Classical Theatre of Harlem a Black Theatre Company? Does It Matter?" *American Theatre* 21.6 (July/August 2004): 26, 88.

Samuels, Allison. "Diary of an Angry Black Man: Isaiah Washington Can't Stop Talking about Being Fired from 'Grey's Anatomy.' Is That a Good Thing for Him to Do?" *Newsweek.* June 28, 2007.

Saxon, Lyle, Edward Dreyer, and Robert Tallant [Louisiana Writers' Project]. *Gumbo Ya-Ya: A Collection of Louisiana Folk Tales.* New York: Bonanza Books, 1945.

Schechner, Richard. *Between Theatre and Anthropology.* Philadelphia, PA: University of Pennsylvania Press, 1985.

Schechter, Patricia A. *Ida B. Wells-Barnett and American Reform, 1880–1930.* Chapel Hill: University of North Carolina Press, 2001.

Scheil, Katherine West. Review of Out of Joint's *Macbeth,* Guthrie Theatre, Minneapolis, MN. *Shakespeare Bulletin* 24.1 (2006): 115–18.

Schonberg, Harold C. "Opera: A Fine 'Macbeth.'" *New York Times.* January 27, 1973.

Scott-Douglass, Amy. *Shakespeare Inside: The Bard Behind Bars*. London: Continuum, 2007.
Scott-Heron, Gil. "H$_2$0 Gate (Watergate) Blues." *Now and Then... The Poems of Gil Scott-Heron*. Edinburgh, Scotland: Payback Press, 2000. 82–85.
"Second Graders Take on *Macbeth*, Halloween." National Public Radio. October 30, 2008. http://www.npr.org/templates/story/story.php?storyId=96344515. Accessed January 27, 2008.
Sedgwick, John. "Inside the Pioneer Fund." *The Bell Curve Debate: History, Documents, Opinions*. Ed. Russell Jacoby and Naomi Glauberman. New York: Times Books, 1995. 144–61.
Seligsohn, Leo. "*Macbeth*, Black or Bloody." *Newsday*. May 11, 1977.
"Shakespeare in American Life: The Astor Place Riot." *Folger Shakespeare Library Online*. http://www.shakespeareinamericanlife.org/stage/onstage/ yesterday/ astor.cfm. Accessed May 10, 2008.
Shange, Ntozake. *Spell #7. Three Pieces*. New York: Penguin, 1982. 1–52.
Shaw, Len. G. "Negro Players Give *Macbeth*." *The Detroit Free Press*. September 15, 1936.
Singer, Paula. Interview with Ralf Coleman. Hatch-Billops Collection, George Mason University. November 24, 1972.
Smalls, Gertrude. "In Answer to Africa's Need for Teachers." *The New York Times Magazine*. March 18, 1962.
Smith, Eric Ledell. *Blacks in Opera: An Encyclopedia of People and Companys*. Jefferson, NC: MacFarland and Company, 1995.
Smurthwaite, Nick. "*Macbeth* in Pidgin." *Asia Times Online* February 2, 1999. http://www.atimes.com/oceania/AB02Ah01.html. Accessed December 4, 2009.
Sollors, Werner. *Amiri Baraka / LeRoi Jones: The Quest for a Populist Modernism*. New York: Columbia University Press, 1978.
Splawn, P. Jane. "'Change the Joker[s] and Slip the Yoke': Boal's Joker System in Ntozake Shange's *for colored girls who considered the rainbow when suicide is enuff* and *spell #7*." *Modern Drama* 41 (1998): 386–98.
Stanfield, Peter. "An Excursion into the Lower Depths: Hollywood, Urban Primitivism, and *St. Louis Blues*, 1929–1937." *Cinema Journal* 41.2 (2002): 84–108.
"'Stevedore' Hits a Snag in Boston." *The New York Times*. September 22, 1935.
Stevenson, Michael. "*The Pianist* and its Contexts: Polanski's Narration of Holocaust Evasion and Survival." *The Cinema of Roman Polanski: Dark Spaces of the World*. Ed. John Orr and Elzbieta Ostrowska. London: Wallflower Press, 2006. 133–145.
Stodard, Nicole. Review of Perseverance Theatre's production of *Macbeth*. *Shakespeare Bulletin* 25.3 (2007): 96–100.
Sturgess, Kim C. *Shakespeare and the American Nation*. Cambridge: Cambridge University Press, 2004.
Swander, Homer. "No Exit for the Dead Body: What to Do with a Scripted Corpse?" *Journal of Dramatic Theory and Criticism* 5.2 (1991): 139–52.
Talty, Stephan. *Mulatto America: At the Crossroads of Black and White Culture: A Social History*. New York: Harper Collins, 2003.
Taylor, Gary. *Buying Whiteness: Race, Culture and Identity from Columbus to Hip Hop*. New York: Palgrave Macmillan, 2005.

Taylor, Gary. "*Macbeth* (adaptation)." *Thomas Middleton and Early Modern Textual Culture: A Companion to the Collective Works*. Ed. Gary Taylor and John Lavagnino. Oxford: Oxford University Press, 2007. 383–98.
Teague, Frances. *Shakespeare and the American Popular Stage*. Cambridge: Cambridge University Press, 2006.
Teish, Luisah. *Carnival of the Spirit: Seasonal Celebrations and Rites of Passage*. New York: Harper & Row, 1994.
Thandeka. *Learning to be White: Money, Race, and God in America*. New York: Continuum Books, 2000.
Thompson, Ayanna. "Practicing a Theory/Theorizing a Practice: An Introduction to Shakespearean Colorblind Casting." *Colorblind Shakespeare: New Perspectives on Race and Performance*. Ed. Ayanna Thompson. New York: Routledge, 2006. 1–24.
Thompson, George A. *A Documentary History of the African Theatre*. Evanston, IL: Northwestern University Press, 1998.
Thompson, Krista A. "Preoccupied with Haiti: The Dream of Diaspora in African American Art, 1915–1942." *American Art* 21.3 (Fall 2007): 74–97.
Thompson, Mark Christian. "Voodoo Fascism: Fascist Ideology in Arna Bontemps's *Drums at Dusk*." *MELUS* 30.3 (2005): 155–77.
Tolson, Melvin B. *"Harlem Gallery" and Other Poems of Melvin B. Tolson*. Ed. Raymond Nelson. Charlottesville: University Press of Virginia, 1999.
Tommasini, Anthony. "Critic's Notebook: The Voice Is What Counts, Not the Color of the Skin." *New York Times*. February 18, 1998.
Torgovnick, Marianna. *Gone Primitive: Savage Intellects, Modern Lives*. Chicago, IL: Chicago University Press, 1990.
Townsend, Irving. "When Duke Records." *The Duke Ellington Reader*. Ed. Mark Tucker. Oxford: Oxford University Press, 1993. 319–23.
Tregle, Joseph G., Jr. "Creoles and Americans." *Creole New Orleans: Race and Americanization*. Ed. Arnold R. Hirsch and Joseph Logsdon. New Orleans: Louisiana State University Press, 1992. 131–88.
Tucker, Mark. "List of Favorites." *The Duke Ellington Reader*. Ed. Mark Tucker. Oxford: Oxford University Press, 1993. 268.
Tucker, Robert. "New 'Macbeth' Presented at Keith's." *Indianapolis Star*. August 26, 1936.
Tucker, Veta Smith. "Purloined Identity: The Racial Metamorphosis of Tituba of Salem Village." *Journal of Black Studies* (March 2000): 624–34.
Turner, Joyce Moore. *Caribbean Crusaders and the Harlem Renaissance*. Urbana: University of Illinois Press, 2005.
Van Deburg, William, ed. *Modern Black Nationalism: From Marcus Garvey to Louis Farrakhan*. New York: New York University Press, 1997.
Van Vechten, Carl, and Langston Hughes. *Remember Me to Harlem: The Letters of Langston Hughes and Carl Van Vechten, 1925–1964*. Ed. Emily Bernard. New York: Knopf, 2001.
Vaughan, Virginia Mason. *Performing Blackness on English Stages, 1500–1800*. Cambridge: Cambridge University Press, 2005.
Vekhter, N. S. "An Old Actor's Memories of Ira Aldridge." *Birzhevye vedomosty* (Stock Gazette). No. 227. May 9, 1903.
Verdi, Giuseppe. *Macbeth*. 1847. Libretto by Francesco Maria Piave and Andrea Maffei.

Verrett, Shirley. Email interview with Christopher Brooks on behalf of Wallace McClain Cheatham. April 21, 2008. Virginia Commonwealth University, Richmond, Virginia.

Verrett, Shirley, and Christopher Brooks. *I Never Walked Alone*. Hoboken, NJ: John Wiley and Sons, 2003.

Walcott, Derek. *A Branch of the Blue Nile*. 1983. *Three Plays*. New York: Farrar, 1986.

Walcott, Derek. *Midsummer*. New York: Farrar, 1984.

Warren, Ebenezer W. *Nellie Norton: or, Southern Slavery and the Bible. A Scriptural Refutation of the Principal Arguments upon which the Abolitionists Rely. A Vindication of Southern Slavery from the Old and New Testaments*. Macon, GA: Burke, Boykin, & Co, 1864.

We Work Again. WPA documentary. 1937. *Treasures From American Film Archives Encore Edition*. 2000.

Weiler, A. H. "Modern Macbeth." *The New York Times*. October 20, 1968.

Welles, Orson. "I've Come to Brazil to Learn." Trans. Tom Pettey. *A Noite*. February 9, 1942. Welles Mss. Collection. Lilly Library. Bloomington, IN.

Welles, Orson. *Macbeth* draft script. Welles Mss. Collection. Lilly Library. Bloomington, IN. 1936.

Welles, Orson, and Peter Bogdanovich. *This is Orson Welles*. New York: HaperCollins, 1992.

White, Shane. *Stories of Freedom in Black New York*. Cambridge, MA: Harvard University Press, 2002.

Whittier, John Greenleaf. "Daniel Neall." 1846. *Whittier's Complete Poetical Works*. Boston, MA: Houghton Mifflin, 1894.

Wick, Nancy. "Shakespeare in Reconstruction: Classic Gets Civil War Setting." *University Week*. February 3, 2005.

Wilberforce, Robert, and Samuel Wilberforce. *The Life of William Wilberforce, By His Sons*. 2nd ed. Vol. 2. London: John Murray, 1839.

Wilders, John. "Introduction." *Macbeth: Shakespeare in Production*. Cambridge: Cambridge University Press, 2004. 1–75.

Wilds, Lillian. "Shakespeare in Southern California." *Shakespeare Quarterly* 33.2 (Autumn 1982): 380–93.

Wilentz, Sean. "On Class and Politics in Jacksonian America." *Reviews in American History* 10.4 (1982): 45–63.

Williams, George W. *History of the Negro Race in America from 1619 to 1880: Negroes as Slaves, as Soldiers, and as Citizens*. New York: G.P. Putnam's Sons, 1882.

Williams, Linda. "When the Woman Looks." *Re-Vision: Essays in Feminist Film Criticism*. Ed. Mary Ann Doane, Patricia Mellencamp, and Linda Williams. Frederick, MD: University Publications of America, 1984. 83–99.

Wilson, August. "The Art of Theater XIV." *Paris Review* 41 (1994): 66–94.

Wilson, August. "Foreword." *Seven Black Plays*. Ed. Chuck Smith. Evanston, IL: Northwestern University Press, 2004. ix–x.

Wilson, August. *The Ground on Which I Stand*. New York: Theatre Communications Work, 1996.

Wilson, August. *King Hedley II*. 1999. New York: Theatre Communications Group, 2005.

Winn, Steven. "From 'Cagney' to Berkeley Rep's 'Macbeth': Eureka Theatre Alums Carl Lumbly and Tony Taccone Team up for Production." *San Francisco Chronicle* February 16, 1997.

Wood, Marcus, ed. *The Poetry of Slavery: An Anglo-American Anthology, 1764–1865*. Oxford: Oxford University Press, 2004.

Wright, Laurence. "*Umabatha*: Global *and* Local." *English Studies in Africa* 47.2 (2004): 97–114.

Yankovic, "Weird Al." "White and Nerdy." *Straight Outta Lynwood*. Volcano, 2006.

Young, Al. "Identities." *Heaven: Collected Poems, 1956–1990*. Berkeley, CA: Creative Arts Books Company, 1992. 100–01.

Žižek, Slavoj. *Welcome to the Desert of the Real!: Five Essays on September 11 and Related Dates*. London: Verso, 2002.

Notes on the Contributors and Editors

Elizabeth Alexander is Professor and Chair of African American Studies at Yale University. She is the author of four books of poems: *The Venus Hottentot*, *Body of Life*, *Antebellum Dream Book*, and *American Sublime*. Her collection of essays on African-American literature, painting, and popular culture, *The Black Interior*, was published in 2004 (Graywolf). The recipient of many awards and honors, Professor Alexander was the inaugural poet for President Barack Obama.

Todd Landon Barnes is a doctoral candidate in the Department of Rhetoric at the University of California, Berkeley, where he is working on "Immanent Shakespeares: Politics, Pedagogy, and Performance." His work explores ethical and political philosophies of immanence within Shakespeare studies, critical pedagogy, digital media production, and performance studies. His essay on post-9/11 performances of *Macbeth* and *Macbush* appeared in *Shakespeare Bulletin*.

John C. Briggs is Professor of English and Director of the University Writing Program at the University of California, Riverside. He is the author of *Francis Bacon and the Rhetoric of Nature* (Harvard, 1989), and has published articles and book chapters on Shakespearean catharsis; *Timon of Athens*; *Othello*; and Bacon and religion. His *Lincoln's Speeches Reconsidered* (Johns Hopkins, 2005) is a close reading of the pre-presidential and presidential speeches.

Richard Burt is Professor of English and Film and Media Studies at the University of Florida, where he specializes in Shakespeare and film, literary theory, and censorship. He is the author three books: *Licensed by Authority: Ben Jonson and the Discourses of Censorship* (Cornell, 1993), *Unspeakable ShaXXXspeares* (St. Martin's, 1998), and *Medieval and Early Modern Film and Media* (Palgrave, 2008), as well as editor or co-editor of seven other book collections.

Brent Butgereit is an undergraduate student at Rhodes College. He served as dramaturge for the 2008 production of *Macbeth* at Hattiloo Theatre in Memphis. He has delivered papers on *Macbeth* at the Associated Colleges of the South's British Studies Symposium and the Rhodes College URCAS: Undergraduate Research and Creative Activity Symposium.

William C. Carroll is Chair of the Department of English at Boston University. His publications include the critical books *The Great Feast of Language in Love's Labour's Lost* (Princeton, 1976), *The Metamorphoses of Shakespearean Comedy* (Princeton, 1985), and *Fat King, Lean Beggar: Representations of Poverty in the Age of Shakespeare* (Cornell, 1996). He has published scholarly editions of *Women Beware Women*, *Macbeth*, and *The Two Gentlemen of Verona*.

Wallace McClain Cheatham is an award-winning composer, organist, scholar, and educator. He is credited with introducing the works of major African-American composers to Midwestern audiences, and has penned a number of scholarly works on race, classical music, and opera. He interviewed many prominent singers for his book, *Dialogues on Opera and the African American Experience* (Rowman & Littlefield, 1997).

Celia R. Daileader is Professor of English at Florida State University, and specializes in Renaissance literature, feminist theory, and critical race studies. She is the author of *Racism, Misogyny, and the* Othello *Myth* (Cambridge, 2005) and *Eroticism on the Renaissance Stage* (Cambridge, 1998), as well as co-editor of *Women & Others: Perspectives on Race, Gender, and Empire* (Palgrave, 2007) and the John Fletcher play *The Tamer Tamed* (Manchester, 2007).

Peter Erickson is Visiting Professor of Humanities (Theatre and Art History) at Williams College, and is working with other scholars to establish the study of race in the field of Renaissance culture. His wider interests are both cross-disciplinary and cross-historical—with strong investments in contemporary culture, as well as the Renaissance. He has published six books, including, most recently, *Citing Shakespeare* (Palgrave, 2007).

José A. Esquea is the artistic director of Teatro LA TEA, where he has directed classical theatre with multiracial casts, including *Hamlet*, *Othello*, and *Macbeth*. Esquea has also directed Shakespeare pieces for schoolchildren in the South Bronx for the Knowledge Project. A graduate of Skidmore College, he began his directing career in 2000 with the short film "A Rose by Any Other Year," which was screened at the 2001 Cannes Film Festival as part of the pavilion for student worldwide film.

Charita Gainey-O'Toole is a doctoral candidate in the Department of African and African American Studies at Harvard University, where she is a Ford Foundation Diversity Fellow. Before entering graduate school, Gainey-O'Toole served as a middle and high school teacher under the New York City Teaching Fellowship. Her interests include poetry, intertextuality, and memory, concentrating on Phillis Wheatley in the African-American literary imagination.

Contributors and Editors

Alexander C. Y. Huang is Assistant Professor of Comparative Literature at Pennsylvania State University. He is the author of *Chinese Shakespeares: Two Centuries of Cultural Exchange* (Columbia, 2009), co-editor (with Charles Ross) of *Shakespeare in Hollywood, Asia, and Cyberspace* (Purdue, 2009), and editor of a film review cluster for *Borrowers and Lenders*. He is also the video curator of an exhibition at the Folger Shakespeare Library (curated by Timothy Billings).

Philip C. Kolin is Professor of English at the University of Southern Mississippi, and has published more than 40 books and 200 scholarly articles. His work on African-American playwrights includes his study *Understanding Adrienne Kennedy* (South Carolina, 2005) and his edited collection *Contemporary African American Women Playwrights* (Routledge, 2007). Kolin is the General Editor for the Routledge Shakespeare Criticism Series.

Douglas Lanier is Professor of English at the University of New Hampshire, and specializes in literary theory, film, and cultural studies. In addition to his influential book *Shakespeare and Modern Popular Culture* (Oxford, 2002), Lanier has published articles on Shakespeare, jazz, Jonson, Milton, the Jacobean masque, and literature pedagogy. He is a recipient of a teaching excellence award from the University of New Hampshire.

Courtney Lehmann is Professor of English and Director of the Pacific Humanities Center at the University of the Pacific. She is the author of *Shakespeare Remains: Theater to Film, Early Modern to Postmodern* (Cornell, 2002), a forthcoming study of female directors of Shakespeare, and, with Lisa S. Starks, editor of two volumes of Shakespeare and film criticism: *The Reel Shakespeare* and *Spectacular Shakespeare* (Fairleigh Dickinson, 2002).

Harry J. Lennix is a stage and screen actor; recent prominent roles include Commander Lock in *The Matrix* series, as well as television roles in *Commander in Chief* and *24*. Among his many Shakespearean credits are Aaron in Julie Taymor's *Titus* (stage and film) and the lead role in a 2007 African-American production of *Macbeth* in Los Angeles. He was part of the first American company to be invited to the Royal Shakespeare Company in 2001.

Bernth Lindfors is Professor Emeritus of English and African Literatures at the University of Texas, Austin. He is the author and editor of dozens of books, and has been studying modern African literature since the field's infancy. Lindfors began the journal *Research in African Literatures*, whose editorship he continued until 1990. He is currently working on Ira Aldridge, with a recent study completed (*The African Roscius*, Rochester, 2007) and another in progress.

Joyce Green MacDonald is Associate Professor of English at the University of Kentucky, and specializes in renaissance drama, restoration drama, and American minstrel performances. She is the author of *Women and Race in Early Modern Texts* (Cambridge, 2002) and the editor of *Race, Ethnicity, and Power in the Renaissance* (Fairleigh Dickinson, 1997), as well as the author of essays in *Theatre Journal*, *Shakespeare Studies*, and many book collections.

Anita Maynard-Losh is the Director of Community Engagement at Arena Stage in Washington, DC. Before Arena, she was Associate Artistic Director at Perseverance Theatre; a faculty member in the Theatre Department at Webster University; and head of the Theatre Department of the University of Alaska Southeast. She directed the Tlingit (Alaska Native) interpretation of *Macbeth*, which received a touring grant from the NEA.

Nick Moschovakis recently edited *Macbeth: New Critical Essays* (Routledge, 2007). His articles on early modern English literature have appeared in *Shakespeare Quarterly* and elsewhere. A writer and editor in Washington, DC, he has taught at institutions including the University of the South and George Washington University, where he is adjunct professor. His book in progress is tentatively titled *Shakespeare's Allusiveness*.

Heather S. Nathans is Associate Professor of Theatre at the University of Maryland. She is the author of *Early American Theatre from the Revolution to Thomas Jefferson* (Cambridge, 2003), *Slavery and Sentiment on the American Stage, 1787–1861* (Cambridge, 2009), and numerous articles on early American theatre. She is coeditor and contributing author for the forthcoming collection *Shakespearean Educations: Power, Citizenship, and Performance* (Delaware, 2009).

Scott L. Newstok is Assistant Professor of English at Rhodes College, where he organized the 2008 symposium on *Macbeth* and African American culture. He has edited *Kenneth Burke on Shakespeare* (Parlor Press, 2007) and has published a study of the rhetoric of English epitaphs, *Quoting Death in Early Modern England* (Palgrave, 2009).

Marguerite Rippy is Associate Professor of Literature at Marymount University, and specializes in race and identity in American performance traditions, including classic Hollywood cinema and the Harlem Renaissance. Rippy has published several essays on adaptation and performance, and is the author of the book *Orson Welles and the Unfinished RKO Projects: A Postmodern Perspective* (Southern Illinois University Press, 2009).

Francesca Royster is Associate Professor of English at DePaul University, and specializes in Shakespeare, film, and black feminism. Royster is the author of *Becoming Cleopatra: The Shifting Image of an Icon* (Palgrave, 2003) as well as essays in *Shakespeare Studies* and *Shakespeare Quarterly*. Her current book project is titled *Performing Eccentricity: Black Popular Music and the Performance of "Strange" in the Post-Soul Era*.

Amy Scott-Douglass is a theatre practitioner and scholar based in Washington, DC. She is the author of *Shakespeare Inside: The Bard Behind Bars* (Continuum, 2007); the "Theatre" section of *Shakespeares after Shakespeare: An Encyclopedia of the Bard in Mass Media and Popular Culture* (Greenwood, 2006); and several essays on early modern women, and film and stage adaptations of Renaissance drama.

Lisa N. Simmons is an actress and producer of independent films, and is the co-founder of "The Color of Film" collaborative in Boston. She is a Master's Candidate in Media Arts at Emerson College. Simmons produced *Hunting in America*, a political drama about racial profiling, moral choices, and Black success in America. She is currently at work on a documentary about the Boston Negro Theatre Unit.

Lenwood Sloan is Director of Heritage Tourism for the Commonwealth of Pennsylvania. He conceived the nationally touring *Macbeth* that recreated the Orson Welles production through an African-American perspective. Sloan has served as an National Endowment for the Humanities program director, has trained with the Joffrey Ballet Company and the Alvin Ailey American Dance Theater, and has been a directing fellow at the Lincoln Center Institute in New York City.

Ayanna Thompson is Associate Professor of English, and an affiliate faculty in Women & Gender Studies and Film & Media Studies at Arizona State University. She is author of *Performing Race and Torture on the Early Modern Stage* (Routledge, 2008), and the editor of *Colorblind Shakespeare: New Perspectives on Race and Performance* (Routledge, 2006) and two special issues of *Shakespeare Bulletin* and *Borrowers and Lenders*. Her new book project, *Passing Strange*, is forthcoming from Oxford.

Index of Passages from *Macbeth*

ACT 1

"When shall we three meet again?" (1.1.1–2), 11, 127
"When the battle's lost and won" (1.1.4), 163
"Fair is foul and foul is fair" (1.1.10), 138, 175, 178, 180, 187, 188, 190
"For brave Macbeth—well he deserves that name!" (1.2.16–23), 26–27, 184
"As sparrows eagles, or the hare the lion!" (1.2.35), 218
"I'll do, I'll do, and I'll do" (1.3.3–9), 59, 91, 208
"tempest-tossed" (1.3.23–25), 36–37
"The weyward [weird] sisters" (1.3.30), 3, 4, 11, 27, 115, 214, 220
"Peace! the charm's wound up" (1.3.35), 87, 94
"You should be women" (1.3.43–45), 11
"imperfect speakers" (1.3.60), 219
"strange intelligence" (1.3.74), 219
"Melted as breath into wind" (1.3.79–80), 167
"borrowed robes" (1.3.107), 4, 54, 168
"Two truths are told" (1.3.126–28), 129
"horrible imaginings" (1.3.137), 235
"nothing is / But what is not" (1.3.140–41), 4
"strange garments" (1.3.143–45), 168
"Let not light see my black and deep desires" (1.4.51), 10fn4, 17, 178, 184, 187
"milk of human kindness" (1.5.15), 36, 205
"that which rather thou dost fear to do" (1.5.22–23), 208
"unsex me here" (1.5.38–41), 32, 216

"And take my milk for gall, you murd'ring ministers" (1.5.45–46), 216
"The future in the instant" (1.5.56), 209
"In every point twice done, and then done double" (1.6.15), 95
"If it were done when 'tis done, then 'twere well" (1.7.1–2), 27, 55, 209
"Bloody instructions which, being taught, return" (1.7.8–10), 165
"Will plead like angels, trumpet-tongued" (1.7.16–20), 36, 37, 67, 153, 210n2
"vaulting ambition" (1.7.27), 217, 235
"To be the same in thine own act and valour" (1.7.39–41), 27, 153
"I dare do all that may become a man" (1.7.46–47), 218
"I have given suck" (1.7.54–58), 28, 216
"screw your courage to the sticking-place" (1.7.60–61), 197

ACT 2

"Is this a dagger which I see before me" (2.1.33–34), 25, 48, 56, 123, 197, 218
"bloody business" (2.1.48), 206
"'Macbeth does murder sleep'" (2.2.34), 205
"To wear a heart so white" (2.2.62–63), 10fn4, 17
"Faith, here's an equivocator" (2.3.7–8), 25, 45, 65, 180, 237
"O horror, horror, horror!" (2.3.59), 106
"Confusion now hath made his masterpiece" (2.3.62–65), 161
"That had a heart to love" (2.3.113–14), 106

ACT 3

"*stand in thy posterity*" (3.1.4), 220
"*As hounds and greyhounds, mongrels, spaniels, curs*" (3.1.94), 72
"*for't must be done tonight*" (3.1.132), 209
"*black Hecate*" (3.2.42), 10fn4, 14, 15, 17
"*night's black agents*" (3.2.54), 10fn4, 17, 178
"*But who did bid thee join with us?*" (3.3.1), 107
"*gory locks*" (3.4.49–50), 4–5, 73
"*And push us from our stools*" (3.4.81), 82
"*What men dare, I dare*" (3.4.98–102), 59, 67
"*secret'st man of blood*" (3.4.125), 218
"*for a wayward son*" (3.5.10–11), 93
"*Shall draw him on to his confusion*" (3.5.26–29), 214
"*Come away, come away*" (3.5.34), 14

ACT 4

"*Double, double, toil and trouble*" (4.1.20–21), 4, 24, 107, 128, 130
"*Liver of blaspheming Jew*" (4.1.26–29), 4
"*Black spirits and white, red spirits and grey*" (4.1.44–45), 13, 17, 166
"*you secret, black, and midnight hags*" (4.1.64), 10fn4, 14, 17, 178
"*I conjure you by that which you profess*" (4.1.66), 206
"*none of woman born*" (4.1.96–97), 19, 183, 191
"*pale-hearted fear*" (4.1.101), 17
"*unfix his earth-bound root*" (4.1.112), 168, 197
"*Whither shall I fly?*" (4.2.73), 107
"*black Macbeth*" (4.3.53–54), 10fn4, 17, 247
"*black scruples*" (4.3.117), 10fn4, 17

"*Alas poor country*" (4.3.165), 106
"*How does my wife?*" (4.3.177), 107

ACT 5

"*When was it last she walked?*" (5.1.2–13), 123
"*Out, damned spot*" (5.1.30), 5, 67, 123, 205
"*Here's the smell of blood still*" (5.1.42), 28, 169
"*What a sigh is there!*" (5.1.44), 124
"*I tell you yet again, Banquo's buried*" (5.1.53), 235
"*To bed, to bed*" (5.1.56), 106
"*What's done cannot be undone*" (5.1.57–58), 98, 208
"*I think, but dare not speak*" (5.1.68–69), 28
"*The devil damn thee black, thou cream-faced loon!*" (5.3.11), 10fn4, 17
"*Towards which, advance the war*" (5.4.16–21), 33
"*banners on the outward wall*" (5.5.1), 35
"*I have almost forgot the taste of fears*" (5.5.9), 35
"*She should have died hereafter*" (5.5.16), 106
"*Tomorrow, and tomorrow, and tomorrow*" (5.5.18), 68, 139
"*Signifying nothing*" (5.5.27), 6, 122
"*At least we'll die with harness on our back*" (5.5.50), 71
"*bear-like I must fight the course*" (5.7.1–2), 71
"*untimely ripped*" (5.10.16), 94, 168
"*palter with us in a double sense*" (5.10.19–20), 40, 219
"*And damned be him that first cries 'Hold, enough!'*" (5.10.33–34), 35, 69, 70, 236
Exeunt fighting. Alarums (5.10sd), 33, 236
Enter Macduff, with Macbeth's head (5.11sd), 236

General Index

Entries in **boldface** indicate illustrations

Abela, Alexander: see *Makibefo*
abolitionists, 23, 25–32, 36, 38, 40–41, 43, 47, 67, 73, 230. *See also* slavery
Actors' Equity Association (AEA), 246
The African Company Presents Richard III (1988, Carlyle Brown), 224, 229–31. *See also* African Grove Theatre; Hewlett, James
African Grove Theatre, 31, 45, 62. See also *The African Company Presents Richard III*; Hewlett, James
Agamben, Giorgio, 184; *homo sacer*, 184–90
Aidoo, Ama Ata, 173
Ainsworth, John, 243
Alaskan (Tlingit) theatre, 6, 127–31, 250
Aldridge, Ira, 6, 31, 45–48, **49**, 50–53, 211–12, 214, 217, 220–21, 221n1, 241
Aldridge, Karen, 252
Alexander, Robert, 135n2, 251
Allied Arts Players (Boston), 79–81
Alonso, Alicia, 102
Amin, Idi, 250–51
amnesia, historical, 6–8, 10
Anderegg, Michael, 94
Anderson, Thomas, 96
Anozie, Nonso, 193
antebellum America, 7, 23–33, 40, 55–56, 60, 63, 67
Aranha, Ray, 246
Arneaux, John A., 242
Ashe Cultural Center (New Orleans), 102, 104
Asian-American theatre, 6, 9, 121–25, 134, 137–39, 196, 246

Association for Theatre in Higher Education (ATHE), 201
Astor Place Riot, 6, 29–31, 57–58, 63. *See also* Forrest, Edwin; Macready, William Charles
Atkinson, Brooks, 88, 236

Bad Breath, or the Crane of Chowder, 55
Baker, Houston A., Jr., 206, 210n3
Baldwin, Alec, 248
Baldwin, James, 223–24, 228, 230–32
Banquo's ghost, 4–5, 25–26, 38, 43n4, 52, 59, 67–68, 75, 119, 138, 205, 210n2, 235. *See also* haunting
Barrax, Gerald, 205
Basset, Angela, 19, 248
Bavan, Yoland, 245
Bearden, Romare, 77, 97
Beaty, Powhatan, 242
Beecher, Henry Ward, 27
Berry, Ralph, 194, 200
Bharadwaj, Vishal: see *Maqbool*
blackface, 167, 214; Laurence Olivier and, 174; minstrelsy and, 55–56, 58, 60–63; Orson Welles and, 83–84, 94; Renaissance drama and, 12–14, 16; "sonic," 162, 164
"blackness," 4–5, 12, 14–19, 83–85, 105, 114, 162, 178, 183, 196, 210
blaxploitation films, 178–79
blood, 4–5, 7, 17, 19, 23, 25–29, 33, 38–42, 65, 169, 180, 191, 206, 213–20
The Bloody Child (1996, dir. Nina Menkes), 183–91, 248
Bloody Streetz (2003, dir. George Barclay), 161

Bogdanovich, Peter, 94
Boisseau, Robin Jackson, 251
Bontemps, Arna, 71–73, 75, 75n8
Booth-Barrett Amateurs (Louisville, KY), 242
The Boston Players, 79–81
Boys, Barry, 245
Braugher, Andre, 247
Breuer, Lee, 92, 247
Briggs, John R., *Shogun Macbeth* (1985), 121–25, 246
Brooks, Gwendolyn, 205
Brooks, Preston, 23
Brown, Carlyle: see *The African Company Presents Richard III*
Brown, John, 26–29, 67
Brown, William Alexander, 230, 231
Brown, Williams Wells, 79
Buchanan, James, 23
Bully, Alwin, 122, 244, 248
Burgess, John, 247
Burroughs, Charles, 242
Burroughs, Eric, 86, 89, 90n1
Burroughs, Norris, 90n1
Burrows, Vinie, 244
Burton, Nefertiti, 247
Bushnell, William, Jr., 246
Butcher, James W., 244, 246
Butler, Buddy, 250

Caitlin, George, 30
Caldwell, L. Scott, 248
Calhoun, John C., 23
Cambridge, Edmund, 96, 246
Campbell, Ken: see *Makbed*
"Caribbean" settings, 6, 48, 83–85, 89, 97, 99, 116–17, 122, 155, 220, 227, 230, 245, 248, 249, 251, 252. *See also* "Haitian" settings
Carlson, Ben, 252
Carroll, Jacqui, 251
Carter, Jack, 84, 88, 92–93, 96, 100, 235, 237, 243
Catherine Market (New York), 167–68
Chamillionaire, 162m, 169
Chan, Christina R., 247
Chapman, Maria Weston, 32
Chappelle, Aleta, 122, 244, 252
Chesnutt, Charles, 207
Chessé, Ralph, 242–43

Chevalier, Maurice, 102
Child, Lydia Maria, 67
Christophe, Henri, 83, 224–25. *See also* "Haitian" settings
Christy's Nigga Songster, 55–56
Civil War, American, 24, 26, 33, 38, 41–42, 66–67, 105, 243, 247, 250
Clark, J. C., 4
"classical" performance, 31, 79, 84, 92–93, 113–14, 119, 134, 251
Cliff, Michelle, 173
Clifton, Lucille, 209–10
Cocteau, Jean, 94–95
Code Noir, 105
Coe, Peter, 245
Coleman, Gary, 162
Coleman, Ralf, 80–83, 243
"colorblind" casting, 8, 113, 193–201. *See also* "non-traditional" casting
Confederacy, Southern, 24, 26, 105–06
Constitution, U.S., 5, 35, 40–41
Cover, Jennifer A., 246
Creoles, New Orleans, 103–09
Crooks, Rich, 248
Cullen, Countee, 71, 84
Cuney-Hare, Maud, 79–81
Cushman, Charlotte, 32
Cwiklik, Frank, 249

Davis, Clinton Turner, 246
Davis, Henrietta Vinton, 242
Dawes, Carroll, 245
Day, Doris, 245
de Grazia, Margreta, 3
Dean, Phillip Hayes, 224–27
Dekker, Thomas, 20n4
Derrida, Jacques, 191
Dessalines, Jean-Jacques, 73–74, 80
dialect, 62, 88, 97, 100n4, 117, 140
Diaspora, African, 7, 107, 113–17, 206, 252
Dodson, Owen, 243, 244
Douglas, Stephen A., 23
Douglass, Frederick, 10, 31, 35–43
Dove, Rita, 208, 210n4
Dowden, Edward, 65
Dowling, Joe, 248
drumming, 9, 19, 73–74, 84–89, 93, 96–97, 103, 109, 127–28, 130, 155–57, 195, 224, 236, 243, 249, 251

Du Bois, W.E.B., 65–68, 73–74, 75n6, 210n2
Dunham, Katherine, 101–02

Easton, William Edgar, 80
Edwards, Gale, 252
Eisenstein, Sergei, 184, 187
Elder, Lonne, III, 211–12, 220–21
Ellington, Duke, 8, 10, **143**, 151–59
Ellis, Maurice, 84, 237, 243
Emeka, Justin, 250
equivocation, 7, 25, 45, 65, 178, 180, 237
Erickson, Peter, 181n1, 181n8, 205, 210n4, 232n2
Esquea, José: see *Macbeth 2029*
Essandoh, Ato, 250
Evans Erville, Victoria: see *MacB: The Macbeth Project*
Ewing, Julia "Cookie," 251
Ewing, Marc: see *Mocbet*

Fairey, Shepard, 10, **233**
Fajer, Marc, 251
Fast and Furious (1931, dir. Forbes Randolph), 243
Federal Theatre Project (FTP), 6–7, 9, 81–82, 84–85, 89, 91–93, 95, 97, 99–104, 235, **239**, 243, 247, 248, 249. *See also* Negro Theatre Unit; Welles, Orson
Feore, Colm, 252
Fields, Julia, 208–09
Flanagan, Hallie, 89
Flash, Grandmaster, 168
Flocabulary [Blake Harrison and Alex Rappaport], 161, 163
Foden, Giles, 251
Ford's Theatre (DC), 237, 242
Forman, Simon, 20n1
Forrest, Edwin, 24, 26–31, 57, 60, 63, 63n3. *See also* Astor Place Riot; Macready, William Charles
Foucault, Michel, 165, 175
France, Richard, 87, 93, 98
Francis, Decima, 247
Freire, Paulo, 164–65
Frothingham, Nathaniel, 244
Furness, Horace Howard, 118–19

Gaffney, Perri, 252
Gaines, Barbara, 252
Garrison, William Lloyd, 23, 25, 40
Garvey, Marcus, 116, 217
Gaylord, E. Keith, 246
Gill, Peter, 247
Glover, Danny, 6, 135n2, 246, 252
Gonzalez, B. W., 249
Goold, Rupert: see *Macbeth* (2008)
Gounod, Charles-François, 4
Granville-Barker, Harley, 117–18
Greeley, Horace, 23
Green, Paul, 80–81
Grey's Anatomy, 196, **197**, 198–99
Gulf War, the first, 184, 186, 188–89; the second, 184, 191

Haden, Kyle, 249
"Haitian" settings, 12–13, 19, 46, 71–75, 80, 83–84, 86–89, 92–93, 95, 102–03, 224–25, 236, 251. *See also* "Caribbean" settings
Hall, Kim F., 174
Hammond, Mona, 245, 247
Hammond, Percy, 89
hands, Lady Macbeth's, 4–5, 26, 33, 123, 213. *See also* stain (of blood)
Harper's Ferry, 25, 28
Harper's Weekly, 23, **24**
Harrison, Richard B., 242
Hatch, James V., 80, 229, 241–43, 245
Hattiloo Theatre Company (Memphis), 3, 251
haunting, 4–5, 7, 12, 25, 38–40, 43n4, 67, 73, 207, 220, 235, 237. *See also* Banquo's ghost
Haven, Reverend Gilbert, 5
Hawai'ian theatre, 6–7, 125, 137–41, 252
Heard, Josephine Delphine Henderson, 205
Hecate (character), 12–15, 17, 3, 86–87, 90, 93, 96–98, 107
Hefner, Hugh, 179–80
Hendricks, Margo, 174
Hewlett, James, 31, 229–30. *See also* African Grove Theatre; *The African Company Presents Richard III*
Hill, Errol G., 80, 229–30, 241–46
Hill, Leslie Pinckney, 71–73

Hinds, Rasheed, 249
hip hop, 161–70
Historically Black Colleges and Universities (HBCUs), 91, 110; Atlanta University, 243; Dillard University, 92, 102, 252; Florida A&M University, 248; Hampton University, 251; Howard University, 104, 244, 246; Lincoln University, 242; Spelman College, 243, 244; Talladega College, 243; Tennessee State University, 250
Holinshed, Raphael, 12, 115
Holland, Agnieszka, 249
Holly, Ellen, 244
hooks, bell, 175, 181n3
The Hop of Fashion (1856), 60–62
Hopkins, Pauline, 79
Horton, Asadata Dafora, 95, 97, 243
Houseman, John, 84, 87, 89, 93, 97, 98, 103, 243
Howard, Terence, 6, 252
Hughes, Langston, 84; *Emperor of Haiti* (1936), 71–75, 75n8; "Note on Commercial Theatre" (1940), 8–9, 205–06, 211–12, 220–21; *Shakespeare in Harlem* (1942), **203**
Hughes, Margaret, 80
Hurston, Zora Neale, 84
Hwang, David Henry, *M Butterfly*, 125
Hyman, Earle, 244–45

"interracial" casting, 13, 17, 93, 193–201, 244
Iraq, 184, 188–90

Jacksonian era, 58–59, 61
Jafferis, Aaron, 161, 163, 252
James, Oscar, 245
"Japanese" settings, 7, 121–25
Jim Crow, 56, 58, 66, 69, 163, 214
Johnson, Javon, 248
Johnson, Wilbert, 247
Johnston, David Claypool, **21**, 33n1
Johnstone, Diane, 252
Jones, David Richard, 246
Jones, Herman, 247
Jones, James Earl, 244
Jonson, Ben, 14–15
Jordan, June, 205

Jordan, Richard, 247
Julia, Raul, 134, 244, 247
Jurasas, Jonas, 246

Kaplan, Cindy, 247
Kauffer, E. McKnight, **203**
Keach, Stacey, 248
Kean, Edmund, 24, 47–48
Kelley, Sean, 249
Kenerly, Kevin, 252
Kennedy, Adrienne, 212–13, 221
Kennedy, Randall, 137
King James I, 4, 176
King James Bible, 117
King, Martin Luther, Jr., 162, 178
King, Woodie, Jr., 91, 96, 246. *See also* New Federal Theatre
The King's Men, 14, 176
Kinney, Arthur, 206–07
Kissoon, Freddie, 243–44
Kissoon, Jeffrey, 245
Kotto, Yaphet, 245
Kurosawa, Akira: see *Throne of Blood*

Lafayette Theatre (New York), 80, 84, 100n1, 102–03, 223
Lamarque, Kimberly, 250
Lambert, Mikel, 246
Landin, Bo: see *Sámi Macbeth*
The Last King of Scotland (2006, dir. Kevin Macdonald), 251
Lathan, Sanaa, 6, 252
Latino theatre, 6, 9, 133–35, 246, 251, 252
Lee, Canada, 100n4, 243
Lee, Gihieh, 161
Lee, Tonia, 161
Lennix, Harry J., 6, 251, 252
Lessig, Lawrence, 166, 168–69
Lhamon, W. T., 16, 56, 62, 167
Lincoln, Abraham, 40, 42–43, 237, 242
Little, Cleavon, 244–45
Lombardo, Greg: see *Macbeth in Manhattan*
Longenhagen, Greg, 248
Loomba, Ania, 16, 174
Lott, Eric, 56, 162, 167
L'Ouverture, Toussaint, 71–72, 224–25, 230

General Index

Ludlow, Noah, 27, 30
Lumbly, Carl, 92, 247, 248

Mac, A Gangsta's Tale (2006, dir. Ayodele Nzinga), 161, 163, 251
MacB: The Macbeth Project (2002, 2008, dir. Victoria Evans Erville), 161, 249
Macbeth (1948, dir. Orson Welles), 93–94
Macbeth (1971, dir. Roman Polanski), 8, 173–81
Macbeth (2008, dir. Rupert Goold), 161
Macbeth 2029 (2008, dir. José Esquea), 111, 133–35, 252
Macbeth in Manhattan (1999, dir. Greg Lombardo), 171, 194–96, 248
Macbeth, Robert, 100n1
Macbeth Travestie (1813, anon.), 58
Macbeth Travestie (1847, W. K. Northall), 58–59
Macdonald, Kevin: see *The Last King of Scotland*
Macon, Peter, 252
Macready, William Charles, 29, 57–58, 63n3, 118. *See also* Astor Place Riot; Forrest, Edwin
Mad Dawg (2004, dir. Greg Salman), 161
Makbed (1999, dir. Ken Campbell), 139, 141n4, 248
Makibefo (1999, dir. Alexander Abela), 249
Mandela, Nelson, 245
Mann, Horace, 43n3
Maqbool (2004, dir. Vishal Bharadwaj), 250
Marienstras, Richard, 95, 100n4
Marshall, Larry, 148n1
Marshall, William, 246
Marvel, Steve, 251
Mason, Luoneal, 79, 81
Mathews, Charles, 60
Maynard-Losh, Anita, 250
McAvoy, James, 251
McCarthy, Mary, 92
McIntosh, Yanna, 252
McKay, Claude, 70–71
McKinney, Chris, 248
McLendon, Rose, 84

Melville, Herman, 29, 57
Menkes, Nina: see *The Bloody Child*
Menkes, Tinka, 183, 185
Mercury Theatre, 89, 92–93
Metamora (1829, John Stone), 57, 60, 63n4
Middleton, Thomas, 6, 11–20, 170n1
Miller, May, 205
minstrelsy, 16, 55–57, 62–63, 151, 154, 162, 166–67, 214–15, 221
Mitchell, William, 58–60
Mitri, Paul T., 138–40, 252
Mocbet (2005, dir. Peres Owino and Marc Ewing), 250
Monson, Lex, 97, 246
Montgomery, Barbara, 246
Morrison, Toni, 115
Msomi, Welcome, 245
Muhammad, Honorable Elijah, 116, 217
multilingualism, 128–29, 137–41
Munday, Anthony, 16
Myrdal, Gunnar, 116

National Endowment for the Arts (NEA), 101, 161, 235–36, 249
National Museum of the American Indian (NMAI), 128–29
National Public Radio (NPR), 10n3
National Spirit Project (NSP), 104–05
Neall, Daniel, 36
Negro Theatre Unit (NTU), 79–84, 91, 93, 243. *See also* Federal Theatre Project (FTP)
New Federal Theatre, 77, 91, 96–98, 246. *See also* King, Woodie, Jr.
Newman, Harry, 246
Newstok, Scott L., 3, 236
Nickelson, Jeffrey, 249–50
Nietzsche, Friedrich, 65–66, 74
Nixon, Richard, 205
"non-traditional" casting, 5–7, 10, 12–13, 91, 95, 99, 115, 130, 145–49, 201, 241–52. *See also* "colorblind" casting
Northall, W. K.: see *Macbeth Travestie* (1847)
Nzinga, Ayodele: see *Mac, A Gangsta's Tale*

General Index

Obama, Barack Hussein, 9–10, 137–38, 235–37, 252
Obama, Michelle, 252
O'Neill, Eugene, 81, 217; *The Emperor Jones*, 81, 85, 87, 90n3, 223–26, 242
Opera Memphis, 3, 145, 148n1
Orman, Roscoe, 91
Owino, Peres: see *Mocbet*
Oz (HBO series), 201–02n2

Pantoja, Adriana, 250
Papp, Joseph, 96, 113, 135n1
Parker, Patricia, 55, 174
Parks, Suzan-Lori, 207, 219–21
Patterson, Rebecca, 249
Perrineau, Harold, **171**, 195, 248
Perry, Felton, 246
Phillips, Josiah, 252
Phillips, Wendell, 230
Polanski, Roman: see *Macbeth* (1971)
Pollock, Max, 95–96
Porter, Cole, 92
post-apocalyptic settings, 6, 133–35, 135n2, 138, 246, 252
Preisser, Alfred, founder of Classical Theatre of Harlem (CTH), 98–99, 249
Price, Leontyne, 147–48, 149n3
primitivism, 9, 84–90, 156–58
prison performances, 199–200

Quarles, Lorenzo, 79, 81
Quinn, Patrice, 251

Rankine, Claudia, 205
Rayne, Stephen, 247
Redlich, Norman, 5
Reed, Ishmael, 207
Reuben, Gloria, 248
Revenge of the Ghetto (2008, dir. "MadskillzMan" [Matt Dacek]), 161–64, 166, 169
Reynolds, Bryan, 175, 178
Rhodes College, 3, 90n3
Richards, Beah, 245
Rivers, Voza, 247
Robeson, Paul, 92, 224–27, 229–32
Robinson, Mary B., 247
Rolle, Esther, 97, 246
Roosevelt, Eleanor, 81

Roosevelt, Franklin Delano, 81
Rowley, William, 20n4
Royal Shakespeare Company (RSC), 113, 193–94, 247
Ruffin, Clyde, 248
Rynders, Isaiah, 30–31, 57, 63. See also Astor Place Riot

Salem witch trials, 209, 210n5
Salman, Greg: see *Mad Dawg*
Sámi Macbeth (2004, dir. Bo Landin and Alex Scherpf), 250
Sands, Diana, 244
Scherpf, Alex: see *Sámi Macbeth*
Schomburg Library (New York), 97, 101, 103
Scot, Reginald, *The Discoverie of Witchcraft* (1584), 13, 17
Scotland, as setting, 6, 82, 93, 97, 103, 117, 134, 138
Scott, Dred, 117
Scott, Sir Walter, 58
Scott-Heron, Gil, 205
"Scottish play," 4, 6, 8, 40, 113, 140, 176, 216, 237
Sedgwick, John, 5
September 11, 2001, 109–10
Shakespeare, William, *Antony and Cleopatra*, 18, 152, 173, 224, 227–28; *As You Like It*, 42, 43n2; *Coriolanus*, 135n1, 249; *Hamlet*, 31, 38–40, 42, 43n2, 52, 55–56, 58, 75, 134, 152, 217, 237, 242; *1 Henry IV*, 43n2, 140; *2 Henry IV*, 43n2; *Henry V*, 140, 152, 153; *2 Henry VI*, 37, 43n2; *Henry VIII*, 43n2; *Julius Caesar*, 24, 43n2, 113, 135n1, 152, 241; *King Lear*, 29, 31, 39, 43n2, 75n3, 193–94, 208, 228, 230, 249; *Love's Labour's Lost*, 18; *The Merchant of Venice*, 17, 40–41, 43n2, 115, 190; *The Merry Wives of Windsor*, 13–14; *A Midsummer Night's Dream*, 139, 152, 157, 200, 210n1; *Much Ado About Nothing*, 18; *Othello*, 3–6, 16–18, 24, 31, 40, 43, 45–48, 50–51, 54, 75, 114–15, 134–35, 152, 153, 155, 158, 174, 199–200, 202n2, 212, 224–27, 229–30, 242; *The Passionate Pilgrim*,

18; *The Rape of Lucrece*, 19; *Richard II*, 249; *Richard III*, 43n2, 45, 48, 61–62, 224, 229–31, 242; *Romeo and Juliet*, 4, 17–18, 43n2, 45, 55, 135, 152, 177, 179, 195, 208, 241; *Sonnets*, 18, 152; *The Taming of the Shrew*, 113, 152, 155; *The Tempest*, 3, 6, 43n2, 127, 185; *Titus Andronicus*, 17, 93, 115, 173–74, 201, 212; *Troilus and Cressida*, 18, 38, 43n2; *The Two Gentlemen of Verona*, 18
Shange, Ntozake, 213–17
Sheridan, Richard Brinsley, 45
Siddons, Sarah, 118–19
Silvera, Frank, 79, 81
slavery, 4–5, 7, 23–32, 36, 38–41, 43, 46–47, 67–68, 73, 79, 117, 188, 210, 250, 251. *See also* abolitionists
Sloan, Lenwood: see *The Vo-Du Macbeth*
Smith, Len Anthony, 245
Smith, Morgan, 241
Spicer, Captain Arthur, 6
Stafford-Clark, Max, 250
stain (of blood), 4–5, 27–28, 67, 205. *See also* hands, Lady Macbeth's
Stallybrass, Peter, 3
Stendhal (Marie-Henri Beyle), 202n2
Stickney, Timothy, 252
Stone, John: see *Metamora*
Stowe, Harriet Beecher, 32
Strange, R. Henri, 242
Strayhorn, Billy, 151–52, 155. *See also* Ellington, Duke
Sumner, Charles, 23
Syer, Fontaine, 250

Taccone, Tony, 248
Taylor, Candace, 249
Taylor, Gary, 12, 15, 176
Terry, Clark, 152, 154, 158. *See also* Ellington, Duke
Thomas, Edna, 84, 93, 243
Thomas, G. Valmont, 247, 249
Thomas, Richard J., 251
Thompson, Ayanna, 194
Throne of Blood (1957, dir. Akira Kurosawa), 121–22, 130, 180, 244, 248
Tolson, Melvin, 68–69

Turner, Nat, 25–28, 127
tyranny, 24, 35, 37–38, 43, 46, 70, 73–74

Underwood, Blair, 6, 252

Vechten, Carl Van, 84
Verdi, Giuseppe, 3, 4, 7, 55, 145–49
Verrett, Shirley, 146–48
Vietnam War, 175, 178, 189
The Vo-Du Macbeth (2001–05, dir. Lenwood Sloan), 99, 101–10
"voodoo," 6–7, 9–10, 11, 19, 72–74, 79, 82, 83–89, 91–92, 94–97, 99, 102–03, 130, 207, 209, 224, 236, 243, 251; *Vodun*, 107, 109, 110n1. *See also* Federal Theatre Project (FTP); Welles, Orson
Vrazel, Ray, 252

Walcott, Derek, 227–29, 231, 248
Walford, Glen, 251
Walker, Sullivan, 245
Wallace, George, 5
Ward, Nick, 247
Warner, Earl, 248
Warren, Ebenezer W., 26
Washington, Isaiah, 196–98
Watergate, 205
We Work Again (1937), 89, 237
Webster, Daniel, 43n3
Webster, John, 178
Welles, Orson, 6–7, 9–12, 74, 82, 83–100, 101–03, 107, 114, 130, 205, 207, 209, 220, 223–25, 235–37, 243, 246. *See also* Federal Theatre Project (FTP); Negro Theatre Unit
Wells, Luther D., 248
Wells-Barnett, Ida B., 70, 74
"weyward," 3–4, 6, 8–10, 11, 27, 115, 183, 206, 214
Whitaker, Forest, 251
"whiteness," 4, 8, 17, 173–81, 183, 187, 188, 190–91
Whitman, Walt, 57
Whittier, John Greenleaf, 36
Wilburforce, William, 57
Williams, Cecelia V., 241
Williams, George Washington, 67–68
Williams, Linda, 185–86

Wilmot, David, 23
Wilson, August, 114–15, 212, 217–19, 221, 231–32
The Witch (Thomas Middleton), 13–19, 170
Wolfe, George C., 248
Works Progress Administration (WPA), 81, 85, 89, 91, 101, 235–37. *See also* Federal Theatre Project (FTP)

Wright, Henry C., 25

Yankovic, "Weird Al," 169
Young, Al, 205–06
YouTube, 10, 161–63

Zaslove, Arne, 122, 244, 247
Žižek, Slavoj, 188, 190–91
Zvonchenko, Walter, 98